Failure Analysis

Jose Luis Otegui

Failure Analysis

Fundamentals and Applications in Mechanical Components

Springer

Jose Luis Otegui
University of Mar del Plata
Mar del Plata
Argentina

ISBN 978-3-319-03909-1 ISBN 978-3-319-03910-7 (eBook)
DOI 10.1007/978-3-319-03910-7
Springer Cham Heidelberg New York Dordrecht London

Library of Congress Control Number: 2013955883

Printed on acid-free paper

Springer is part of Springer Science+Business Media (www.springer.com)

Preface

When one reads a book, and cannot stop reading, it means that the book is pleasant and that, somehow, generates the expectation of the reader to continue reading compulsively until the end. The same happens with a play or a film. If the spectator begins to feel uncomfortable in his armchair, it is a signal that the play/film is decaying in interest. By the opposite, nobody moves, even in uncomfortable armchairs, and a spectator will keep his position until the end.

This happens with Jose Luis' book. Those of us that love Engineering and the resolution of its problems find in this work the right synthesis of what we always have looked for in books of the specialty. As mentioned in the prologue, this book answers why a failure occurred that jeopardized the integrity of a component or piece; the conditions that led to this failure and, finally, what we must do so that it does not repeat.

This book has been written by an education professional, also a successful researcher in this area as well as other areas of engineering, who has permanently performed consulting activities. In this book, the author pours all his experience in the prediction of damage, fitness for service, and reliability in concrete problems of professional life.

Therefore, this work that comes to light today covers all the expectations that are looked for to satisfy the high demand of specialized human resources in the selection, application, optimal operation, and service of industrial pieces and components. It will allow, in addition, forging highly enabled professionals, apt to work in projects in different economic sectors. I take for granted the success of this contribution by my colleague and friend of so many years.

Argentina, 2013 Raul H. Conde

Prologue

Back in the twelfth century, Bernard of Chartres said once that we are like dwarfs on the shoulders of giants. We can see beyond them, not because of any physical distinction, but because we are lifted up by their great height. Recapturing the idea, the famous seventeenth century English physicist Isaac Newton wrote in a letter to the scientist Robert Hooke: "if I have seen more is because I am on top of the shoulders of giants." We all are stepping on shoulders of giants that preceded us, in our continuous missions for whatever there is to be obtained beyond the horizon. The Holy Grail of engineering is the perfect design, something that always works exactly as envisioned and which never requires improvement. By all means, if we could obtain it, the perfect design would never fail.

In his seventeenth book, Henry Petroski explores something that he had already explored before: the inevitability of failure and the role it has had in the advance of technology. His last work is "To Forgive Design: Understanding Failure" (Harvard University Press). In his preface, the author defines this book as a sequel to his first book "To Engineer Is Human: The Role of Failure in Successful Design," published in 1985 and still available. The following fragment belongs to Chap. 2, "Things happen" or "Forgiving design":

It would not have to surprise us that failures exist. After all, the structures, the machines and the systems of the modern world can be exceedingly complicated in their design and operation. And the people who conceive, design, construct and interact with these complexities without a doubt can fail. Sometimes they use a defective logic, they transpose digits in a numerical calculation inadvertently, fit a bolt a screw of more or of less, they read a dial erroneously or they push hastily when they must throw. Also they can fail in concentrating themselves, anticipating and communicating at critical moments. In other occasions, the accidents can be due to people stopping being honest, ethical or professional.

For some reason, accidents happen and they invariably come from or lead to a failure by something or somebody. In fact, which would have to surprise us is not that failures happen, but that they do not happen more often. When they happen in our field, we tend to defend ourselves against accusations; we try to derive the failure. Failures are too often attributed to the things we design, we do, we sell and we operate, and not to the people who design them, do them, sell them and operate them.

In this book, we will start from a basis that is not shared by most other authors in the field: all causes related to a failure are exclusively human. From the philosophical point of view, this concept is clear. Still, those clearly physical causes, as is the case of a defective material, can be considered the result of a human error, although this error was committed at another moment (sometimes decades before) and in another company or country. Therefore, it is outside our analysis, and we can simply analyze the physical cause, the defect in the material, and forget that it is the result of a previous human fault.

The phrase "human error" commonly used in the journalistic media is tautological: only Humans have reasoning and freedom to choose between diverse alternatives, thus we are the only ones "enabled" to commit an error. Natural disasters (earthquakes, flooding, etc.), also called "Acts of God," cannot be considered as a human error. But the consequences of these natural events on the component that has failed are indeed due to human errors. Somebody did not consider adequately the probability of such event, or its intensity, and he/she did not make the necessary adjustments to the design and the construction of the component or structures.

But the previous statement does not mean that a person has to take the blame or be the cause of a failure. To look for guilt leads to the opposite direction of finding solutions. This will be seen specifically when we deal with the techniques of Root Cause Analysis. An axiom for lawyers, particularly for those involved in oral trials when they interrogate witnesses, these being friendly or hostile for the interest of his cause, is: never ask something if not sure of what the answer will be. This attitude, taken to other fields of the human activity, assures that the participation of a person will generate very little value added to the communitarian knowledge.

This phrase is attributed to Albert Einstein: "The man finds God behind each door that science manages to open. Chance does not exist; God does not play dices. The important thing is not to stop asking questions." Probably this has been said in different contexts. A failure analyst will have to assure that each member of the team makes at least all the necessary questions to arrive at true answers, that are not known a priori, and not regarding if these might not be considered suitable.

This book approaches failures of structural elements, that is, those components whose basic mission is to sustain mechanical loads. From this we can conclude that failure causes will be related to the incapacity of the components and materials to support operative loads, at some time of their life. We will not speak here of failures in functional components, where the fundamental utility of the component is referred to a specific capacity. Thus it is the case of failures in electronic and optical components, measurement and control systems, etc.

Construction technologies and use of structural components were refined by trial and error over thousands of years until the first age of metals. The use of metals was also refined by trial and error. The science of metals (and structural materials in general) is really as recent as the twentieth century. Of course, the use and engineering of these materials are much older. As components and systems become more complex, so do the mechanisms by which they fail. And because of this, understanding the causes of failures has never been more relevant. The modern methods for the analysis of real causes of failures respond to the need to interpret:

Why failures occur;
How failures occur;
How to avoid recurrence.

We live the way we do, and not another, because of the strengths, weaknesses, working costs, and difficulties of the materials and components we use. This book is not only for engineers and scientists, although most of the technicians in these disciplines would benefit from reading. Its real benefit is for the "lay" curious who want to know more, those who may face circumstantially the consequences of failures of systems or mechanical components. This book should also be useful for students, future scientists, or engineers.

One of the objectives of the professional activity of the author and collaborators has been the consolidation of experimental and numerical tools to address scientific research related to the main propagation mechanisms of mechanical damage in industrial components. The authors is a professor in the area of Mechanical Engineering, at the National University of Mar del Plata (UNMdP, Argentina). Dr. Otegui and some partners specialize in research, with activity in the Mechanics of Materials Division of INTEMA (Research Institute of Materials Science and Technology). This institute is funded by CONICET (National Scientific and Technical Research of Argentina) and is based in the Faculty of Engineering UNMdP. Contributors are from GIE SA, prestigious Engineering Consultancy firm with 20 years of experience in integrity assurance and life extension in the energy industries and hydrocarbon transportation and processing.

Experience in the evaluation of piping, vessels, rotating machinery, and other industrial components, allows the author to formalize this text, suitable for use as a professional reference book for those interested in how best to respond to failures and other incidents of mechanical origin. This text is intended as a self-contained source for those with different technical grades, who face the challenge of learning from the experience, often bitterly and at a great cost, of equipment failure, with the intention of reducing its likelihood of recurrence.

Acknowledgments

This book is dedicated to my most loved ones my wife Alejandra and my daughters Milagros and Victoria.

As it usually happens to us, I feel I should have spent more time with them. I wish I had known Alejandra years before. My daughters matured so soon! The rhythms of life are OK, I am not complaining. On the contrary, I find myself in a difficult situation for agnostics: I would like to thank, but I am not sure to whom. In any case, as my mother Chela says, I thank whoever is currently in charge of God's business in this world, recognizing that He probably does not exist. I thank Chela for helping me discover this and other secrets.

This book is the result of many training activities, in-house courses for companies, and particularly of lectures given during the last 10 years for Executive Corporate Training (pcernich@criteriocapacitacion.com). The author is grateful to Mr. Pablo Cernich, for his enthusiasm and professionalism in the pursuit of technical and pedagogical improvements in these conferences.

I want to thank my friends and mentors, among them, Luis A. de Vedia (UNSAM), Raul Conde and Juan Belmonte (UNMdP), Patricia Frontini (CONICET), and Carlos Manfredi (GIE SA). With large differences in temperament, they have shown me the way to applied technological research with a social sense. I also thank Carolina Miranda for her efforts in correcting and editing the book, her talent and bonhomie.

The participation of colleagues and friends in the writing of this book is noteworthy. They contributed their opinions, efforts, and original material based on their own research:

- Mr. Aníbal Marquez amarques@fi.mdp.edu.ar, Professor and Researcher at INTEMA (UNMdP-CONICET). Aníbal is also consultant at GIE SA, leader in Failure Analysis of Rotating Machinery. He authored Chap. 8 of this book.
- Mr. Pablo G. Fazzini fazzini@giemdp.com.ar
- Mr. Ariel Ibarra Pino aaibarra@fi.mdp.edu.ar
- Ms. Janine Booman, PE booman@giemdp.com.ar
- Mr. Kunert Hernán Kunert@giemdp.com.ar
- Ms. Paola Venturino venturino@giemdp.com.ar
- Mr. Jeremías De Bona debona@giemdp.com.ar

We have made reasonable efforts to publish this book on reliable data and information, but the author and publisher cannot assume responsibility for the validity of all the material or the consequences of its use. At the end of each chapter there is a mention of all material reproduced in this publication. If we have not recognized any copyright, it is requested to let us know, so that we can rectify in any future reprint.

Contents

1 Introduction 1
- 1.1 Historical Perspective 1
- 1.2 Current Technology is Conditioned by our History 2
- 1.3 The Danger of Myths and the Public 4
- 1.4 Defects in Components, Ductile and Brittle Materials 6
- 1.5 The Industrial Revolution and Failures in Pressure Components 7
- 1.6 The Advent of Fracture Mechanics 9
- 1.7 Scope of the Failure Analysis 10
  - 1.7.1 Example 1.A Expert Analysis After a Traffic Accident 13
- 1.8 Concluding Remarks 15
- References 18

2 Tools for Preliminary Analysis of a Mechanical Failure 19
- 2.1 Methodologies for Field Investigation After a Failure 19
- 2.2 Collecting Data and History 21
- 2.3 Visual Inspection Techniques and Field Photography 21
- 2.4 How to Detect the Site of Initiation of Mechanical Failure 25
  - 2.4.1 Fractographic 25
  - 2.4.2 Example 2.A Identification of Failure Origin 28
  - 2.4.3 Initiation Site of a Fracture 29
  - 2.4.4 Example 2.B Detection of Previous In-Service Damage 32
- 2.5 Failure of Threaded and Rotating Elements 33
  - 2.5.1 Example 2.C Failure of a Bolted Structure 36
- 2.6 Extraction and Storage of Samples 39
- 2.7 Inspection by NDT Techniques 42
- 2.8 Organization of Work Teams 48
- References 51

3 Tools for the Microscopic Analysis of a Mechanical Failure 53
3.1 Microstructural and Metallographic Characterizations of Metallic Parts 53
3.2 Cutting and Preparation of Samples 54
3.3 The Polycrystalline Nature of Metals 58
3.4 Microscopic Examination of the Structure of Metals 60
3.4.1 Example 3.A Metallographic Analysis of a Welded Pipe 65
3.5 Microscopic Examination of Fracture Surfaces 66
3.5.1 Example 3.B Fractographic Analysis of a Thick-Walled Tube 68
3.6 Identification of Subcritical Growth Surfaces 71
3.6.1 Example 3.C Failure of a Flexible Connection 73
3.7 High Magnification Tools, Electron Microscopy 75
3.7.1 Example 3.D SEM Analysis 76
3.8 Extra-High Definition Fractographic Analyses, Nano Devices 80
References 83

4 Mechanisms of Damage and Failure 85
4.1 Introduction 85
4.2 Failure Mechanisms 87
4.3 Mechanisms of In-service Damage 89
4.4 Corrosion 90
4.4.1 Example 4.A Corrosion in Exchanger Tubes 96
4.5 Propagation of Cracks, Fatigue 100
4.5.1 Example 4.B Fatigue Failure in a Pipeline 100
4.6 Hydrogen Damage 104
4.7 Damage by Prolonged Exposure to High Temperature 106
4.7.1 Example 4.C Fire in a Hydrocarbon Furnace 107
4.8 Environment Assisted Cracking 111
4.8.1 Example 4.D Cracks in a Heat Exchanger Tube Plate 114
4.9 Discontinuities or Defects Introduced During Manufacture 116
References 120

5 Damage Resistance Tests of Materials 121
5.1 Tensile Testing 122
5.1.1 Example 5.A A Stainless Steel Tube at High Temperature 125
5.2 Ductile–Brittle Transition Temperature 126
5.3 Experimental Determination of Fracture Toughness 129
5.4 Spectrometry Analysis and Hardness Tests 132

5.5 Experimental Measurement of Residual Stresses 135
5.6 Experimental Determination of the Resistance to In-service Damage 137
5.7 Testing of Fatigue Life 138
5.8 Techniques for Corrosion Monitoring 140
5.9 Estimation of SCC Propagation Rates and Threshold Stress 142
5.10 Creep Resistance and Monitoring of In-service Damage 144
References 148

6 Modeling Tools Applied to the Analysis of Mechanical Failures 149
6.1 Introduction 149
6.2 Numerical Modeling Tools 152
6.2.1 Example 6.A Modeling of Soil-Pipeline Interaction 155
6.3 Criteria for Modeling Pressure Components 158
6.3.1 Example 6.B Stress Analysis of Damper Vessel 161
6.4 Stress Analysis of Cracked Components 162
6.5 Calculation of the Load Required for Brittle Fracture 164
6.6 Calculation of the Conditions for Ductile Fracture 168
6.6.1 Example 6.C Calculation of Applied Stress at a Pre-existing Defect 170
6.7 Mechanical Modeling of Longitudinal Cracks in Pipes 172
6.8 Mechanical Modeling of Fatigue Crack Propagation 174
6.8.1 Example 6.D Modeling of Fatigue Crack Growth in Compressor 177
6.9 Determination of Fracture Toughness in Post-failure Samples 178
References 180

7 Root Cause Analyses 183
7.1 Introduction 183
7.2 The Explosion of the Challenger Space Shuttle 184
7.3 Methodology for Root Cause Analysis 187
7.3.1 Example 7.A Is a Blown Fuse a Failure? 189
7.4 The Cause–Effect Tree 189
7.5 Data Collection: Interviews and Statements 191
7.5.1 Example 7.B Causal Tree in a “Hot Tap” Repair . . . 193
7.6 Data Collection: Documents and Records 198
7.7 Rebuttal, Iteration, Coincidence, and Causality 199
7.7.1 Example 7.C RCA of Failure in Directional River Crossing 202
7.8 Levels in Failure Analyses 209

7.9 Alternative Formats for a Failure Investigation 212
7.9.1 Example 7.D RCA of Fire in a Heater at a Petrochemical Plant 214
References 217

8 Damage and Failure Mechanisms in Machinery 219
8.1 Definition of Failure in Machinery 219
8.2 Modes of Failure in Shafts 220
8.2.1 Example 8.A Failure of Shaft in a Hydraulic Pump 220
8.3 Failures of Bearings 221
8.3.1 Example 8.B Failure of Bearing in A.C. Generator 225
8.4 Failure in Sliding or Plain Bearings 227
8.4.1 Example 8.C Failure of Sliding Bearing in a Large Gas Engine 228
8.5 Failure of Transmission Elements: Gears and Couplings 230
8.5.1 Gears 230
8.5.2 Couplings 233
8.5.3 Example 8.D Failure of a Flexible Coupling 234
8.6 Failure of Fasteners, Bolts, and Other Threaded Elements . . . 237
8.6.1 Example 8.E Failure of an Allen Type Fastening Bolt in a Compressor Piston Head 241
8.7 Characteristic Failures in Turbo Machines 244
8.7.1 Steam Turbines 244
8.7.2 Example 8.F Failure in the Rotor of a Steam Turbine 246
8.7.3 Gas Turbines 247
References 250

9 Failure Mitigation and Extension of Service Life 251
9.1 Introduction and Historical Perspective 251
9.2 Failure Mode and Effect Assessment 253
9.3 Procedures for Assessing Fitness for Service 255
9.4 Evaluation of General and Local Loss of Thickness 257
9.5 Evaluation of Geometrical Discontinuities and Cracks 259
9.6 The Leak Before Break Criterion for Pressurized Components 264
9.7 Monitoring Damage and Stresses During Service 264
9.8 Strength Tests Using Instrumented Indentation 270
9.9 Monitoring of Machines 271
9.9.1 Vibration Analysis 273
9.9.2 Lubricant Analyses 274
9.9.3 Acoustic Emission Analysis 275
9.9.4 Analysis by Infrared Thermography 276

9.10 Estimation of Residual Life Against Damage Mechanisms 276
9.10.1 Example 9.A Fatigue Life Prediction in a Pipeline. . . . 278
References 281

10 Consequences of a Failure Analysis 283
10.1 Writing Reports After a Failure Analysis. . . . 283
10.2 Failure Analysis in Potentially Litigious Cases 285
10.3 Learning or not Learning from Accidents, Here the Question. . . . 288
10.4 On Human Error. . . . 291
10.5 Human Errors at Different Levels, the Chernobyl Case 293
10.5.1 The Worst Nuclear Disaster 293
10.5.2 Levels in the Chain of Human Errors 295
10.6 Business Impositions and Political Decision-Making. . . . 297
10.6.1 Example 10.A Incident on a Water Pumping Plant 298
10.7 The Rules are Relaxed: The Case of the Macondo Oil Well 301
10.8 Underestimating Consequences of Natural Events, the Fukushima Event. . . . 305
10.9 Management of Old Plants and Equipment 307
10.10 Causes and Consequences of the Most Famous Accident. . . . 310
References 313

About the Author

Dr. José Luis Otegui graduated in 1980 as the Mechanical Engineer UNMdP, Argentina. In 1988 he was awarded the academic degree of Doctor of Philosophy in Mechanical Engineering at the University of Waterloo, Canada. Since then he has been dedicated to the development of technologies in the area of mechanical modeling of metallic materials for structural use, and their application for the prediction of damage in industrial equipment. His articles are regularly published in scientific journals and conferences related to forensic engineering, pressure vessels, fatigue and fracture. He is currently Full Professor at the Faculty of Engineering of the University of Mar del Plata, and Researcher at CONICET (National Counsel for Scientific and Technological Research of Argentina).

Chapter 1
Introduction

1.1 Historical Perspective

Technology has always been a risky business, but quantifying that risk is relatively a new phenomenon in the world of engineering and management. No matter what the technology, our best estimates of their success tend to be too optimistic. The history of mechanical constructions dates back to engine. The most successful in this field were the Romans. Roads and aqueducts developed by them were one of the pillars of strengthening their hegemony in the "known world" of the time. One of their most notable developments is the arch bridge, see Fig. 1.1a. It is based on transferring vertical loads due to vehicle weight by compression between the stones (Fig. 1.1b) to the embedding in pillar bases. Some historians take it as true that one of the keys to the success of this design was based on a practice, normal then as it is now: the pre-operational load test. Once the bridge was built, and before its entry into operation, a test was carried out with a dead weight much larger than the maximum expected in service. The peculiarity was that, in such event, the responsible for its design and construction should remain under the bridge… This system would ensure, on the one hand, a careful design and construction, and on the other, avoidance of repeating mistakes by incompetent engineers.

The need to reduce costs and longer spans between supports, and the availability of new materials and construction methods led to new designs, now based in tensile-loaded members and not only to compressive load transfer, see for example Fig. 1.2. This type of steel structure, which we call truss, was alongside the steam engine, one of the pillars of the European industrial revolution from the late eighteenth century. By then, Sir Isaac Newton had already laid the foundations of mechanics by introducing previous calculation as an auxiliary tool to the "genius," and reducing the importance of trial and error in the evolution of the designs.

Designs circumstantially exceeded capabilities of load calculation and prediction of material properties. In the period of greatest growth of the industrial revolution, there was no single week without a train accident attributable to defects in

J. L. Otegui, *Failure Analysis*, DOI: 10.1007/978-3-319-03910-7_1,

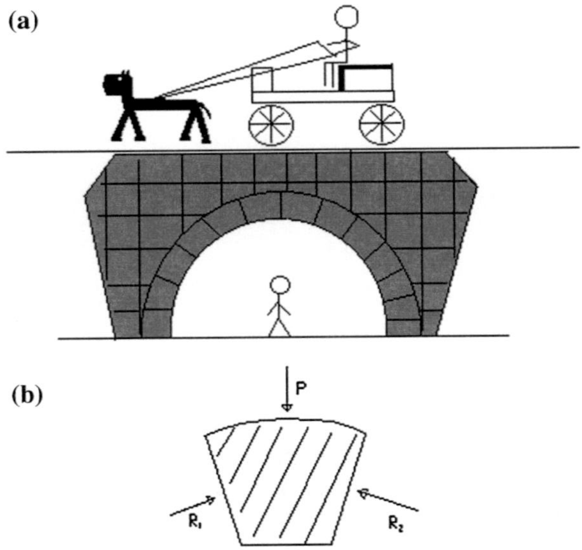

Fig. 1.1 **a** The arch bridge from Roman times, **b** load transfer by compression

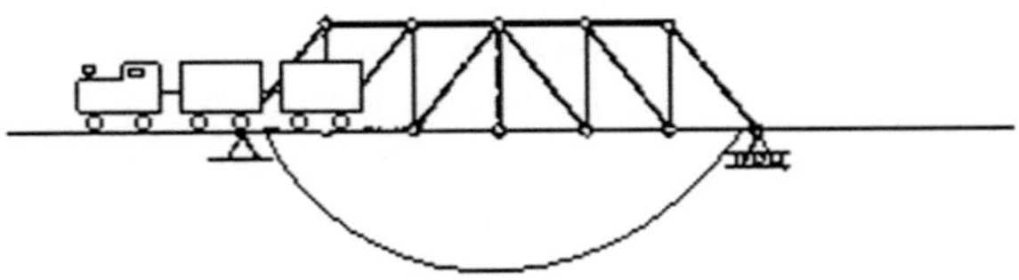

Fig. 1.2 The steel truss bridge and the industrial revolution

design or materials reported by British newspapers. The search for increasingly efficient structures generated the development of increasingly powerful design methods and materials with improved strength and reduced weight and cost.

1.2 Current Technology is Conditioned by our History

One of the most resounding failures, and thus more useful in the field of failure analysis, was the loss of the space shuttle Challenger. Failure analysis revealed not only deficiencies in design, but also serious organizational shortcomings. Understanding these shortcomings led, among other things, to the development of fault tree techniques (events, previous defects, and overcome barriers) for the determination of root causes. We will return to these issues in Chap. 7.

The American Space Shuttle used two SRB (Solid Rocket Booster) tanks that were manufactured by Thiokol in Utah. These long, thin tanks had to be built into sections. As discussed in some detail in Chap. 7, the failure of a joint was responsible for the Challenger disaster. Engineers perhaps would have preferred sections of different dimensions, but had size limitations due to railway tunnels through which the tanks would be transported, and they designed them in accordance to railway dimensions.

The gauge of railroads in the U.S. is 4 feet, 8.5 inches. Why did they use that size? Because it was the width of British railways and, as American trains were built by the British, this measure was taken as a matter of compatibility. Why did the British use this measure? Because British companies that built cars were the same who built the wagons before the train came up and used the same elements used to manufacture floats and wagons. Why did floats have that wheel span (4 feet, 8.5 inches)? Because the distance between the wheels should fit the old streets of Europe, which had exactly that width. And why did the streets measure that way? Because these streets were opened by the Roman Empire during its conquests, and were based on the ancient Roman chariot. Why did the Romans have chariots of that width? Because they accommodated the buttocks of two horses side by side.

It can be concluded that the Challenger disaster, one of the most shocking failures, affecting one of the most advanced machines of global engineering design and technology was greatly influenced by the size of the buttocks of Roman horses. This is just an example, but can be extended to almost all our technologies.

Let us see another example, a common accident defined as the "curse" that makes a toast that falls off the table impact the floor "almost always" with the jam (or shortening) down, the worst-case scenario. The impact that this issue has had on the internet is remarkable. There are two most frequently proposed causes:

- The evil of inanimate things (Murphy's Law): this cause, also called fate, acts of God, etc., will be discussed elsewhere in this book (do not bet that this cause is real in many cases!)
- The weight of the layer of butter or jam is larger than that of the underlying bread (here we see a serious attempt to physical explanation).

Does this topic have any engineering explanation? Sure! But it is not the weight of the butter layer, as might be expected. Experiments have been done, and verified that "face down" is indeed the most likely way for a toast to fall. The phenomenon has been analyzed with engineering tools; the mechanics of the toast path has been modeled. Apparently, due to the mass and moment of inertia of the toast (that is, its resistance to move in linear and angular paths) and conditions of friction with the table, it is seen that during the time the toast takes to fall off the table into the floor, the toast tends to give an average half turn. This is determined by the height of the table, if higher or lower, the path of the falling toast would be different. But the height of the table is the result of the evolution of the human species, its status, and its height (Fig. 1.3).

Dr. Matthews showed that the phenomenon of the falling toast has deeper implications than might be expected. If tables were 3 m high, the problem would disappear because the toast would have enough time to complete a full rotation. Why are tables as high as they are? For human convenience. Why are humans the

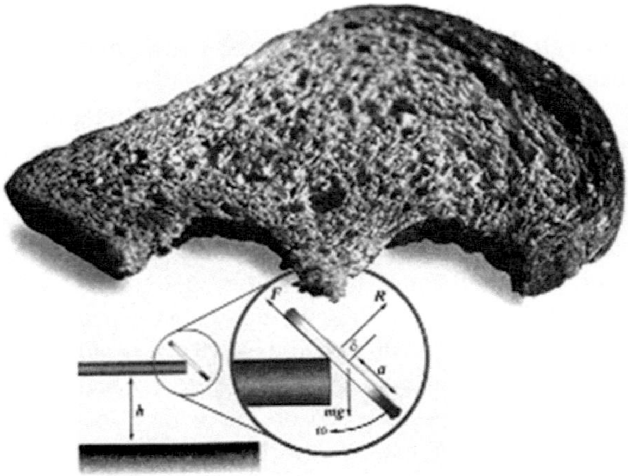

Fig. 1.3 Physics of the falling toast

height they are? Using a simple chemical bonding model of the human constitution, Dr. Matthews showed that there is a limit of about 3 m for the safe height of a cylindrical bipedal such as humans. Above that height, a simple fall would cause sufficient kinetic energy to break the bones of the skull. This limit sets a maximum of 1.5 m high tables, which is not sufficient to prevent the butter side falling toast.

The thesis of Dr. Matthews argues that the toast falls butter side down because the universe is as it is. The formula that gives the maximum height of humans contains three so-called "fundamental constants of the universe." The fine structure constant determines the strength of the chemical bonds in the skull, the universal gravitational constant determines the strength of gravity, and the Bohr radius indicates the size of the atoms that make up the body. The values of these three constants were fixed in the design of our universe moments after the Big Bang.

But the height of the table and the fall of a toast are also consequences of our cultural evolution. In cultures where it is customary not to use chairs, but eating in a squatting position, tables are lower and this problem would probably not occur as frequently. If everyone did the same, the problem of jam stuck to the floor would not exist, as there are many other problems that we have not heard of… yet. No doubt that as we move forward in building new technologies, new failures will appear, which we will have to learn to not repeat them.

1.3 The Danger of Myths and the Public

In any project, large or small, it is expected that each engineer's work is consistent and transparent so that another engineer can review his hypotheses following inadvertent errors, logic, and computation. This is the epitome of team play, and is the giving and receiving of concepts and calculations among engineers working on a project making it a success. Of course, occasionally there are lapses of logic, mistakes that result in faulty design, which may or may not lead to immediate failure. If the project is a building, a poorly designed beam or a column could show up during construction, under a well-trained eye from the field engineer. Then the design could be sent back to the drawing board, where the error can be solved. Unfortunately, not all errors are detected, either in the design office or in the construction site, and those that are not detected can eventually lead to failures.

Long-term success, either in a space shuttle program or in the design and construction of simpler components, tends to lead to change or complacency, and both can lead to a failure. As an engineer, "every success sows the seeds of failure and the success gives you great confidence". When we are confident and secure, satisfied that we did everything correctly because there have been no failures, we also tend to be inattentive and careless.

The general reaction of people in front of the occurrence of a negative event, such as a breakdown or accident, is seeking an explanation (the simpler, the better) that allows rationalizing the loss. There are different theories about the origin of Murphy's Law and the details of how it was originally made. Between 1947 and 1949, the USA at Edwards Air Force Base developed a plan called MX98, designed to test human resistance to G-forces during rapid deceleration. Initial tests used a rocket on rails with a set of end brakes and a humanoid doll tied to a chair in the sled. The following tests were made with a person, Captain Stapp. Edward Murphy proposed using electric strain gauges attached to Stapp's harness to measure the force exerted on them by the rapid deceleration. Murphy's assistant cabled the harness and made a first test using a chimpanzee. Sensors gave a reading of zero. Then he realized that there was a mistake in the installation: each sensor was wired backwards. Then Murphy made his famous statement. According to George Nichols, another engineer who was present, Murphy, frustrated, blamed his assistant, saying: "If that person has a way of making a mistake, he will." The "Murphy's Law" was later condensed to "If it can happen, it will happen" by other team members. Other versions claim that the original sentence by Murphy was something like "If there is more than one way to do a job and one of them will end in disaster, someone will do it that way."

The phrase came to light when Stapp was asked why no one was injured during the rocket testing. Stapp replied that it was because they took into account Murphy's Law. He cited the law and translated it into that it was important to consider all the possibilities before making a test. In 1952, the phrase was changed to "Everything that can go wrong, will" in a section of the book "The Butcher: The Ascent of Yerupajá" by John Sack, and in the 1955 book by Lloyd Mallan: "Men, Rockets and Space Rats." The current quotation for the law is "What can go wrong will go wrong," that was never uttered by Edward Murphy. This variant, often known as Finagle's law captures the general tendency to emphasize the negative things that happen in life.

The explanation that the weight of the jam layer on the bread is what justifies its face-down fall to the floor brings us to the discussion of the MYTH. This is one of the most contentious issues after the discovery of a breakdown or accident: for any unfortunate event, there is an explanation that is simple, clear, logical… and false. What leads us to the first law concerning preliminary reports from an expert failure assessment: the first version of the causes, particularly if they involve people to blame, will be indelible in the minds of general public. Any opposite result, arising from engineering analysis, no matter how it is later published, will not replace the initial idea.

1.4 Defects in Components, Ductile and Brittle Materials

One factor in the number of unexpected failures in the nineteenth century was the increase in the use of metals instead of the previously used building materials (wood, brick, stone, etc.), for which design patterns, service history, and maintenance procedures were well known. The use of metal (most notably iron and steel) in construction rose from only 20 % at the beginning of the industrial revolution, to about 80 % in the early twentieth century. This served as an engine for overall development of engineering science, with the building of large structures, pressure vessels, and boilers with increasing pressures and temperatures. It also led to the development of materials science, the emergence of new materials, better steel production technologies, etc. New subjects appeared, such as metallography, new materials, nondestructive testing, preoperational testing, etc.

In the sixteenth century, Leonardo Da Vinci did the first historical observation related to what we call today "fracture mechanics." He noted that the longer the rope, the lower its strength. He called it "the inverse proportionality of rope strength and length."

Galileo (1638) rejected the proposed size effect by Leonardo and speculated on the effect of size on large animal bones, calling its bulkiness the "weakness of the giants." Half a century later, Mariotte (1686) contributed, based on extensive experiments, with a very important idea. He noted that "a long rope and a short rope support the same weight, but it is more likely to have a long rope with a defective section that will break early." He suggested (linguistic model) that the greater the length, the greater likelihood of larger defects in the material, which weaken the rope (no rope is stronger than its weakest section). We now know the influence of defects. At present, it is possible to "see" and "measure" defects by Non Destructive Test (NDT) techniques (X-rays, ultrasonics, magnetic particles, dye penetrant). Fracture mechanics allows us to determine how these defects affect the strength of the component (ropes, pipes, vessels, structural elements, etc.). At that time, there were no tools to validate the influence of these small defects, the same way microorganisms could not be even imagined.

Subsequently, not much occurred until Griffith (1921) demonstrated experimentally that the mechanical resistance of glass fibers increased from 42,300 to 491,000 psi when reducing the diameter from 0.0042 to 0.00013." He noted that "the weakness of isotropic solids… is attributable to the presence of discontinuities or faults." The issue was not mentioned in the impressive "Strength of

Materials" text by Timoshenko in 1953. The classical theories of elasticity with an endurance limit show no size effect.

England and Germany had key roles during the second industrial revolution in the nineteenth century. An important part of the research efforts in these countries was devoted to avoid, reduce and control unexpected failures. In 1836, the French author Stendhal mentioned a serious problem related to fatigue damage in his novel "Memoires d'un Touriste." The mysterious breakdowns increased between 1840 and 1860. Accidents were catastrophic, without warning. We now know those failures were due to brittle fracture, i.e., failures were not preceded by plastic deformation which could give a perceptible warning. The novel "No Highway" by Nevil Shute was the first fiction depicting fatigue failure of an aircraft (1948).

The situation was serious when it was noticed that machine components could fail at stress levels well below the critical fracture stress. This required the presence of cyclical fluctuations of the load. It was observed (although the reason was not understood until later) that a crack would initiate in certain locations and start to grow, first slowly and then more rapidly, until finally breaking the component apart, often in a brittle mode. A new type of failure, which is now called fatigue, was identified. As fatigue was recognized as the growth of a crack across a component, new tests were developed to evaluate this form of degradation in samples with various types of grooves.

1.5 The Industrial Revolution and Failures in Pressure Components

The industrial revolution was characterized by steam generation. In those days, the designers and builders of boilers had only little prior knowledge as there were no design and construction codes to guide them in their efforts to manufacture a steam boiler that could operate safely. Human knowledge about these systems was inadequate and this was demonstrated by the many boiler explosions occurred.

April 28, 1865, at the end of the American Civil War, 2,021 Union prisoners of war were released from Confederate prison camps in Vicksburg, Mississippi. The transport that carried them home, a steamboat called "Sultana," sailed down the Mississippi River when suffered an explosion due to a failure in the boilers, 7 miles north of Memphis. The boat was completely destroyed and 1,547 passengers died. In 1894, there was another spectacular explosion and fire. Twenty-seven boilers exploded in quick succession in a coal mine near Shamokin, Pennsylvania, destroying the place and claiming the lives of six people (Fig. 1.4).

Boiler explosions continued going on. In a period of 10 years, from 1895 to 1905, there were 3,612 boiler explosions, with an average of one per day. The loss of life was enormous; about 7,600 people were victims of these tragedies. In 1905, the R. B. Grover Shoe Company Plant in Brockton, Massachusetts, was destroyed leaving as a result 58 deaths and 117 injured. A year later in Lynn, Massachusetts,

Fig. 1.4 "SULTANA" steamboat explosion, 28 April, 1865

Fig. 1.5 Grover Shoe Company Plant

another night explosion in a factory boiler injured three people and caused a loss of \$500.000 (Figs. 1.5 and 1.6).

The problem could be clearly defined: steam boilers, although considered valuable power sources, were not safe. There should be an engineering solution to protect society and prevent disastrous explosions. A Uniform Inspection Code was published in the USA in 1884, which is considered the first standard issued by the American Society of Mechanical Engineers (ASME). In August, 1907, the first effective legislation on boiler design in the United States was established by the Board of Boiler Rules in the state of Massachusetts. In the following 4 years, other states and cities enacted similar legislation (Fig. 1.7).

The first ASME Boiler Code was issued on February 13, 1915. Six additional sections were added the following 11 years. The first rules for pressure vessels were released in 1925. This publication was entitled Section VIII "Rules for the Construction of Unfired Pressure Vessels" (Fig. 1.8).

Fig. 1.6 Boiler explosion, Grover Shoe Company Plant

Fig. 1.7 Remains of Grover Shoe Company boiler (found hundreds of meters away from the plant)

1.6 The Advent of Fracture Mechanics

Fracture mechanics is a discipline that relates material toughness, flaw size, and stress level. A Brittle Fracture is a process associated with the collapse of a structure or component involving very little plastic deformation, leading to catastrophic failure. Ductile fracture is, on the other hand, a process of collapse of a component or structure that is associated with excessive plastic deformation.

The evolution of tests related to fracture mechanics can be divided into four periods. The first is related to the early developments up to the time of standardized testing procedures, as will be seen in Chap. 3. The second was the scene of the brittle fracture, a period beginning in the 1950s. With the analysis of the

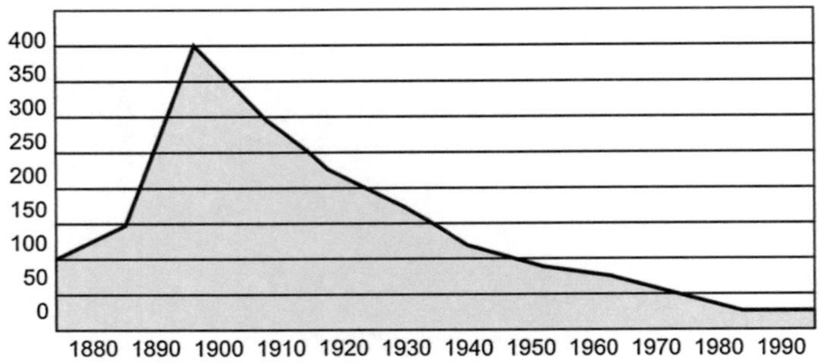

Fig. 1.8 Number of explosions in the USA 1880–1990

fractures occurred in the Liberty ships during World War II and other similar cases, the concept of transition temperature was introduced, characteristic of this period. The fourth period began in the early 1980s, characterized by the development of fracture mechanics as we know it today, including correlations between absorbed energy in Charpy tests and other fracture mechanics tests.

The case of the Liberty Series ships is very interesting. At the beginning of the Second World War (1939–1945), the German army had a strong technological supremacy, which was largely counteracted by their U.S. rivals with mass production of large numbers of machines affected to war. German submarines were decimating the American freighter fleet; many were sunk even before leaving the Gulf of Mexico. Freighters of the type called Liberty were built in large numbers. Different sections were built inside the country, then ships were assembled in ports and sea launched. Entirely welded hulls were used for the first time. Some split open after launching into the water. Figure 1.9 shows the hull of one of these ships, seen from the pier, with a circumferential fracture breaking the hull in its middle. The hull is resting on the bottom bow and stern, with its central area elevated.

1.7 Scope of the Failure Analysis

According to the objective of the company operating the failed equipment, a subsequent failure investigation may include different levels of analysis and different purposes. Here are the most common ways to address the problem:

Condition Report

In this study, all abnormal conditions in the part or component that may have been caused by the main event, or may have contributed to its occurrence, are reported. The report includes a detailed survey for the condition of all parts of interest, documented with high-resolution images that can be used for further investigative purposes. All damages, malfunctions or nonconformities are identified, typifying each and identifying their most likely causes. NDT can be performed in situ.

Fig. 1.9 The broken hull of a Liberty ship seen from the pier

Failure analysis

At this level of analysis, the investigation focuses in identifying the immediate cause or causes of the failure. All causes of failure had been said to have a single source: human error. However, in terms relevant to an operating unit, it is common to identify three types of causes that require different actions:

- Physical causes: problems in fluid composition, defects in materials, extreme environmental conditions, etc.
- Human causes: improper installation, design errors, errors in the implementation of procedures, etc. They are related to staff training and adequacy of facilities.
- Latent causes: lack of procedures, lack of training. These are conditions that increase the probability or consequences of an eventual failure.

Each of these causes requires a particular methodology for analysis. The tools used for the analysis of the physical causes will be addressed in Chaps. 2, 3, 4 and 5. The latent and human causes are part of broader aspects, and will be addressed in the final chapters of this book.

A failure analysis of mechanical elements typically includes:

1. The evaluation of the photographs and other records and recollection of field data.
2. Defining fractographic and metallographic characteristics of the failure:
    - Surface and dimensional characterizations
    - Microstructural characterizations.
3. Chemical and mechanical characterization of materials.
4. Mechanical modeling of the formation of the initial damage and the final break.
5. Engineering critical analysis
6. Specific meetings of experts to discuss results.

Root Cause Analysis
Due to its scope and specificity, this methodology, often called RCA, deserves a more complete discussion and therefore will be treated in depth in Chap. 7. In short, a typical study includes:

1. Completing the cause and effect (causal) tree.
2. Collection and review of background and field data.
3. Definition of fractographic and metallographic characteristics of the failure.
4. Experimental tests for characterization of materials and preexisting defects (mechanical and chemical characterizations).
5. Mechanical modeling of the formation of the initial damage and the final break.
6. Engineering Critical Assessment (ECA) and Definition of Root Causes: In an iterative process, all alternatives in the causal tree are verified or eliminated through investigation and experimental data. The assessment concludes with the description of the causes of the failure, identifying the events, conditions and exceeded barriers that really influenced its occurrence. The findings are discussed with the technical manager in charge of the project.
7. Specific meetings, which are usually defined as brainstorming.

Expert Analysis
This branch of activity, also called forensic engineering, is significant when it is carried out in the context of actual or potential litigation. Unlike the RCA, which seeks to find all causes, in this case the study aims to find all legal and technical backgrounds that favor the interests of one or more of the parties involved in a harmful incident. This analysis shares most of the techniques with the other levels, but experts must interact with other agents: the legal department, insurance companies involved in the incident, insurance adjusters, regulatory or legal state agencies, etc. The expert analysis usually involves:

1. Evaluation of photographs and other operating history and field data collection.
2. Defining fractographic and metallographic characteristics of the failure.
3. Chemical and mechanical characterization of materials.
4. Mechanical modeling of the formation of the initial damage and the final break.
5. Engineering critical assessment
6. Advice to managerial, legal, and technical staff in relevant negotiations.

As we will see at the end of the book, any failure analysis carries a high degree of confidentiality, but this is particularly important for forensic analysis in potentially litigious contexts. A particular branch of forensic analysis involves cases where the definition of responsibilities, and particularly the economic costs of a failure, involves legal trials. In these cases, it is common to find experts who handle legal and confrontational aspects, interact with media and orient public opinion, etc., in what is known as "expert witnessing." Most of the skills and techniques involved in this discipline are beyond the scope of this book.

1.7.1 Example 1.A Expert Analysis After a Traffic Accident

The expert or forensic analysis described in this example was carried out after a collision occurred between a truck and a bus in a two lane road, which caused 15 deaths among bus passengers. The client in this case was the insurance company of the truck. In a previous analysis by the local police of the road conditions, performed immediately after the crash, evidence emerged proving that the accident was caused by a mechanical failure in the truck, and that the crash had occurred in the lane corresponding to the bus.

The truck involved is of the type tractor and trailer. The tractor (or chassis) includes the engine, cabin, and a small cargo box, while the trailer has a front steering axe and two back axes, and is pushed through a lance or "A-frame." This lance is connected to a mooring hoop in the tractor, by a buttonhole bolted to the lance. Control over the trailer can be difficult for the driver, especially when, as is usually the case, the weight of the trailer is much greater than the weight of the tractor.

Several marks on the pavement in the bus lane (right in Fig. 1.A1) revealed interesting aspects of the accident. These marks correspond fundamentally to the lance of the trailer and to diverse parts of the chassis of both vehicles during the collision. Two areas are observed: 1–5 mm deep semicircular scratches, that begin in the truck's circulation lane, are due to the haulage of the lance onto the pavement (see insert to the left in Fig. 1.A1). Figure 1.A2 shows a sequence of three steps in the truck path to collision, defined after the analysis.

The photographic composition in Fig. 1.A3 summarizes the results of the fractographic investigation. Figure 1.A3 (a) shows the broken buttonhole and bolt assembly. The eye is connected to the mooring ring on the tractor. On the left, the bolt is seen broken in its thread (b) and corresponding nut (c), found in the field adjacent to the road, about 10 yards from the bridge shown in Fig. 1.A2.

The details and results of the fractographic, metallographic, and mechanical studies and numerical modeling of load transfer and crack propagation are beyond the scope of this example. After the break, the trailer remained hooked to the tractor by safety chains, but started to deviate to the left, invading the opposite lane (Fig. 1.A2). If the truck could have continued its way a few more feet, the tubes of the trailer air brakes would have disengaged, and the trailer would have stopped.

The accident was due to the failure of the bolt assembly, which in turn was due to the propagation of a fatigue crack (see arrow in Fig. 1.A3c). Evidence of premature wear was found in the bolt (Fig. 1.A3d) due to friction against the walls of the housing at the lance. The bolt was subjected to fast degradation due to wear and cyclic loads, augmented by excessive clearances in the load transfer train between tractor and trailer, after only 1 year of use.

The major responsibility of the accident is referred to a mechanical fault in the truck. However, expert analysis gave the client some weapons to defend his position. First, it had been subjected to mandatory inspections few months earlier. Inspection methods were not sufficient to detect any wear or damage, which allows

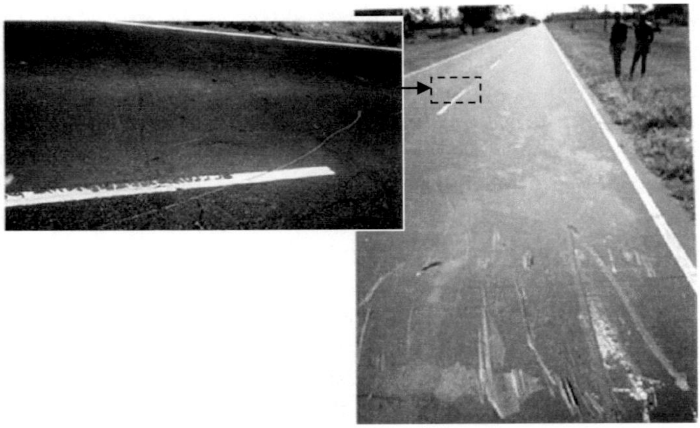

Fig. 1.A1 Marks in the road

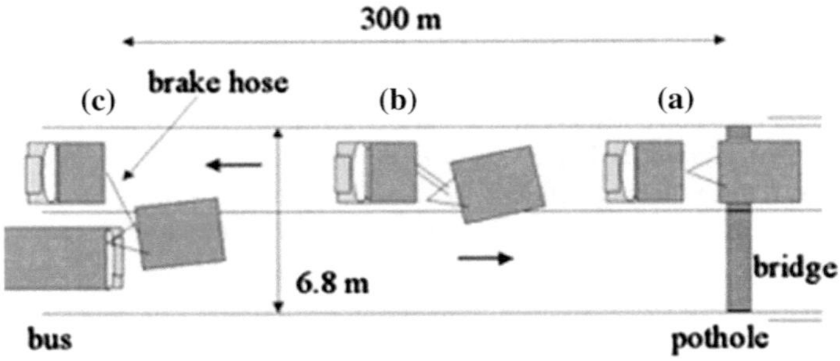

Fig. 1.A2 Schematic path of tractor and trailer up to crash with bus: **a** Bolt failure, **b** Trailer pushed by chains, **c** Impact

defining them as a cause of the incident. Another factor that contributed to the accident is related to inadequate road conditions. Figure 1.A1 indicates that at the bridge exit (first event) there was a hole or bump in the road. There was a mismatch between the structure of the bridge and the asphalt coating of the road. This was the strong point for the defense of the client; the position in the bank where the bolt was found confirms its breakage occurred there.

Regardless of previous damage in the bolt assembly, high impact loads caused by the pothole caused the instantaneous fracture of the bolt. At that time, the route was under concession to a private company under a tolling arrangement. It was incumbent upon this company to maintain the road. A week after the accident, the pothole at the exit of the bridge was repaired. The outcome of this litigation is beyond the scope of this example.

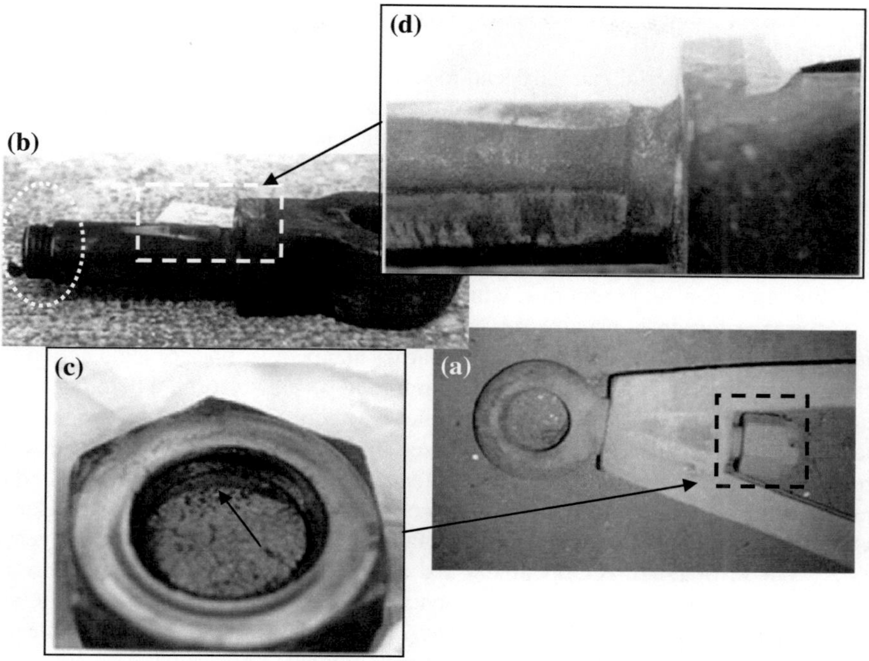

Fig. 1.A3 Fractographic analysis: **a** Bolt assembly, **b** Failed bolt, **c** Nut, **d** Friction marks

1.8 Concluding Remarks

To establish proper communication between legal and engineering parts, it is worth clarifying the meaning of some terms. By "discontinuity" we mean an imperfection (such as those introduced during welding). If the discontinuity exceeds the limits defined in quality requirements (code, standard or agreed specification) it is called a defect and must be removed or repaired. A defect raises the probability of occurrence of a failure, and failure is the occurrence of an event that prevents further normal operation. Thus, for example, a slag inclusion trapped in a weld is a discontinuity. If it exceeds a certain size, defined by quality standards, it becomes a defect.

All cracks are usually considered defects and rejected. Not so with other discontinuities, such as porosity and slag inclusions in welds, distortions in bars or plates, etc. The term "deviation" should be understood as anything that deviates from the specification indicated in Manufacturing. Normally, defects are a not sure cause for failure. The likelihood of failure increases with the size of the defects and with the loads or pressures applied to the material during operation. When a defect is very important in size or is associated with other defects, then failure may occur. Defects mostly generate a reduction of the section sustaining the load, but when they are sharp (planar or crack-like) defects also generate what is called a stress concentration.

Defects in a material can only lead to failure if they are subject to high enough stresses. These stresses may come from thermal distortions during fabrication and welding (called residual stress) but are also generated by the operating conditions of the equipment (due to loads or pressure). Mechanical stresses are concentrated at the apex of acute notches or cracks. In ductile materials, defect propagation usually generates a localized plastic collapse, which in the case of pressurized components would lead to leaks. In the case of brittle materials, fracture can occur quickly, and span large sections, which usually generates much larger losses.

A service failure is usually very detrimental to the operator, as costs include repair plus lost income, damage to third parties and people, fines or legal action for breach to customers, loss of data etc. Therefore, tests are performed before starting normal operation of a new or repaired piece of equipment. Pre-operational load tests are a necessity because it is recognized that manufacturing processes or weld repairs are not perfect, and often leave defects that can sometimes go undetected in nondestructive testing. If the defects do not generate pre-operational test failures, experience suggests that there will be no failures in service for a number of years.

A Code or Standard is a document that sets mandatory compliance rules in the country of origin. Contractual requirements may state that the parties agree to the use of standards from other countries. Thus, the ASME Boiler and Pressure Vessel Code is a U.S. standard that is commonly used in many countries, especially in the Americas. However, this rule applies only to manufacturing new equipment. Standard procedures are not mandatory when repairing old equipment and, in many cases, some of its provisions do not apply to repairs.

For example, when repairing old technology welding materials, which may also be previously degraded by use, post weld heat treatment is not always possible. These repairs require welding materials and procedures different from those used in the original manufacture. The repair is often performed on the assembled equipment, with which problems of access and residual stresses become added difficulties. For example, when welding the mid section of a tube with free ends, the tube will be free to shrink to accommodate thermal contraction during cooling of the weld. But, if tube ends are already welded to fixed parts, the tube cannot contract and then high axial tensile stresses develop in the tube (and the weld).

A common failure is the result of an “unfortunate” combination of defects and errors. For example, leaks in a boiler tube are found to be produced by the association of undercuts in the inner surface of the tube wall, combined with slag inclusions near the outer surface. This joining of previous defects produced a marked reduction of section and a high stress concentration. Applied axial tensile stress in the tube is due to the sum of the residual stress introduced during girth welding the opposite end of the tube and that due to operation (internal pressure).

For those wishing to understand the physical aspects of the problem of strength and failure of materials, and have no previous scientific knowledge, some other titles are easy to read:

- "The New Science of Strong Materials or Why You Do not Fall Through the Floor" James Edward Gordon. ISBN 0691023808. This is a very readable treatise on materials science that highlights the strength of the chemical and physical bonds, crystalline structure, and cracks. The author presents physical and chemical problems, and how they were solved. He also shows that the strength of materials is influenced by their shape and microstructure, as well as environment and applied loads.
- "Structures: Or Why Things Do not Fall Down," also by Professor Gordon. This book covers the same material but in greater depth and with more illustrations.
- "Why Things Break: Understanding the World by the Way it Comes Apart," Mark E. Eberhart. ISBN 1400048834. "Why objects break" versus "When objects are broken" defines the author's practical interest in addressing the issue of fracture design. The author is interested in the reasons of fracture, which is ultimately a property of interatomic bonds. Mr. Eberhart begins by defining fracture as the first technology used by man: the breaking of rocks to form useful tools. This technology was refined by trial and error for about a million years to the first age of metals; then the use of metals also advanced by trial and error through copper, bronze, iron and steel.

The development of topics related to failure analysis and prevention in pressurized components can be found in a previous publication by the author: "Piping and Pressure Vessels" (ISBN 978-987-1371-18-1), particularly in Volume 2: "Damage in Service."

Many companies are reluctant to publish reports of accidents. This attitude can be understood, but Kletz (1991) argues that the publication of information that can prevent other accidents is a moral obligation. We recommend subscription to "Engineering Failure Analysis" (Elsevier Sc, ISSN 1350-6307). This journal specializes in the resolution of practical and innovative developments in failure analysis techniques; its technical articles mostly relate to failures of structural elements.

Some international organizations offer online expertise. ASM International (formerly American Society for Materials) offers the ASM Center for Failure Analysis to help technicians, consultants, and engineers on technical issues related to correct and prevent failures. With more than 1,000 cases, the new online hub is designed to help find and understand materials related cases and failure modes present in an ongoing investigation. The ASM Failure Analysis Center is based on collections of case histories published in ASM journals, teaching materials, and reference books. Covering more than 700 specific material types, each case is classified by material type, industry, and application involved. Professionals who require more specific help regarding failure analysis may contact any of the organizations specializing in this field.

References

1. H. Petroski, *To Forgive Design: Understanding Failure* (Harvard University Press, Cambridge, 2012) p. 410
2. R. Matthews, Tumbling toast, murphy's law and the fundamental constants. Eur. J. Phys. **16**(4), 172–176 (1995)
3. Mechanical Engineering Magazine, March 2012, vol. 134 No. 3
4. ME. Eberhart, *Why Things Break: Understanding the World by the Way it Comes Apart.* ISBN 1400048834
5. T.A. Kletz, *Lessons from disaster—how organizations have no memory and accidents recur* (Institution of Chemical Engineers, Rugby, 1993)
6. Fallas mecánicas: Distintos niveles de análisis. www.giemdp.com.ar
7. J.E. Gordon, *The New Science of Strong Materials or Why You Don't Fall Through the Floor*, (Princeton University Press, Princeton, 1984). ISBN 0691023808
8. J.L. Otegui, Binder C.A. s/supuesto múltiple Homicidio culposo en accidente de tránsito. Juzgado No. 1, (Corrientes, Argentina, 1994)
9. J.S. Baker, L.B. Fricke, *The Traffic-Accident Investigation Manual*, 6th edn. (Northwestern University Traffic Institute, USA, 1986)
10. J.L. Otegui, Fatigue Damage Leads to a Serious Traffic Accident. Eng. Fail. Anal. **9,** 109–122 (2002). ISSN 1350–6307
11. T.A. Kletz, *Learning From Accidents*, 2nd ed. (Butterworth–Heinemann Ltd, Oxford, 1994). ISBN 0 7506 1952 X
12. J.L. Otegui, E. Rubertis, Cañerías y Recipientes de Presión, EUDEM, Editorial U.N.M.d.P. ISBN 978-987-1371-18-1. 2008

Chapter 2
Tools for Preliminary Analysis of a Mechanical Failure

2.1 Methodologies for Field Investigation After a Failure

The aim of this chapter is to define guidelines for the preliminary analysis of a mechanical failure, in order to facilitate subsequent investigations and determine the causes of the failure. This chapter includes the definition of the investigation team, the removal and storage of failed parts and evidence, and the definition of guidelines for the failure analysis to be performed. The success of a post-failure investigation, especially in the case of mechanical breaks involving large energy (shocks, explosions, etc.) depends critically on the quality of the evidence collected at the site. The objectives of this section are:

- To list the main activities after an accident or explosion, and to indicate how to carry out a systematic investigation of causes.
- To discuss how the damage can be evaluated, and to understand the chain of events.

After a major accident, the task force, team, or investigation committee will be in charge of analyzing the accident. For smaller events, this activity is often done internally by the company responsible. Figure 2.1 shows the main objectives of this work. As shown in the figure, the coordination function is important since some of these activities will have different priorities or objectives. For example, repair of the failed equipment and damage documentation might be contradictory, if not coordinated. Evidence of failure mode may be lost during cleanup. After an accident, a common reaction is to start cleaning without documenting damage. It is necessary, therefore, that a person chairing the investigation committee coordinates all activities.

Investigating the cause of the accident is usually time- and resource-consuming. The purpose of the site assessment is to obtain the information necessary to reconstruct the events backwards, from observed damage and eyewitness recounts. Qualified personnel are required to perform these analyses. Baker et al. [3] recommend the participation of experts immediately after the accident. Otherwise we might miss most useful damage indicators. Damage documentation must begin

J. L. Otegui, *Failure Analysis*, DOI: 10.1007/978-3-319-03910-7_2,

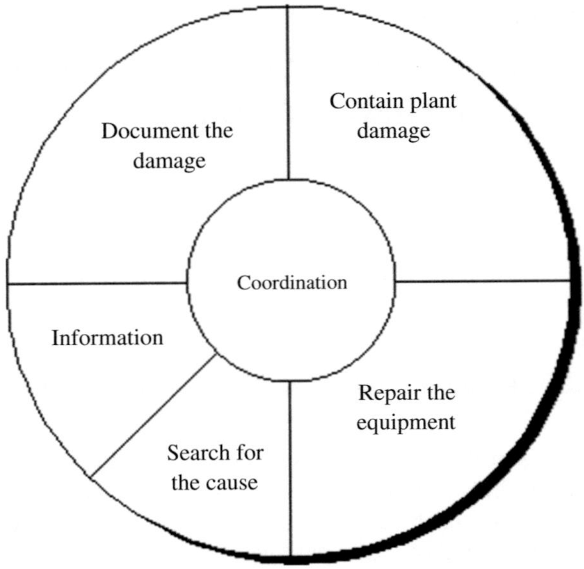

Fig. 2.1 Main objectives of investigation

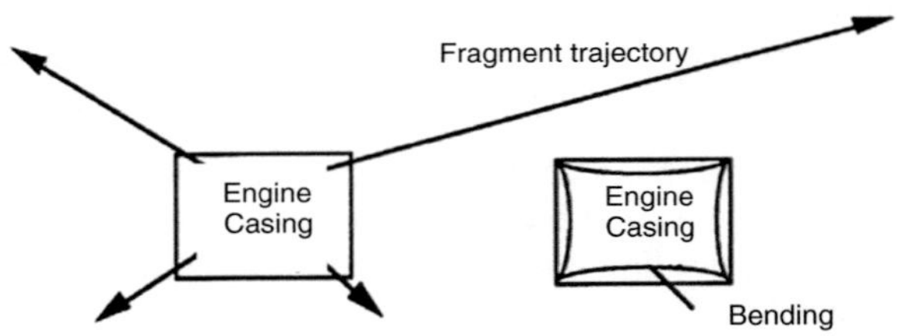

Fig. 2.2 Fragments and structural deformations provide valuable information

immediately and should be done by a specific expert in failure analysis (structural response, combustion turbines, etc.).

In many cases, the methods are based on a posteriori analysis of photographs taken at the site, of the general area and of specific damage, so a professional photographer should be used. Making systematic records of locations and directions of all pictures taken is required. A map of fragments, with original and final positions, should also be organized. Fragments are usually a good indicator of where the initial failure occurred and its magnitude. Figure 2.2 shows the trajectories of four parts of motor covers of a real case. A motor housing flew 15 m from its original position. The fragments of the motor housing "a" support the hypothesis that fuel gas entered the engine housing and the failure began as an explosion inside that housing.

The deformations and deflections of structures are also indicators of damage. Inward deflections in the motor housing "b" indicate loading from the outside. Typically, the deflection of pipes, panels, and other elements as well as the direction of broken glass can be used to estimate origin and load intensities during an explosion [3].

2.2 Collecting Data and History

The field representative must collect all the evidence of previous failures in the same area, failures of the same type, on-site inspections of the failure, etc. He should collect additional information if the case so requires, such as weather reports, flooding, or earthquake reports, etc. The field representative shall collect the historical operating data of the component or equipment, specifically at the time of the failure occurrence. He is in charge of interviewing operators, nearby residents, and any person who can provide information on the failure, and collect information on the extraction of samples and features of the failure zone, depending on the type of failure. The representative needs to know the environment at the initiation of a fire or explosion and/or what features of the environment helped it spread. It is important to make a sketch of the terrain and failure zone.

The field representative tentatively determines what type of failure caused the event and associated damage mechanisms identified in the preliminary integrity report. He should search databases for the same type of failures, with the aim of finding a correlation. The responsible for field inspections will develop a preliminary report detailing what he relieved, location plans, and operating conditions at the time of the failure, survey of external failure conditions, and other data of interest to define a preliminary overview of the event.

Usually it is a senior representative, e.g., a plant integrity chief, who will define the need and determine what types of failure analyses are undertaken. The field representative must then assume the responsibility to interact with the company performing this failure analysis, coordinating, and providing the necessary information, as well as with the various departments within the company.

2.3 Visual Inspection Techniques and Field Photography

Visual inspection is a test method based on the detection of specific elements using basically the human eye and the experience of the inspector. Usually help tools such as magnifying glasses, cameras, meters, borescopes, camcorders, etc., are required. To carry out a visual inspection, good lighting must be ensured; at least 1,000 lux can be taken as a reference value. Light sources may be needed in addition to those existing at the site. Also, the area and the components where the test is performed should be properly cleaned. All instruments used to enhance appreciation must be in good condition.

Visual inspection is usually the first test and one of the most useful ones, as it is a low-cost test it does not require sophisticated equipment and is complemented by all other test methods. Findings must be documented to avoid losing validity. Within the limitations of the method, it can be mentioned that only surface defects can be detected and that their detection requires a “trained eye.”

Examination of fractured surfaces allows for a lot of information regarding the origin and causes of a broken component when subjected to mechanical loads. During the preliminary analysis, usually at the site or somewhere protected within the same premises where the failure occurred, visual assessments are made with the eventual help of a magnifying glass; magnification varies between **X**1 and **X**10. With this evaluation, it is possible to analyze damage, possible deformations, cracks, and other defects, and most importantly the failure initiation site. Qualitative visual descriptions are to be performed of the various areas of the fracture surface and other sites with signs of damage of any kind (corrosion, wear, fatigue, manufacturing defects, etc.). In some cases, it is necessary to resort to other methods of nondestructive evaluation, which are briefly discussed at the end of this chapter.

The field investigator must have knowledge and experience in photographic techniques. The core variables of a camera are diaphragm aperture and exposure time; their combination gets the right balance of light in each exposure. The diaphragm is a part of the objective that limits the amount of light entering the camera. It works like the iris of the human eye, opening and closing to allow more or less light. If the diaphragm aperture is increased (more light), the exposure time (time that the film or sensor is exposed to light) has to be reduced, otherwise the picture gets too bright (overexposed). If aperture is reduced (less light), the exposure must be extended, otherwise the picture gets dark (underexposed). The aperture is measured in f-numbers. The basic f number scale is f1, f1.4, f2, f2.8, f4, f5.6, f8, f11 and f16. The smaller the number indicates larger aperture. In each f-step, the brightness is reduced by half. F11, for example, has twice the brightness than f16 and half than f8.

The shutter limits the time that the beam of light entering the camera exposes the film or sensor. Usually, exposure times range from seconds (for very poor lighting conditions) to milliseconds (for very fast pictures). Most usual time, in seconds, are: … 4, 2, 1, 1/2, 1/4, 1/8, 1/15, 1/30, 1/60, 1/125, 1/250, 1/500, 1/1,000 and 1/2,000. Both automatic and manual cameras can pick a particular opening and get an exposure time estimate, and vice versa (Fig. 2.3).

In failure analysis, gently shooting and camera support are very important. Many times it is necessary to provide another source of light. If exposure times are greater than 1/15, a tripod must be used. When a photograph is completely blurred, it means that the shutter speed was too slow for the way the camera was held. When a portion of the image is blurred and another is not, no proper focus was achieved (Fig. 2.4a).

The "macro" function is essential to get good details of a failed piece. It allows focusing very close to the target; in some cameras as close as 1 cm. Focal lengths under 30 cm are highly recommended. When seeking to obtain photographs of fracture surfaces, it must be understood that not necessarily all the fracture is in a same plane. Therefore, good depth of field must be obtained, whenever possible (Fig. 2.4b). Another important aspect is the variation in the direction of the light source. Figure 2.5 shows the same fracture surface illuminated from different

Fig. 2.3 Shutter opening and focus depth

Fig. 2.4 **a** Bad depth of field **b** Good depth of field

angles; note how each picture shows different things. Believe it or not, some may be relevant in the future, even though at this time it would not seem so.

To photograph fractures, flash is not our friend since it "flattens" the image. Shoot under the sun or against a strong light source. Contradicting the usual rules for artistic or landscape photography, it is better to point against the light (angle of incidence).

Flat surfaces, such as metallographic samples, go well with both artificial and natural light. The depth of field is no longer a problem, just concentrate on the frame. If shooting with artificial light, whites should be compensated. Very shiny surfaces do not usually give much information on the photos; it is always easier if the surface is opaque, as in the case of a metallography revealed with a chemical etch, such as Nital, because there are fewer glares. If photographing a shiny surface is required, place the light source so as to eliminate glare into the lens (Fig. 2.6).

In summary, one can say that modern digital cameras have simplified the failure expert's activity, but they do make no miracles. Before taking the pictures, one

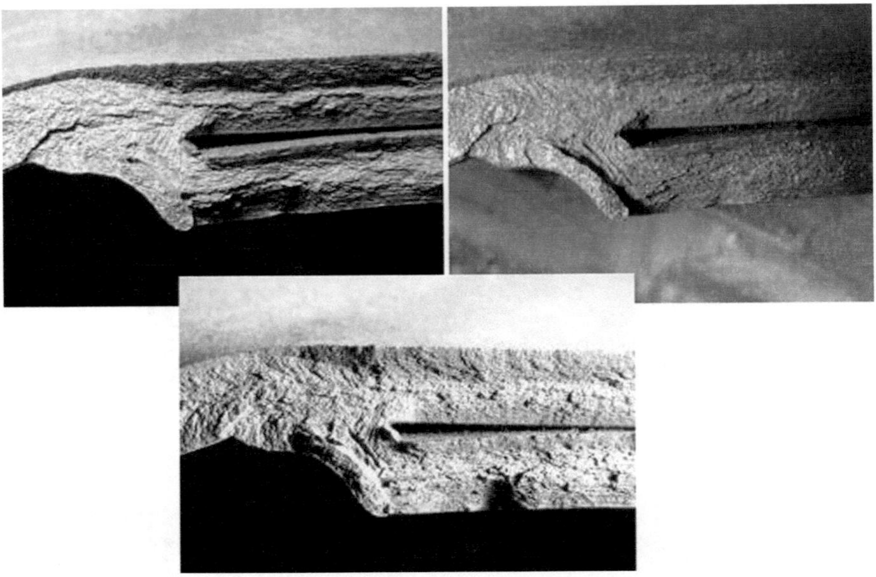

Fig. 2.5 Crack illuminated from different angles

Fig. 2.6 Macrophotograph of a metallographic section

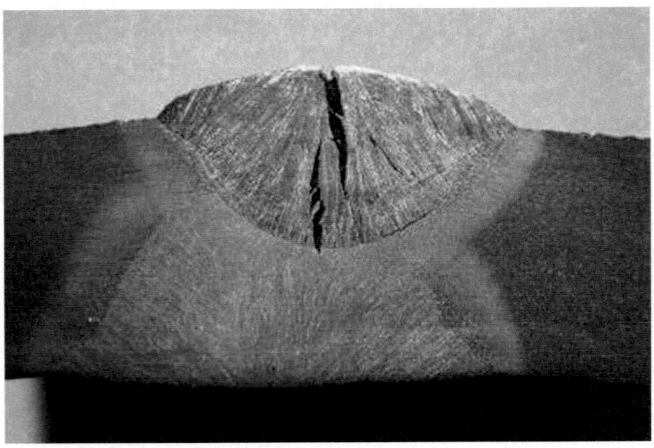

must know what to display and search the position of the light source and the eye to see what is wanted to see. Then the camera will see the same as the operator. The appeal of the new high-resolution cameras (5 MP or more) is that in a well-taken photo more details can be later appreciated than those seen with the naked eye, through electronic zoom and a computer program. As with any photographic activity, it is always recommended to take lots of pictures with slight variations, and then choose the most suitable.

2.4 How to Detect the Site of Initiation of Mechanical Failure

2.4.1 Fractographic

By fractographic analysis or characterization, we mean the inspection of the surfaces of fractures and failures. In the preliminary in field stage, this can only be done to a low magnification (up to approx. **X**20). In laboratory, with optical techniques and scanning electron microscopy, it is possible to reach much higher magnifications. Preliminary fractographic analysis helps to define:

- Size of critical defect at failure
- Mode of flaw propagation
- Failure initiation site and pre-existing defects
- Mode of in-service propagation of pre-existing defects.

Let us see some examples of typical cases of fracture of components subjected to mechanical loads. We can define three types of fracture surface, which can be differentiated at this stage of preliminary inspection:

- Brittle fracture: instantaneous event at final failure (hundredths of a second).
- Ductile fracture: although its propagation can take from a few hundredths to a few seconds, it is also defined as an instantaneous failure event.
- Previous propagation: also called stable or subcritical propagation or growth, it occurs for some time during service life of the component, due to specific conditions of operation (cyclic loading, temperature, aggressive environments, etc.).

Table 2.1 summarizes the characteristics of ductile and brittle fracture, and their manifestations in a visual inspection of the fracture surface. Figure 2.7 shows typical ductile fractures in laboratory samples. Figure 2.7a shows a slant fracture at 45° to the surface of the component; this is typical in structural steel plates. In modern high strength steel pipes and pressure vessels, discontinuities are frequently produced by controlled thermal–mechanical treatment, which is discussed in some detail later. In these cases it may occur that the ductile fracture surface shows a series of flakes, as shown in Fig. 2.7b.

Another feature of ductile fracture is that it occurs after the section suffers a significant amount of plastic deformation. Plasticity usually results in a reduction of section, as shown in Fig. 2.8.

Brittle fracture of metals at the macroscopic level is characterized by a generally flat surface, perpendicular to the direction of maximum stress, with no signs of previous plastic deformation. If the fracture surface is clean enough to allow inspection with a magnifying glass, often a faceted surface with small bright planes can be defined, Fig. 2.9.

Table 2.1 Characteristics of ductile and brittle fracture

Ductile	Brittle
Metal surfaces are usually opaque with fibrous appearance	Metal surfaces are normally bright with granular appearance
Characterized by metal tear accompanied by plastic deformation	Characterized by rapid propagation and no plastic deformation

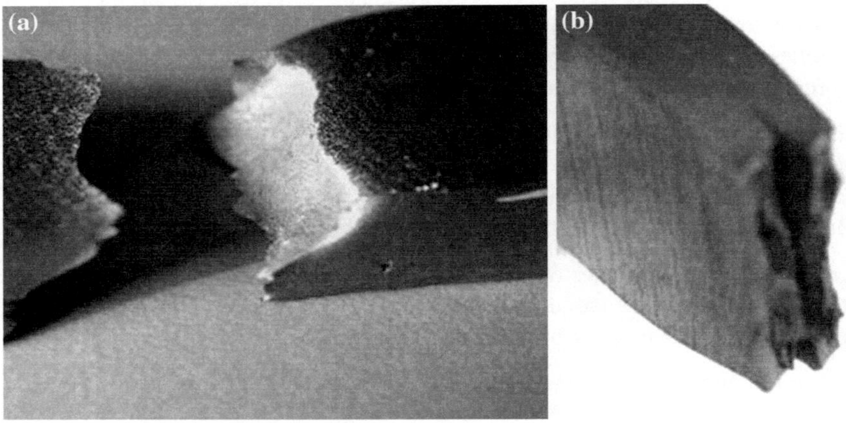

Fig. 2.7 **a** Fracture at 45° from the surface **b** Series of flakes at 45°

Fig. 2.8 Reduction of section previous to break in a welded joint (*shaded*)

At the macro level, a characteristic of brittle fracture is the formation of chevron marks. Figure 2.9 shows an example of such "chevrons"; these are arrow or V-shaped marks in the fracture surface. In brittle fractures these arrows point to the initiation site.

The reason for the formation of these chevrons is as follows. In areas close to the surfaces of the piece, the front of the crack propagates in a plane stress field, while in the middle of the thickness the crack front grows in an area of high tensile

Fig. 2.9 "Chevrons" in a brittle fracture, and shear lips near plate surfaces

stress triaxiality. As will be seen in Chap. 6, a triaxial stress field hinders plastic deformation, so that crack growth tends to occur in a more brittle condition. Therefore, crack propagation tends to be delayed near the component surfaces, which leads to a V-shaped crack front, instead of a flat front. Small variations in the successive planes of the crack path through different microstructures in the material produce steps in the crack surface which are normal to the crack front. These form the characteristic chevron marks, the tips of the arrows located near the middle of the thickness. When the stress field is not fully tensile but has a bending component, stresses near one surface are larger than in the other. This makes the two branches of the chevron marks not to be of the same length.

Frequently, in the vicinity of the component surfaces a change in the restriction to plastic flow reduces the degree of stress triaxiality. This change in conditions (from plane deformation to plane strain, which will also be discussed in more detail in Chap. 6), in many cases causes a change in propagation mode. Although the fracture is brittle in most of the thickness, two thin layers of ductile fracture are formed adjacent to the component surfaces. These layers are characterized by forming 45° slanted lips, called shear lips. These are also related to ductile growth, which is controlled by maximum shear stresses.

Such shear lips can be clearly seen in Fig. 2.9, as two opaque narrow bands at the top and bottom of the figure. The width of the shear lips indicates the ductility or brittleness of the material. The shear lips in Fig. 2.5, for example, cover 75 % of the fracture surface, and are much larger than those in Fig. 2.9. The use of this information is discussed in greater detail in other sections.

A common feature is that a fracture begins in a brittle mode in areas with previous defects and/or stress concentrations. When the fracture reaches a certain size, the driving force is enough to keep it growing, even if the material is ductile. The mechanical conditions for his phenomenon will be seen in some detail in Chap. 6. Thus, the transition between brittle fracture and ductile fracture is a first indication of the zone where the failure began. For example, Fig. 2.10 shows a circumferential fracture occurred in a girth weld of a pipe, under the effect of axial loads. Most of the fracture is ductile (Fig. 2.10a), with steps at 45° and lateral shrinkage. But at initiation (Fig. 2.10b), the fracture is through the thickness

Fig. 2.10 **a** Ductile fracture of a pipeline **b** Close-up of the through-thickness fracture

(at 90° from the surface of the tube), without lateral contraction. Also seen here are parallel marks advancing from the outer surface of the tube, like sea marks on a sandy beach. We will later discuss in more detail about these "beach marks."

2.4.2 Example 2.A Identification of Failure Origin

Figure 2.A1 shows the casing of a centrifugal pump from a liquid hydrocarbon (oil) pumping station. Circumferential fracture was propagated by the stresses due to internal pressure. The arrows indicate how the crack propagated, as defined by the fractographic analysis. In this case, there has been an unusual situation: the

Fig. 2.A1 Cracked casing of a centrifugal pump

propagation of the fracture was such that a window opened. The right (final) part of the window broke due to bending; the area that became a "hinge" is highlighted by an ellipse in the figure.

Figure 2.A2 shows the chevron marks in the fracture zone shown at the top of Fig. 2.A1. From the analysis of the chevrons in Fig. 2.A1, it was possible to reconstruct the failure occurred in the pump casing, and the definition of its origin (central circle in the Figure). Note that the tips of the chevrons are not located in the middle of the thickness, but in the third closest to the inner surface. This asymmetry is associated with the bending stress field generated by the outward force produced by fluid pressure acting against the inner surface of the "window."

2.4.3 Initiation Site of a Fracture

Any fracture resulting in instant failure that starts at a given site can be called failure initiation. We have seen that the most important part of the effort during the preliminary analysis is referred to detecting failure initiation. This site is defined by one or more of the following factors, which will be discussed in detail in other sections of this book:

- Stress concentration.
- Material weakness.
- Previous defects (manufacturing or grown during the previous service).

Initiation of fast fracture is often the result of a damage mechanism developed during service. In this case, we speak of three stages in the fracture surface:

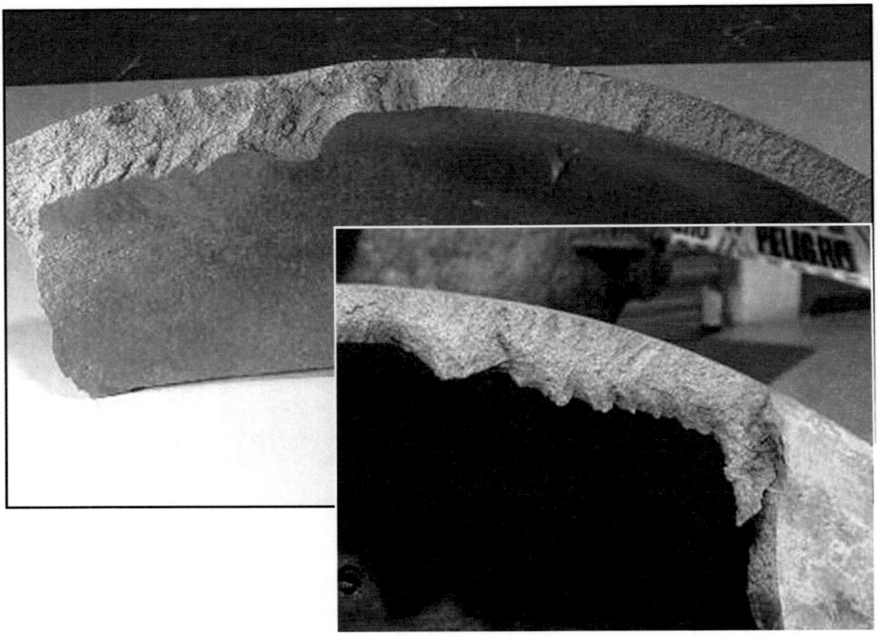

Fig. 2.A2 Chevrons in different parts of the fracture surface

- Initiation of previous damage.
- Propagation of previous damage.
- Final failure.

Figure 2.11 shows an example of the initiation of a crack from a stress concentration due to cyclic loading during service; which is called fatigue and is also discussed later in greater detail. It is a welded joint; weld reinforcement is at the bottom of the figure. Cracks propagated from defects in the material of the two welds, identified as A and B. Direction of crack growth is indicated by black arrows. When cracks reached a certain size, they joined (coalesced), and continued to propagate as a single crack. The dotted line indicates both crack fronts at the time of coalescing. The dashed line indicates the crack front when the component finally fractured; direction of fast fracture is indicated by white arrows.

What are the clues that indicate initiation and prior in-service propagation? Many are on the fracture surface. First, the slow (stable or subcritical) service propagation typically generates a rather smooth fracture surface, while fast propagation tends to form rough fracture surfaces. As stable propagation occurs during a certain time, operating conditions frequently change during crack growth, thus changing propagation rate. These different growth rates result in variations in the appearance of the fracture surface. The boundaries between different surfaces

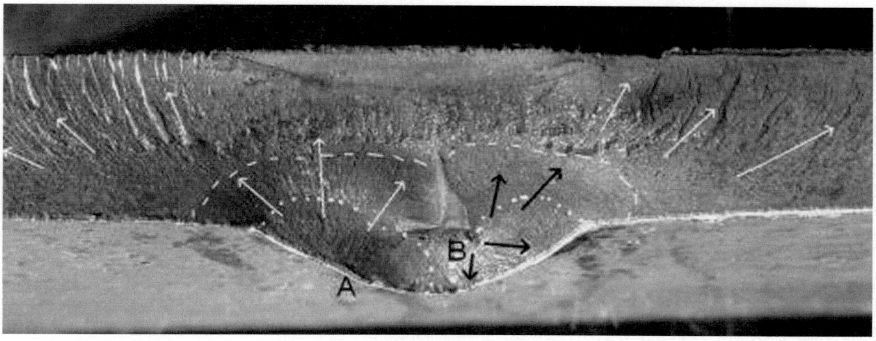

Fig. 2.11 Crack initiations from a stress concentrator

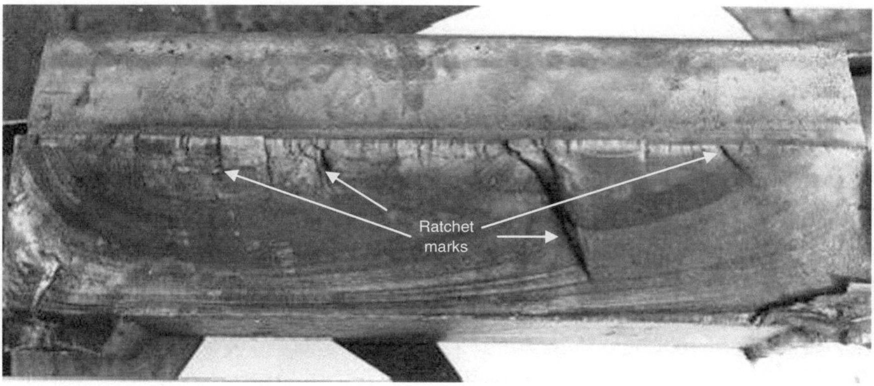

Fig. 2.12 Fatigue crack growth

indicate the shape of the crack front at each time. These limits are those already defined as beach marks, and are often detectable during visual inspection.

If the service propagation occurs from multiple initiation sites, cracks initiate in different planes, which eventually merge. This leaves between them wedge-shaped marks (ratchet marks). The analysis of beach marks and ratchet marks identify crack initiation sites.

As an example, Fig. 2.12 shows a fatigue crack grown from the surface of the component, in the upper part of the figure. Beach marks are clearly visible even when the cracks are very small. Ratchet marks in the vertical direction are indicated with white arrows.

Figure 2.13 shows another fatigue crack, also grown from the surface (upper part). Beach marks are visible in most of the fracture surface shown.

When the crack reached the dark mark there was an increase in growth rate, presumably due to an increase in operating load. Beach marks become less clear,

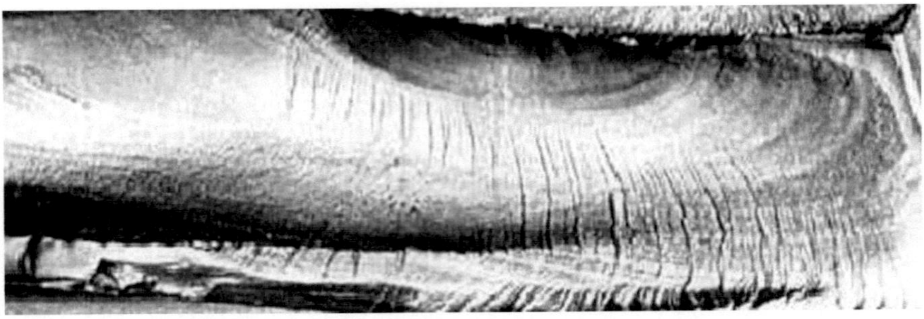

Fig. 2.13 Fatigue crack grown from the *upper* surface

and other marks which are mutually parallel are visible in the vertical direction. These are called river marks, and also frequent in fast fractures. These river marks indicate different planes of crack growth, which occur along certain crystallographic planes (also discussed in a later chapter), and indicate the direction of fracture.

2.4.4 Example 2.B Detection of Previous In-Service Damage

An instant fracture of a cylinder liner of a piston compressor at a petrochemical plant is depicted in Fig. 2.B1. Figure 2.B1a shows the characteristic chevron marks and shear lips of a brittle fast fracture. Chevron marks indicate initiation at the left of the Figure, and point to the location of a threaded hole. This hole hosts one of the bolts used in positioning of the cylinder during maintenance. Figure 2.B1b shows the smooth crack surface, typical of subcritical propagation, that is to say, it occurred during a service period. The analysis of beach marks and ratchet marks allow identifying the site of crack initiation at the bottom of the second thread in the hole, Fig. 2.B1c.

Higher magnification allowed observing signs of plastic deformation within the thread. Apparently, during maintenance activities at the last plant stop, a bolt was used that did not correspond exactly with the geometry of the thread. As a result, interference generated between the threads damaged the threaded hole. Many months later, the cyclic stresses from the compressor ended up propagating a fatigue crack, which grew as a quarter circle. When the size of this crack was of the order of the hole depth, a critical condition was reached: the crack became unstable and propagated instantaneously, causing the catastrophic failure of the component.

Fig. 2.B1 **a** Brittle propagation with chevron marks and shear lips **b** Smooth crack surface **c** Crack initiation site

2.5 Failure of Threaded and Rotating Elements

Due to their large share in the occurrence of failures in mechanical components, some attention will be paid to the visual analysis of threaded (screws, bolts, and nuts) and rotating (shafts and axles) elements.

The main function of a screw is to transfer the load. Threaded components come in many types, dependent upon design requirements and environment in which the fastener is to be used. Threaded elements frequently fail due to high stresses at the root of the thread, which is a geometrical stress raiser. Small machining errors, surface defects caused by corrosion or mechanical damage (as we have already seen in our previous example) are sufficient to initiate cracks, often propagated by fatigue.

Surface damage by fretting may be the result of small movements between adjacent surfaces. Atmospheric corrosion, galvanic and crevice corrosion, stress corrosion, and hydrogen embrittlement are frequent contributors to the failure of these elements. Those will be discussed in other chapters.

One aspect that is not minor in the case of failures involving bolts and nuts is recovering failed parts. Many times the remains are scattered, and then it becomes difficult to define where each recovered piece was located before failure. It is common for a failed component to involve many threaded fasteners; it is important to identify those screws that caused the failure and separate them from those who

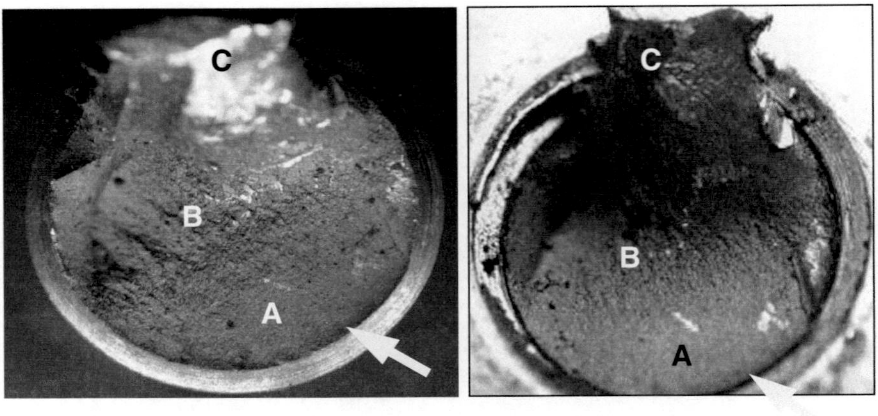

Fig. 2.14 Typical failure of a screw

may have failed due to overload after the failure of the others. That is, separate the elements causing the failure from the elements that failed as a result of it.

American Codes ASTM and SAE require that heads of threaded fasteners are identification marked with its grade (strength). These markings allow quick verification of proper selection by visual inspection. A visual confirmation of grade and verification of specification is the kickoff of any investigation. Then the areas in the fracture surface related to modes of initiation and propagation should be investigated, similarly to what has already been discussed. Figure 2.14 shows a typical failure of a screw, and we find:

a: initiation area, opaque and flat, without visible beach marks.
b: a fibrous area with river marks indicating direction of propagation of the fracture.
c: a narrow shear lip corresponding to the final fracture due to ductile overload.

An example of a threaded bolt which failed by fatigue is shown in Fig. 2.15. Here almost 100 % of the fracture surface is subcritical growth. The insert shows in some detail the initiation of cracks in the bottom of the thread. Conversely, the overload failure of a bolt is indicated by a 100 % fracture surface corresponding to ductile tearing. Also in these cases it is possible to distinguish some features related to the load that caused the failure, and identify signs of crack growth due to tension or to shear.

The relationship between the areas of cracking and ductile tearing is related to the relationship between the amplitude of cyclic loading and the maximum load achieved during service. Surfaces of final fracture in threaded elements of high strength, low ductility steel show brittle fracture, with indications such as river marks, as shown in the example of Fig. 2.16. Two initiation sites (A and B) are seen at the top of the figure, separated by a ratchet mark; the black line underscores

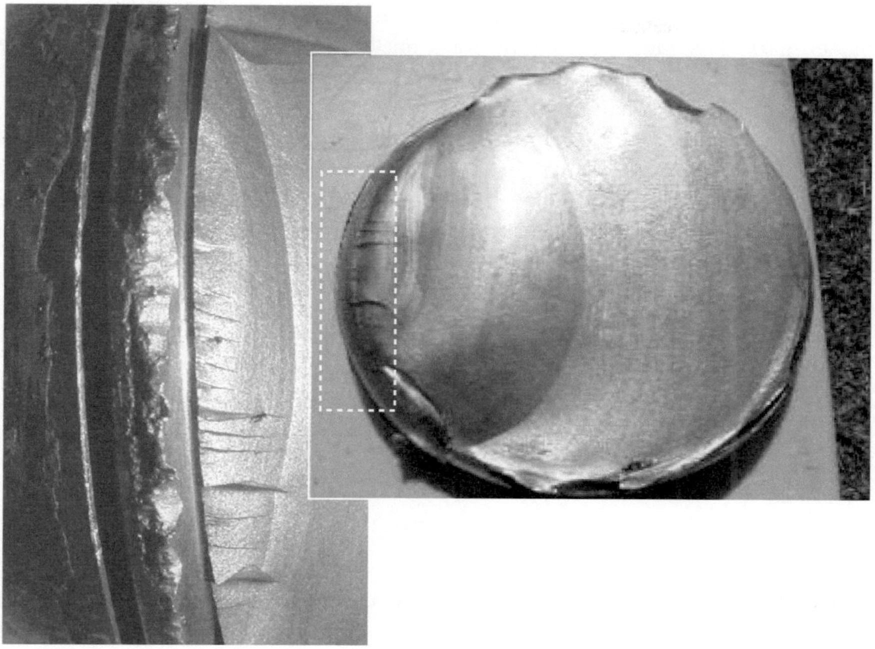

Fig. 2.15 Fatigue failure of threaded bolt

Fig. 2.16 Initiation sites separated by a ratchet mark

the beach mark at the time of final breakage. Note that this coincides with the ratchet mark. This indicates that the crack unstabilized at the time the two fatigue cracks coalesced, which is consistent with fracture mechanics concepts for criticality of cracks (see Chap. 6).

Shafts work in a wide range of conditions, including corrosive environments, at very high or very low temperatures, and may experience a range of loading conditions: tension, compression, bending, torsion, or a combination of these conditions. In addition, they may experience vibratory stress. The most common causes of failure are:

- Abrasive wear: material detaches from a solid surface due to hard particles or protuberances on the sliding surface. Examples of abrasive materials are sand, dirt, metal particles, and other debris in the lubricant.
- Fatigue: commonly begins at a stress concentration, as do also other cracking mechanisms. Typical features of these stress concentrators are corners, keyways, gutters, forged shapes, weld defects, nicks, cracks, pitting corrosion, clearances, and bends.
- Misalignment: may be made after a repair; resulting vibrations often cause fatigue failure of the shaft.

Each cause of failure is associated with certain characteristics of both the location of the failure and its fractographic surfaces. Figure 2.17 shows some characteristics associated with the shape of the beach marks and the relationship between the areas of subcritical propagation and final fracture.

2.5.1 Example 2.C Failure of a Bolted Structure

In this example we analyze the failure of a crane used in port activities, which failed by rupture of the anchor bolts to the base. The principal interest is aimed at determining the causes of the incident, if it is due to gradual wear or deterioration resulting from overload in operating conditions, if it is related to construction deficiencies or material, or installation deficiencies. Figure 2.C1 shows the jib crane, mounted on a base. Detachment occurred in the highlighted area at the bolted joint, which allows rotation of the structure along the base. The ring is composed of two rows of 44 pins each, Fig. 2.C2. In this case, the plume fell by the failure of the bolts that secure the inner bearing top. These bolts are the only connection between the rotating loader arm and the rest of the base.

In loaders and cranes which are mounted on a rolling base, the most common failure modes are failure of the bolts and bearing track wear failures. With this preliminary information and failed items collected in the field, previous to calculations and analysis of the material, it was possible to obtain some important conclusions regarding the cause of the fall of the crane. Broken bolts were found at the level of threads due to tensile overload (Fig. 2.C3). The position of these bolts

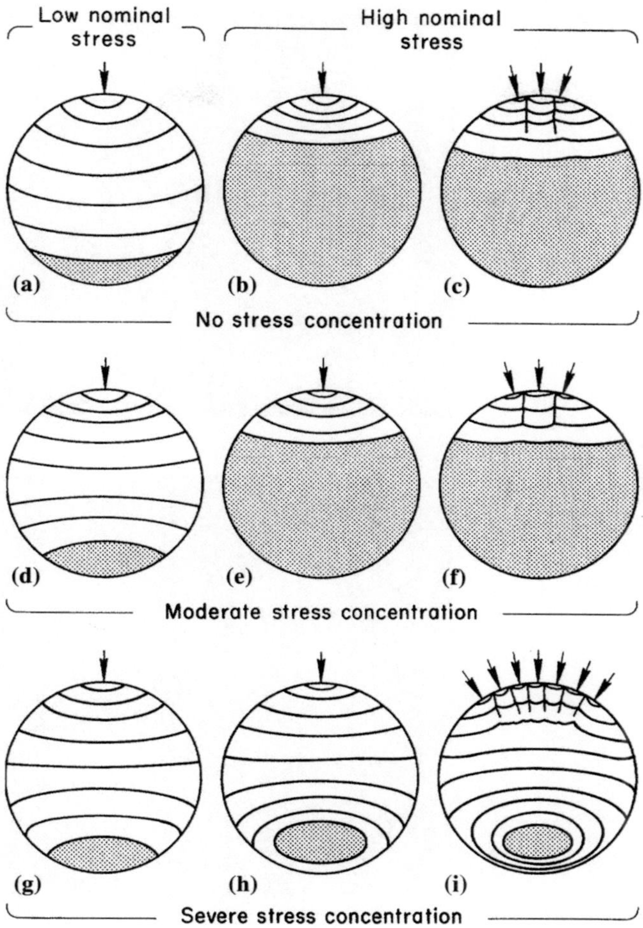

Fig. 2.17 Fatigue crack growth of rods (from ASM materials handbook)

confirmed that the ultimate catastrophic event occurred when the arm dropped, first forward, and after a few minutes later laterally, to its final position.

Other bolts presented worn out threads, both in rod and nuts, Fig. 2.C4. This failure can be progressive in nature and no obvious cumulative damage be detected until final failure. Nut threads were sliced at their roots, as the bolt material is stronger than the nut. This is normal, intended to redistribute the stresses in the fillets.

Some bolts were also found with obvious signs of fatigue; this is normally associated with a deficiency in preloading (Fig. 2.C5). Black arrows indicate microcracks at initiation sites at the bottom of the thread, white arrows indicate fatigue propagation until final failure.

Fig. 2.C1 Jib crane

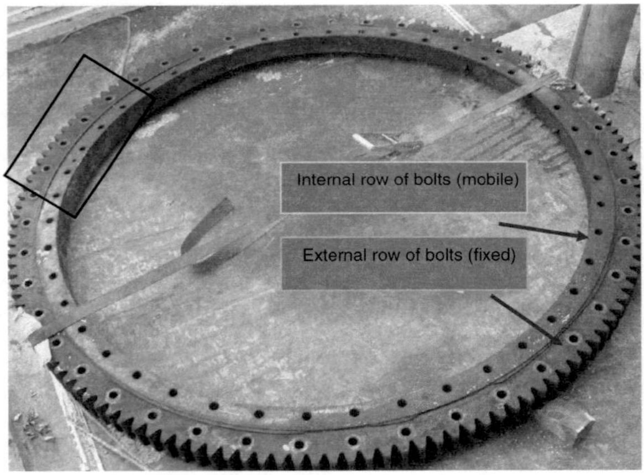

Fig. 2.C2 Ring composition

Fig. 2.C3 Cut bolts at the level of threads

Fig. 2.C4 Worn out threads

These bolts were located in the crown area diametrically opposite to the direction of the loading arm and were the first to fail, thus increasing the load on the rest of the fastening system. Due to lack of proper tool access, verification of torque applied to the bolts of the inner row was performed using an impact wrench, making it impossible to apply the torque accurately (Fig. 2.C6).

Figure 2.C7 shows an outline of the damage to the bolts around the crown, viewed from below. The distribution of damage clearly identifies what were the bolts whose rupture was the cause of the failure, and what were broken during the collapse of the structure.

The requirement to control tightness in the crown was in the maintenance documentation, but through two indirect references and without the importance that such control has. Currently, this check is required on a biannual basis for certification of offshore equipment and, in the future, this criterion should be applied to this equipment.

2.6 Extraction and Storage of Samples

The collection of evidence must be carried out independently while repairs are being made to restore service without affecting these repairs. Ideally, there should be a protocol for hurried working conditions, which are never the same as the previous ones. A protocol would avoid methodological errors that might prejudice the outcome of the analysis. Here are some things to keep in mind:

Prior to removing the pieces

- Collect the evidence in the area that may have contributed to the failure: state of coatings, component identification, welds, foreign bodies, etc.
- Evidence of the position of the failure and the different components with respect to physical references, using graphic recordings (drawings, video, and/or photos).

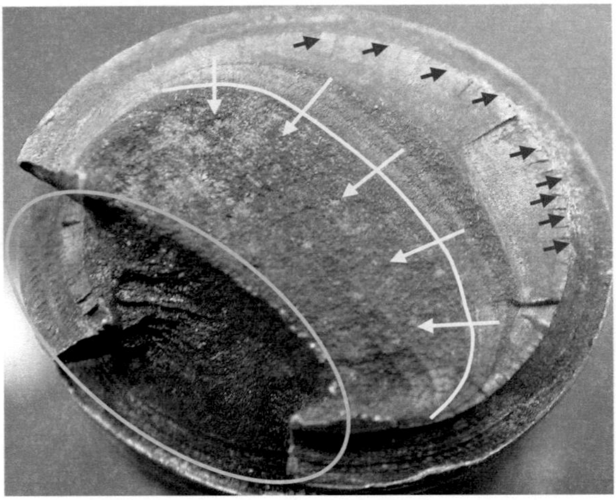

Fig. 2.C5 Bolts with clear evidence of fatigue

Fig. 2.C6 Excessive wear increased clearance between balls and bearing track

- Perform the appropriate markings to identify the location where each piece was removed.
- Photographically record all marks made.

Removal of pieces

- Prevent degradation of the sample during cutting, for example by overheating of the fracture zone when grinding or cutting with thermal processes (plasma, oxy-acetylene, etc.). Make cuts at an adequate distance zone and remove samples of both the failure zone and in non-failed areas.
- Perform cold cutting. If the distance available for the cut is not enough or would be dangerous to make flame cutting, grinding and coolant should be used to keep the area at low temperature.

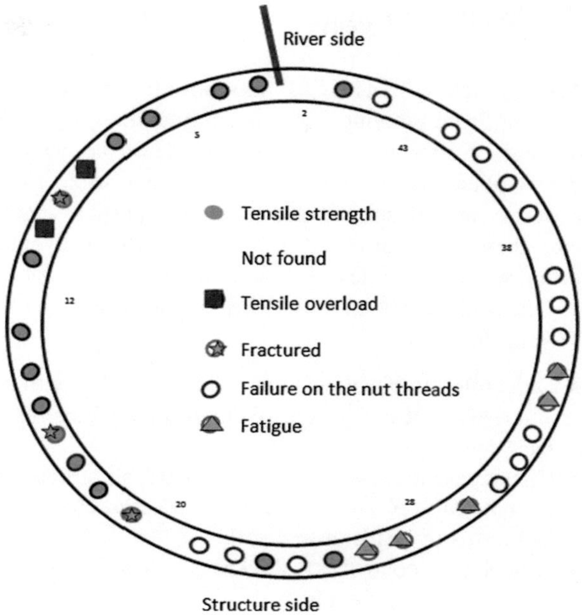

Fig. 2.C7 Scheme of damage in the bolted ring

- Transport samples so that fracture surfaces are not subjected to the components' own weight.
- While moving large pieces, use preferably nylon slings at all times avoiding the use of steel cables that may accidentally lean on the fracture surface and deform evidence.

Conservation of samples

- Do not clean the samples by any mechanical or manual means; avoid erasing important evidence such as corrosion and surface cracks, among others.
- If the part is wet or with a certain level of humidity, spray a jet of alcohol or kerosene to the fracture surface to displace the water and then air-dry.
- Waterproof the fracture area with transparent lacquer or varnish spray, cover with kitchen plastic film, or apply lubricating grease if no lacquer or varnish is available. Cover the fracture surface with corrugated cardboard, expanded polystyrene, or paper in order to avoid damaging the piece at the time of transportation.
- Store samples in a suitable location, which does not change the characteristics of the failure.
- If the failure surfaces are separated, DO NOT reassemble them together. This will generate mechanical damage by crushing at a microstructural level, and valuable information might be lost.

2.7 Inspection by NDT Techniques

Nondestructive Tests (NDT) are those that allow an assessment of the equipment or component without compromising its integrity. This means that the tests do not damage or interfere with the future use of the parts inspected. NDT are based on the application of physical phenomena such as electromagnetic, ultrasonic, acoustic, and elastic waves, subatomic particle emission, capillarity, absorption, etc. There are a variety of NDT, each with its advantages and disadvantages; they differ in the principles of operation and the type of application.

The wide application of NDT methods in materials is summarized in the following three groups:

- Defectology: this allows the detection of discontinuities, evaluation of corrosion and deterioration caused by environmental agents, determination of stress, and leak detection.
- Characterization: evaluation of chemical, structural, and mechanical properties of materials, physical properties (elastic, electrical, and electromagnetic), and isotherms in heat transfer path.
- Metrology: control of thickness, local dimensional measurements, coating thickness, filling levels.

An indication is produced by a disturbance in a signal, detected through a nondestructive testing method. In this sense, the information may be false, relevant, or not relevant. A discontinuity can be defined as the lack of homogeneity or interruption in the normal physical structure of a material, or a deficiency in the normal configuration of a piece, part, or component. The discontinuities can be classified according to:

- location: surface, subsurface, and internal (or embedded),
- origin: inherent to manufacture, process, service, etc.,
- morphology: planar and volumetric.

The generation of corresponding records is as important as testing. These should include test conditions, data on geometry, and properties of the components under study, and an appropriate record of the characteristics of the discontinuities detected.

When using such NDT techniques for integrity assurance in the manufacturing steps or in-service inspection, discontinuities may be relevant or nonrelevant. A defect is considered when the size, shape, or location of the indication exceeds acceptable limits established by the Code, Standard, or applicable specification.

The use of NDT techniques for the preliminary evaluation of a failure becomes important when looking for evidence of damage to other parts of the components, apart from the fracture zones. The detection of these secondary defects is often very useful; rarely a failure due to a defect has only one initiation site. Often what ends up causing the failure is the most critical of a number of previous defects or discontinuities of similar origin. But nonpropagated defects are rarely detectable by visual inspection. In case of suspicion of their existence, it is required to apply evaluation techniques sensitive enough to find and evaluate them, or to rule out their existence.

The method of magnetic particles (MP) detects surface and occasionally subsurface (near surface) discontinuities in ferromagnetic materials. When placing the poles of a magnet on the surface to evaluate, flow lines are formed by the magnetic field generated between the north and south poles. These power lines are altered by the presence of discontinuities, causing leakage of the magnetic field. The magnetic particles will be directed according to the flow lines and will give an indication of the discontinuity. Typical defects found are fatigue cracks and stress corrosion cracking (SCC).

Magnetic particle testing can be dry or wet. In the latter case, the particles are applied in suspension in a liquid (water or kerosene). The test requires white light with a minimum intensity of 500 lux to ensure adequate sensitivity for the evaluation of indications. Also, the wet test can be performed with fluorescent particles with ultraviolet light from mercury arc lamps. In this case, it is necessary to obscure the working area and light will be at a maximum of 1,000 mW.

To determine whether a defect at a failure origin is repeated elsewhere, it is common to test the rest of the component. This is particularly common in welds. Figure 2.18 shows the position of the magnetizing yoke for cracks parallel to the weld joint.

Figure 2.19 shows an example of application of the magnetic particle technique to the verification of an SCC failure of a buried pipeline. SCC is characterized by the initiation of colonies of microcracks on the outer surface of a buried pipeline (as shown in the middle part of the figure). Propagation and coalescence of some of these cracks may cause longitudinal fracture (burst) of the tube (upper Figure), or in rare cases a leak (bottom of Figure). It is unlikely to obtain sufficient field magnification to detect crack colonies visually, especially when they cover large areas of the surface of the component. Figure 2.20 shows a colony of this type, as revealed by the black over white MP technique (a thin white coating and black particles).

The liquid penetrant testing reveals surface discontinuities by absorbing contrasting dye in ferrous and nonferrous metals, and other nonporous materials. It is standardized by ASTM E-165, which classifies fluorescent (ultraviolet visible) and colored dyes, which are also classified by the method of removal. The principle is very simple. First a liquid is dispersed on the clean surface of the material and allowed to penetrate the discontinuities. The liquid is then removed leaving the surface dry; part of the liquid will be retained within cracks or other defects. Finally a developer is applied, which absorbs the liquid housed in the cracks, indicating their presence.

Figure 2.21 shows an example of the technique applied to detect the cause of a failure in a threaded rod. The figure shows the threaded housing in the steel body of a reciprocating compressor. Cracks become apparent, showing a process of

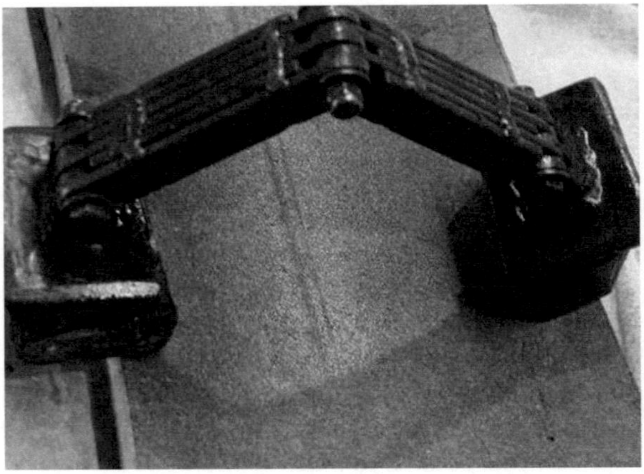

Fig. 2.18 Magnetizing yoke

tearing in layers parallel to the seat surface, indicative of high tensile loads in the connection. Note that in this case, even when the material is ferromagnetic, there is no way to magnetize the piece in the thickness direction. Although the method is fast and inexpensive, it only detects surface defects, has a lower resolution than magnetic particles, and can lead to misinterpretations if not applied correctly.

The ultrasonic test (US) can be classified into two types: Thickness Measurement and Failure Detection. The ultrasonic inspection method involves sending ultrasonic waves through the material with the aid of a coupling material. Sound travels through the material losing some of its energy when reflected at each interface. The reflected waves are picked up by a transducer (receiver) and then analyzed. In the case of Thickness Measurement, sound velocity in the piece is known, so that the time taken for the signal to be released back from the component is measured, when reflected by the other wall surface.

In the case of failure detection, the method uses the same principle to detect the presence of discontinuities within the material. The big difference is that the equipment measures intensity of emitted and received sound. With these data, a reflected wave can be analyzed, indicating position and size of the defect. Normal defects that can be identified and measured include laminations, cracks, pores, inclusions, lack of penetration, lack of fusion, etc.

Some of the advantages of the method:

- The high sensitivity of the method makes it possible to detect small discontinuities.
- US only requires access to one surface of the material.
- Excellent resolution to determine size and location of the defect.
- US does not require special personal safety items.

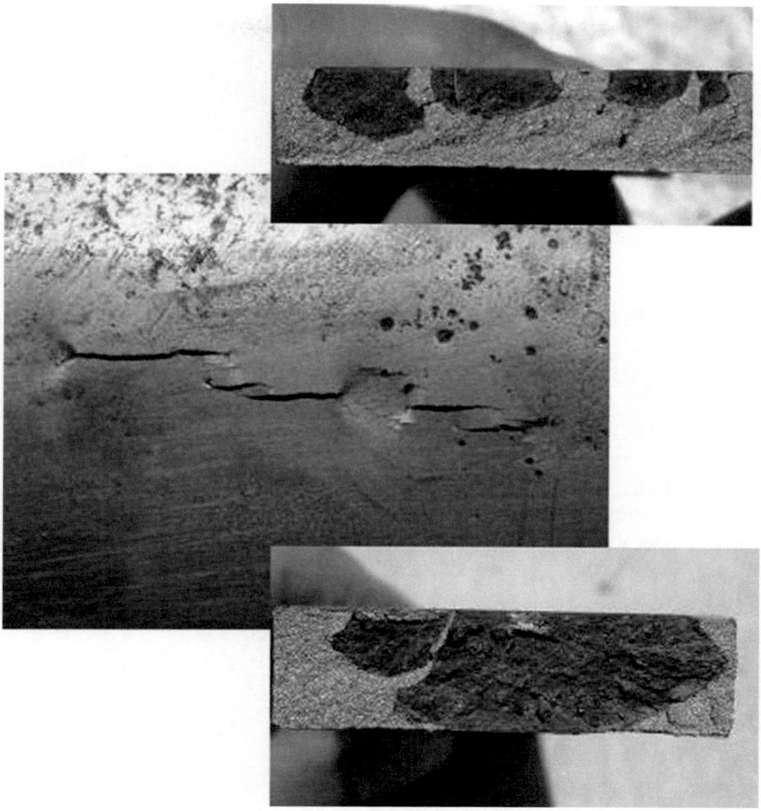

Fig. 2.19 Magnetic particles reveal surface cracks (Center)

In contrast, high training is required for planning and operation, and application of a coupling medium. Rough parts or parts with complicated, small, very thin, or inhomogeneous shapes are very difficult to inspect. In addition, records are interpreted only by competent personnel.

US testing must be in accordance with a written procedure. Each procedure must include at least the information referred to welding and/or material being tested, thickness, surface preparation, and final cleaning. It must also incorporate data of the equipment and coupling used, type of probe (straight or angled) and size of the transducer, angles and modes of wave propagation in the material, frequency, instrument type, calibration description, directions and magnitude of waves, test data to be recorded, and recording method (manual or automatic).

Figure 2.22 shows an example of the use of US for detecting angular secondary defects in a circumferential weld of a large diameter pipe. A fundamental aspect of the test is equipment checking and calibration. To verify proper operation of the

Fig. 2.20 Colony of microcracks

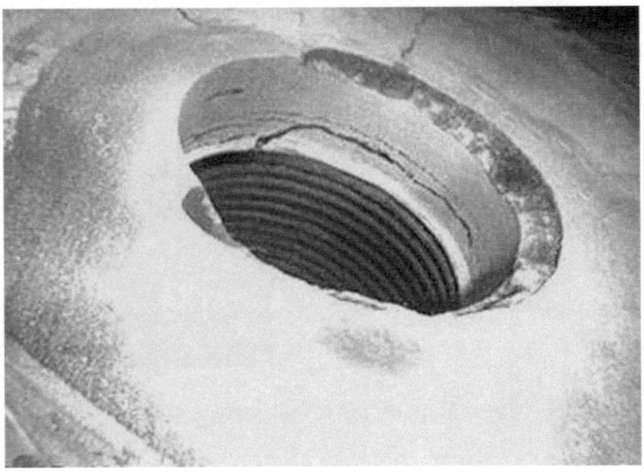

Fig. 2.21 Application of liquid penetrant to threaded bore

test, the equipment must be calibrated at the beginning and end of each test, when examiners change and at any time a malfunction is suspected.

Eddy Current (EC) is an electromagnetic leak testing in which small currents are induced in the material and any changes in the flow of these currents are recorded. These changes are due to inhomogeneities in the material. The basic principle of operation is to pass an alternating current through a coil located in the vicinity of the part to be analyzed. The so-created alternating magnetic field generates a small circulation of eddy currents in the piece. This current in turn generates its own magnetic field, which interacts with the detection coil. The coil then captures disorders caused by discontinuities through an electrical circuit.

The interpretation is based primarily on the analysis of the amplitude and phase of voltage changes in the detection coil by the influence of variations in the parameters of interest of the tested part. The evaluation of the indications is based on comparisons with calibration samples, specially prepared for each test case,

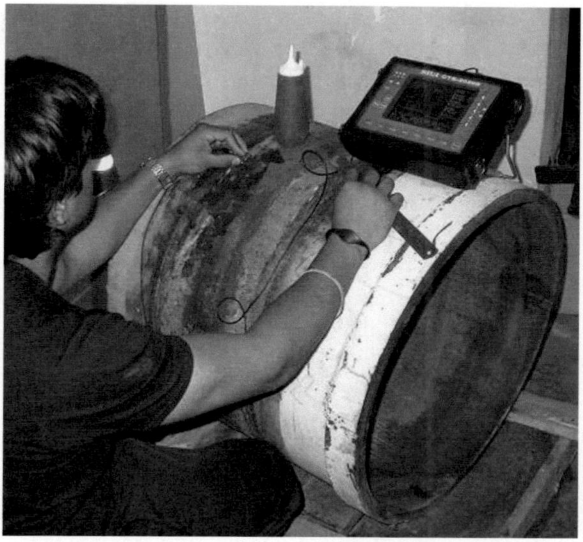

Fig. 2.22 Ultrasonic test with angular probe

according to established rules and procedures. The test is very versatile and very useful in pipelines. It also quickly identifies the defects location and generates a lot of information, but sometimes what is important is difficult to distinguish; it requires highly specialized interpretation of results. EC creates a magnetic field in ferromagnetic materials that can mask some of the results.

Other NDT methods widely used in manufacturing and inspection are not applicable to the preliminary inspection of failures. Such is the case of industrial radiography, which uses radiation to penetrate the material and reveal information. In acoustic emission (AE) testing, transient elastic waves are generated by the rapid release of energy from localized sources, such as plastic deformation or crack propagation. Elastic waves move through the solid to the surface, where they are detected by sensors. This method gets information about the existence and location of possible sources. But the test requires that the component is loaded and therefore in-service, with varying operating conditions. After a failure, the equipment or component is normally out of service.

Field metallography and replica testing involve copying the metallographic microstructure of a material (metals, ceramics, etc.) without destroying the part to be examined. Then the "copy" can be seen under a microscope. It is very useful when it is not possible to extract a sample to examine in the laboratory. The main instrument for the realization of a metallographic analysis is the metallographic microscope, with which a sample can be inspected with magnifications between X50 and X200. For this test, mirror polishing of the piece is required; and various methods and elements are used to etch the microstructure. The information that a metallographic examination can provide is varied: microstructural degradation, quality of material, different phases, determination of alloy types, etc.

Hardness is a measure of the resistance to indentation and friction wear. There are several ways to measure hardness. Most tests use a small indenter device that penetrates the material; indenters can be in the shape of a small ball or a needle. Depending on the type of tip used and the range of loads applied, different scales are defined, suitable for different ranges of hardness: Brinell hardness (HB), Rockwell (HRC–HRB), surface Rockwell, Webster, and Vickers (HV). The interest in the determination of hardness in carbon steels lies in the correlation between hardness and mechanical strength. In cases where getting the equipment out of service for Destructive Testing is not practical, hardness testing is very convenient. An improved variant is the instrumented indentation method, which will be discussed in a later chapter.

2.8 Organization of Work Teams

The areas of the company involved in a failure are the same three areas sharing technical responsibilities: engineering, maintenance, and operations. They may have other names, but the roles are well defined. Operations (production, for example) is the heart of the company, the area that gives meaning to the organization, it is the "client" of the other areas. The engineering area is responsible for the design and construction of infrastructure. The maintenance area (also called Integrity, etc.) is responsible for the normal function of this infrastructure. It is also common that external companies are contracted for short or long periods or for specific events or components, and therefore also become involved in these three fields.

In many cases, Integrity Management is included within the Engineering Department; sometimes, Integrity Management is independent or is included within the Maintenance Department. External consultants ("experts") and contractors are also usually involved in the process of surveying information and failure analysis.

During normal operation of a plant, the three mentioned areas (those that built the equipment, those that use it and those that keep it working properly) often have complex relationships. In the case of a failure, the relationship between these groups becomes even more complex. The staffs of these "subcompanies" that coexist within the company tend to make common cause against those responsible for the other areas. To put it simply:

- For "engineering," the most likely cause of failure is an operational error (it is common to hear that components are ran over 100 %, to increase profitability, or to overcome production losses from other causes).
- For operation and maintenance, the most likely cause is an error in the design or construction of the defective component.

- For engineering and operations, the most likely cause is an error in maintenance; perhaps an oversight or a wrongly reassembled component after a routine inspection, etc.

To complete, we find insurance companies and adjusters. Adjusters have a stake in the outcome of a failure analysis, since in some cases the result defines who will be responsible for the costs arising from the failure, and to what amount.

These emotional attitudes must be dismantled and be replaced by a professional and proactive attitude toward failure analysis. As responsibilities and roles to emergencies such as fires are defined, it is also advisable to set (before the occurrence of a failure) the responsibilities for each of the managements. Here, as an example, the responsibilities defined by a natural gas transport company for leak events and other failures in their pipelines and associated components:

Integrity Management

- Communicate to operations and maintenance managers to take actions based on the preliminary failure analysis.
- Provide the necessary resources or outsource for further analysis.

Head of Integrity

- Coordinate as necessary with maintenance and gas transport areas for the activities needed to obtain the required information to avoid occurrence of the incident.
- Define the type of analysis and tests to be performed, as well as the vendor for the services according to the need.

Integrity Supervisor

- Collect information from mechanical, operational, gas control, inspections, and other relevant sources to determine the possible(s) mechanism(s) of damage.
- Develop a detailed drawing with the location of the failure.
- Prepare a preliminary integrity analysis.
- Inform those responsible for field maintenance of criteria for proper identification, extraction, preservation, and transfer of the sample(s).
- Perform coordination and monitoring of the transfer of the samples from the area of the event to warehouses or maintenance base.
- Perform a preliminary integrity analysis report for closing the event.

Operations Management

- Coordinate with maintenance all integrity activities and samples removal.
- Maintain a continuous fluid communication with maintenance personnel involved in these tasks.

Maintenance Head

- Coordinate with the integrity area and transport management all activities for sample removal.
- Set the required implementation of surveys, collection, and transfer of samples, following Integrity recommendations.

Responsible for Site Maintenance

- Provide the necessary elements for taking samples, photographic survey, and other instruments for the analysis.
- Ensure the preservation, identification, and transfer of the samples to the maintenance base.
- Draw up records for data mining activities and preservation of samples (including all necessary field tests), ensuring delivery on time.

In this scheme, the Integrity Supervisor has the central role in organizing the failure analysis. Elsewhere in this section, this person is defined as responsible in the field. Once preliminary expert activity is finished, he is to develop a preliminary report detailing what was relieved, location plans, and operating conditions at the time of the failure, survey of external conditions, and other data of interest. This report is delivered to management with an overview of the event.

In this gas transport company, the responsibilities of the Integrity Supervisor are defined in a protocol or internal procedure, applicable to the entire company:

- Collect the whole failure data in the area, failures of the same type, on-site inspections.
- Collect additional reports if the case so requires, such as weather reports, reports of floods, earthquakes, etc.
- Collect historical information from operation of the line and specifically the moment of failure.
- Interviewing or request that the interview is made to operators, nearby residents, and any person who can provide information on the failure.
- Collect information, samples, and characteristics of the failure zone, depending on the failure occurred. He needs to know environmental characteristics that helped failure.
- Make a sketch of the terrain and failure zone.
- Determine what type of previous defects caused the event, and identify associated damage mechanisms in a preliminary integrity report.
- Search in databases for the same type of failures, if any, with the aim of finding a correlation.
- Interact with the company performing the failure analysis and the other departments of the company, to coordinate and provide the necessary information.

The Head of Integrity, the supervisor's immediate superior, defines the needs and determines what types of failure analysis should be performed. It is also responsible for defining the scope of the analysis, depending on the type of failure and its consequences for the organization. There is a range of possibilities, some were already mentioned in Chap. 1, and others are discussed in greater detail in later chapters.

References

1. R. Matthews, Tumbling toast Murphy's law and the fundamental constants. Eur. J. Phys. **16**(4), 172–176 (1995)
2. D. Bjerketvedt, J.R. Bakke, *Gas Explosion Handbook*, Christian Michelsen Research AS programme "Gas Safety 1990–1992" (GSP90-92), (CMR, Bergen, 1993)
3. W.E. Baker et al., Methods for calculating the effect of explosions from damage indicators (1983)
4. P. Fazzini, J.L Otegui, C. Manfredi, Field & laboratory study in detection and prevention of SCC in buried pipelines. DUCTOS 2001, (Rio de Janeiro, Brasil, 2001)
5. J.L Otegui, E. Rubertis: Cañerías y Recipientes de Presión EUDEM, Editorial U.N.M.d.P. ISBN 978-987-1371-18-1. 2008
6. Expert assessment and definition of root causes of failures in camisea 14" pipeline, Techint Eng. Constr. (2006)
7. *ASM materials handbook.* Vol. 11, The Materials Information Society, USA
8. C. Manfredi, J.L. Otegui, Failures by SCC in buried pipelines, Eng. Failure Anal. **9**(5) 495–509 (2002), (ISSN 1350-6307)
9. Lineamientos para Análisis de Falla – Procedimiento, Compañía Operadora de Gas del Amazonas, Perú, (2011)
10. Report GIE 9401-04-10: Análisis de Falla de Equipo Cargador de Buques, (2010)
11. A. Marquez, P. Venturino, J.L. Otegui, Common Root Causes in Recent Failures of Cranes, Eng. Failure Anal. (2013) (ISSN 1350-630. EFA-D-13-00316)

Chapter 3
Tools for the Microscopic Analysis of a Mechanical Failure

3.1 Microstructural and Metallographic Characterizations of Metallic Parts

We saw in the previous chapter that the examination of the fracture surfaces is a powerful tool for failure analysis of mechanical elements. In this chapter, we will discuss techniques for fractographic analysis that allow to reach large magnifications, thereby observing microscopic details which are often vital for understanding the metallurgical causes of failure. The use of large magnification requires that analyzed surfaces are free of contamination resulting from exposure to the environment after the failure. This is impossible in most cases. Ambient humidity generates a surface deposit of oxides in most metals which are then very difficult to remove. If pollution is reduced we can live with it; that is why preservation and packaging of samples was so emphasized in the previous chapter.

In this chapter, we deal with two relevant activities in microscopic analysis. First, we will discuss the analysis of fracture surfaces, primary or secondary, natural or manufactured in the laboratory to repeat the event. This way it is possible to define the growth mechanisms of defects, the interaction between them and the presence of "secondary" defects or discontinuities. Secondary defects are those that did not participate in the final failure, but can give a hint on the origin of the main defect(s).

Fractographic techniques are an essential tool in the correlation of microstructural characteristics with the properties of the materials and morphology of defect growth. Observations are made by optical microscopy and scanning electron microscopy (SEM) to determine the state of the material and/or determine conditions and modes of propagation of defects that could have contributed to the occurrence of the failure.

Second, we will discuss metallographic studies. For determining the interaction between defects and microstructural characteristics, metallographic samples are removed and prepared. A selected sample of the piece to be analyzed is usually called a specimen; and must be "representative." The sample should be as close as possible to the source of the failure or whatever is wanted to identify. The sample

J. L. Otegui, *Failure Analysis*, DOI: 10.1007/978-3-319-03910-7_3,

may have any shape; typical dimensions are equivalent to a parallelepiped of 5–15 mm sides, approximately. These samples are cut from the material, then polished and etched with proper reagents to reveal their microstructure. For example, 2–10 % Nitric acid in ethylic alcohol is called Nital, and is much used when examining steel samples. Other etchants and etching procedures are required when inspecting different materials and microstructural features. There are recommended procedures for these studies, for example by ASTM standards.

Metallography is the discipline that studies the microscopic structural characteristics of a metal or an alloy. Undoubtedly, the microscope is the most important tool for a metallurgist. It is possible to determine grain size, shape, and distribution of various phases and inclusions, which have big effect on the mechanical properties of the metal. The microstructure reveals the mechanical and thermal treatment of the material and, under a given set of conditions, can predict its expected behavior.

Metallographic examinations are destructive tests. When going into increasing magnifications, it is difficult to maintain a correlation between what is being watched and the piece as a whole. Therefore, it is previously required to perform a photographic record of all zones of interest. It is very useful to make a drawing of the sample to indicate photographs taken and magnifications used. Parameters to be set are balance of whites and zoom mode. It is convenient to keep zooming always to the maximum, so that to maintain a valid reference magnification.

3.2 Cutting and Preparation of Samples

Experience has shown that success in the microscopic study depends greatly on the care taken to prepare the specimen. The most expensive microscope does not reveal the structure of a sample that has been poorly prepared. The procedure followed in the preparation of a sample is comparatively simple and requires a technique that develops only after practice.

Before cleaning the sample, a photographic record is recommended to monitor cleaning, in order to keep information which might later become important. When preparing a fracture surface or a material with relatively open cracks, all dirt in the notches must be removed. Otherwise it can be released during polishing and scratch the specimen. Sometimes, however, it is desired to observe these deposits. There are different cleaning methods, they are ordered in Fig. 3.1 by increasing severity.

For the preservation of the sample it is generally sufficient to wrap it in a tissue. If the sample has not been completed or analysis must be continued, it should be covered with a layer of light oil, or wrapped in plastic film. In all cases, samples should be kept in a sealed plastic container with fresh silica-gel to reduce moisture in the atmosphere. Samples must also be properly labeled with the name of the project, client, or case.

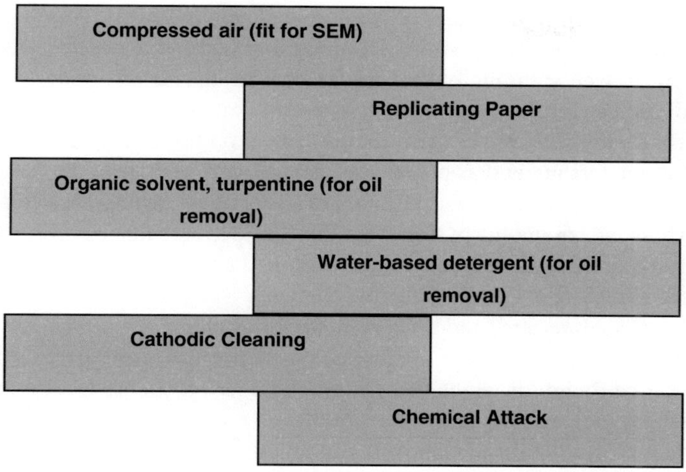

Fig. 3.1 Cleaning methods

The sample must be selected from the right area and proper orientation. For instance, distortion of the microstructure ("banding", as we will see it later) is important, it may occur that a cross-section do not show the elongated grains, only a slice parallel to the rolling direction adequately reveals the elongated grains. Sometimes it takes more than one sample. Usually, welding is examined by a "cross-section," that is, a normal cut to the weld.

Soft materials (hardness less than 35 Rc) can be sectioned by sawing, but the hardest materials are to be cut with abrasive tools. Metal cutting saws with abrasive blades and coolant flow are the tools used for this purpose. The sample should not be overheated, no matter whether it is hard or soft. Grain structures can be altered by high temperature.

Small or awkwardly shaped samples should be mounted to provide intermediate and final polishing. Wire, rods, small samples of metal foil, thin sections, etc., must be mounted in a suitable material or attached rigidly mounted on a mechanism. Thermoset plastics are often used, conformed around the sample by heat and pressure. Bakelite is the most widely thermoset resin used for mounting samples.

Different polishing methods include electrolytic, rotating, or vibrating polishing. The sample is polished on successively finer sheets of sandpaper. The most common procedure involves: first grinding the sample surface in a belt sander, continued by handheld abrasive paper of various grades, from 240 to 600 (the larger the number, the finer the abrasive particles). Usually, the intermediate polishing operations are made with dry emery abrasives, but in certain cases, such as the preparation of soft materials, silicon carbide abrasive may be used. Compared with emery paper silicon carbide is faster. As the backing film is resin-based, it can be used with a lubricant, which prevents overheating of the sample, minimizing damage when the metals are soft. Lubricant also provides a flushing action to clean the products removed from the sample surface, so that the paper is not soiled.

The time spent in polishing and its success depends largely on the care taken during the previous polishing steps. The final approach to a streak-free surface is obtained by a wet spinning wheel covered with a special cloth loaded with abrasive particles, carefully selected in size and composition. While many materials will do a satisfactory job, there seems to be a preference for aluminum oxide when polishing ferrous and copper-based alloys, and cerium oxide for polishing aluminum, magnesium, and their alloys. Other abrasives which are often used for final polishing are diamond paste, chromium oxide, and magnesium oxide. The diamond paste is the most effective when chemical analysis of the polished surface is expected, since it leaves no residue of elements which may then be confused as alloys (typically aluminum and silicon in steel samples).

Several polishing cloths are available, synthetic cloths for general polishing purposes can be found, of which Gamal and the micro cloth are most widely used. Each polishing step should be in the normal direction to the above. A properly polished sample would allow identifying nonmetallic inclusions and other defects. Always suspect of cloth cleanness. Do not clean the sample above the cloth. If the site is left, clean cloth on return, you never know who might have come in between. Do not use the same cloth for copper alloys with other alloys, or vice versa.

The purpose of etching is to make the structural characteristics of the metal or alloy visible. The process should be such that different parts of the microstructure are clearly identified. This is achieved by subjecting the polished surface to the chemical action of an appropriate reagent or etchant. Etchants are mixtures of inorganic or organic acids and alkalis dissolved in alcohol, water or other solvents. Table 3.1 lists the most common etchants for steel. Samples are etched by dipping facedown long enough in the solution. An optional method is to apply the etchant with an eye dropper. The etching action is stopped by placing the sample under a stream of water. The sample is dried with isopropyl alcohol and a hair dryer to finish. It is important not to rub the polished and etched sample with cloth or fingers.

The etching time should be monitored very carefully. If the etching time is too short, the sample will be under etched and grain boundaries and other features will be faded and indistinct when observed under the microscope. If the etching time is too long, the sample is over etched and can be very dark, showing unusual colors. It is advisable to etch in stages. Different microstructural characteristics are observed with varying degrees of etching.

One aspect of the metallographic preparation techniques relevant to failure analysis is the reset of opposing surfaces of a fracture, to obtain a complete view of the interaction between the fracture and microstructural characteristics. The cutting of the samples is usually done in separate parts that are often distorted by the consequences of failure. The sites must be carefully selected to ensure the most accurate matching. The reset can be done by placing the two pieces in the same inclusion, or subsequently matching photographic images.

Such is the case of Fig. 3.2, which corresponds to the point of initiation of a fracture at the root of a weld. It shows two different magnifications. The detail of

Table 3.1 Most common etchants for steels

Metal	Reactive	Composition	Observations
Iron and carbon steel	Nital 5–60 s	2–5 % nitric acid in methanol	Obscures the pearlite Differentiates martensite from pearlite Reveals ferrite grain boundaries Defines depth of surface treatment in nitrided steels
	Picral 5–120 s	4 g of picric acid in 100 ml of methanol	For carbon and low alloy steels. As good as Nital to reveal ferrite grain boundaries. Time 5–120 s
	Hydrochloric acid and picric acid	5 g hydrochloric acid and 1 g picric acid in 100 ml of methanol	Reveals austenitic grains in quenched and tempered steels

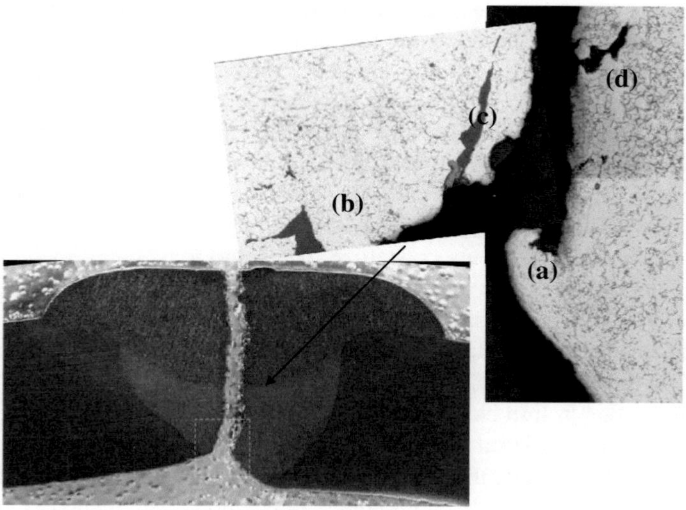

Fig. 3.2 Initiation point of a fracture in a weld (X20). Detail (X200): **a** Protrusion, **b**, **c**, **d** Elongated inclusions

the area of defect initiation (square) is shown at higher magnification (X200). Here, it is shown how microstructural morphologies correlate with fracture surfaces. The point indicated at (a) corresponds to a protrusion which was probably plastically deformed after failure. Defects (b), (c) and (d) are elongated inclusions with sharp edges, probably due to segregation in the columnar grain boundaries of the weld. A defect like this was probably the origin of the fracture, note how it is aligned with the trace of the fracture.

3.3 The Polycrystalline Nature of Metals

Dr. Mario Quinteiro used to say to his graduate students: "The prehistoric man was a tireless seeker of stones, and after several strokes, made his tools. With each stroke a piece of the stone jumped broken and the stone was shaped as a sharp knife or an ax. But some stones did not break but deformed, flattened, stretched. They were not brittle but ductile. Thus Neolithic man discovered metals for its most characteristic property."

Long after men discovered these materials transmitted heat very well and made copper and iron pots; and much later, in the late eighteenth century, men found metals as the best conductors of electricity. From the Bronze and the Iron Ages, one can say that history is the history of metals. Of the 92 naturally occurring elements that appear in the periodic table, 70 are metals. Plasticity and electrical conduction are two properties that are blended with the metal itself.

These two properties are related to the metallic crystalline structure and the link that binds the atoms together. Solids are characterized by a regular arrangement of the particles, atoms, ions, or molecules constituting the element or compound. We can define a spatial arrangement that we call a unit cell of an element, which repeats indefinitely in a space network. Placing the ions, atoms, or molecules at the vertices of the unit cell, 14 different types of spatial networks can be built, which in turn can be reduced to 7 types of solids, base for the 7 crystallographic systems in which all chemical elements crystallize. If a solution or liquid metal is cooled, the solid obtained takes one of these crystal forms.

Metal atoms, separated from their peripheral electrons, are located in the network nodes and form a cluster of electrons; the valence electrons circulate at orbital levels furthest from the atom cores. This is the metallic bond. Metal ions in the crystals are all equivalent; this in turn explains plasticity because electrons can change places without modifying the crystal structure. An electric potential makes electrons formed this way to freely circulate, thus causing electric current metallic conductivity.

Metals crystallize primarily in a body-centered cubic system and in a hexagonal compact system. The compactness of the atomic packing explains the high density of metals, and partly by equivalent crystalline parameters, their contribution to crystal plasticity. Some chemicals, including some metals, crystallize in two or more crystalline systems. This is called allotropy and explains the behavior of some metals, such as iron, which can change some properties under temperature variations or heat treatments.

When the crystal lattice of a metal gets penetrated by an atom of another chemical element, metallic or not, this atom goes into solid solution in the original metal. These atoms can penetrate in two different ways. One is by a substitution of

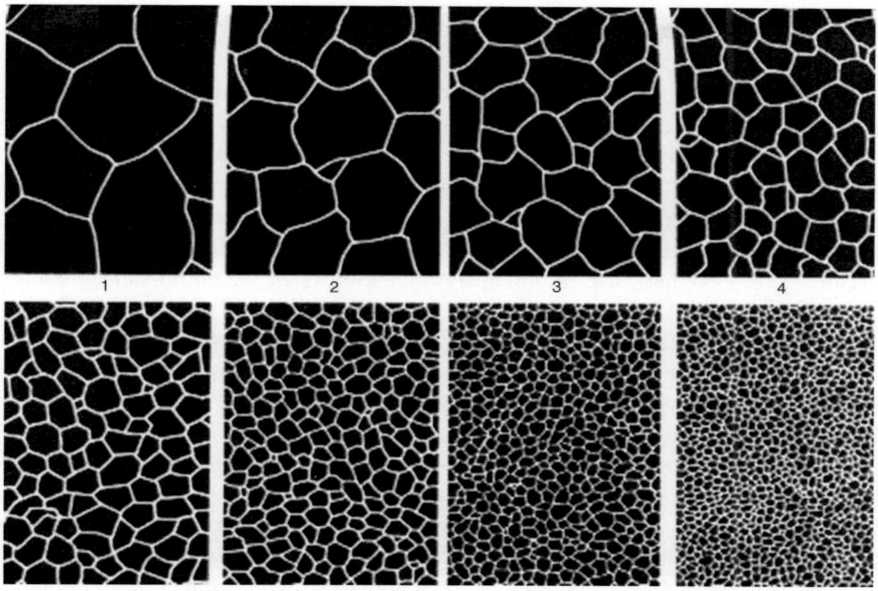

Fig. 3.3 Grain sizes defined by the number of grains per square inch at 100X (ASTM E 112 Standard Guide for Grain Size Measurement)

one or more atoms for one or more solvent solute atoms, which causes a substitution solid solution. In other cases the atoms of the element penetrate into the interstices of the lattice. These atoms are sufficiently smaller than the lattice parameters; thus an interstitial alloy is formed.

The grain size has a remarkable effect on the mechanical properties of the metal. Grain growth effects caused by heat treatment are easily predictable. Temperature, alloying elements, and heating time affect grain size in metals. It is generally preferable a small ("fine") grain size rather than a large one. Fine grain metals have higher tensile strength and hardness, and are less distorted during hardening, but also are less susceptible to cracking. However, in steels, coarse grain increases hardenability, which is often desirable for carburizing steel and also when the steel is subjected to long cold working processes.

A wide range of grain sizes may occur in a given steel. All metals undergo grain growth at high temperatures. However, there are some steels which can attain relatively high temperatures (about 1800 °F or 982 °C) with very little grain growth. But as the temperature increases, there is a rapid growth of grain. These steels are known as fine grain steels. Grain size is determined by the number of grains per square inch when viewed at a microscope with 100X magnification. Figure 3.3 is a chart that represents the actual size of grains as they appear at that magnification. In structural steels, which undergo solid phase transformations, grain size is usually specified as the austenitic grain size (prior to phase transformation).

3.4 Microscopic Examination of the Structure of Metals

Different alloying, operating conditions, and heat treatments generate different structures. Metals and alloys are polycrystalline, i.e., they consist of crystals commonly called grains. The size, shape, and configuration of the grains within a metal or alloy are a function of the way in which the metal is produced and how it is used. It is common for metals and alloys to contain pores, slag inclusions, cracks, or other discontinuities in the grains or along the edges. As mentioned before, if they compromise the integrity or reliability of the component, they are called defects.

Many of these features are sometimes visible only with a microscope (optical or electronic) once the surface has been mirror-polished. This is known as-polished examination. To reveal the grain structure, however, one must resort to contrasting by selective etching. This is known as etched condition examination. Metal characteristics, grain size, and carbon content for example, may be determined by studying the micrograph. The analysis of the microstructure of materials helps determining if the component has been processed correctly. It is critical to determine the reliability of the product and to determine why the material failed. This introduction to metallography is applicable to structural integrity assessment, failure analysis, and determination of residual life of metal components.

A tremendous range of tools are available to support failure analysis with representation and analysis of high resolution images. Each has its advantages, which are often complementary. Successful and efficient analysis requires an understanding of the technical differences and the choice of the tool that best fits the requirements of a particular application.

The reflected light metallurgical microscope is by far the most used. It is similar to those used for other purposes, except that it contains an illumination system within the lens system to provide vertical light. Some microscopes also have a crosshair and micrometer scale to measure the enlarged picture. A reticule containing grain sizes at 100X magnification, used to compare and measure relative grain sizes is also used. Polarizing filters are used in the illumination system to reduce gloss and improve the definition of grain structures. The magnifying power of the microscope is determined by multiplying magnifications of objective lens and eyepiece. Therefore, a 40X objective lens with 12.5X eyepiece would enlarge the image to 500X (500 times).

Inverted stage microscopes offer a more modern design. In this instrument the sample is placed face down on the platen. An inverted stage microscope coupled with a video camera and a CCTV monitor is often used in failure analyses. Optical microscopes can give good images until 500X, with optical magnifications up to 1000X but then some resolution is lost. The image is projected on an opaque screen. Many metallographic instruments have the ability to produce photomicrographs; otherwise adapters are available to obtain photographs in most microscopes.

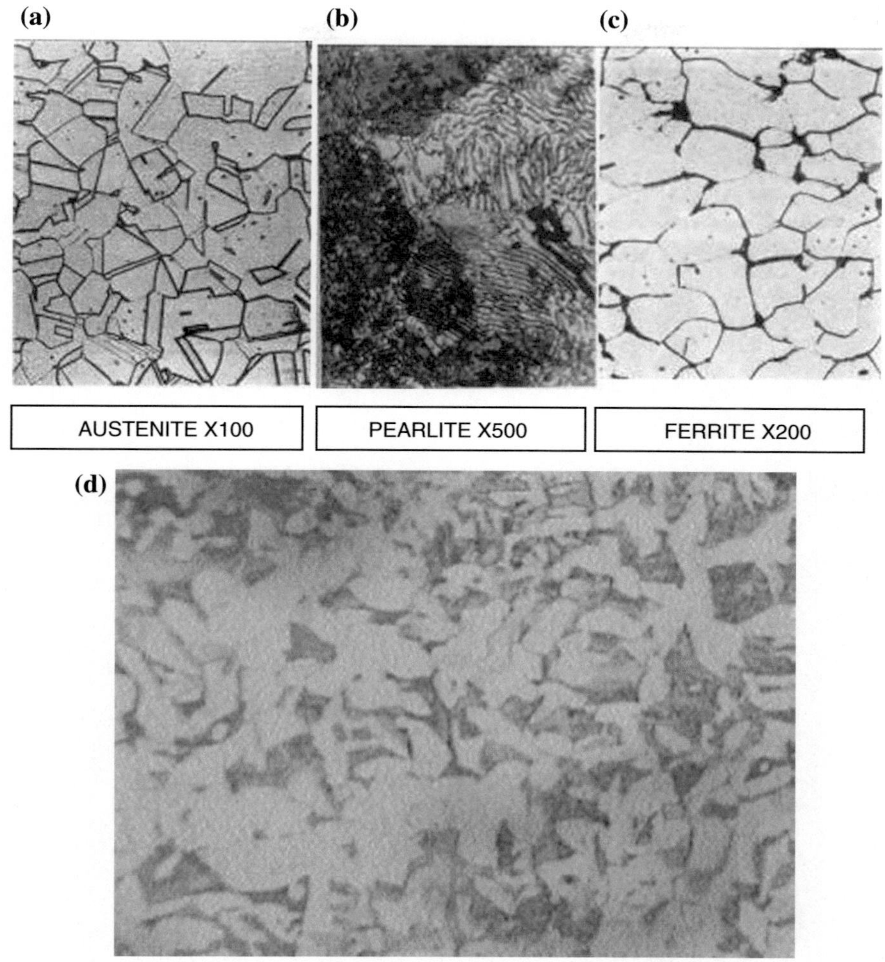

Fig. 3.4 Typical microstructures in structural steels. **a** Austenite X100 **b** Pearlite X500 **c** Ferrite X200 **d** Metallographic sample of a structural steel (X200)

Here are a few characteristics of steel structures used in mechanical components. Figure 3.4a, b, c shows, from left to right, the phases making up the microstructures of typical structural carbon steels obtained by low cooling rate thermal treatment (annealed or normalized). Figure 3.4d shows an acetate metallographic replica of a polished and etched (Nital etchant) structural steel. This Figure shows the continuous Ferrite matrix (clear) with islands of a second phase, the Pearlite, which in normal conditions presents the lamellar internal structure shown in Fig. 3.3b (at larger magnification). The lamellae are ferrite (white) and cementite (dark).

(a) (b) (c)

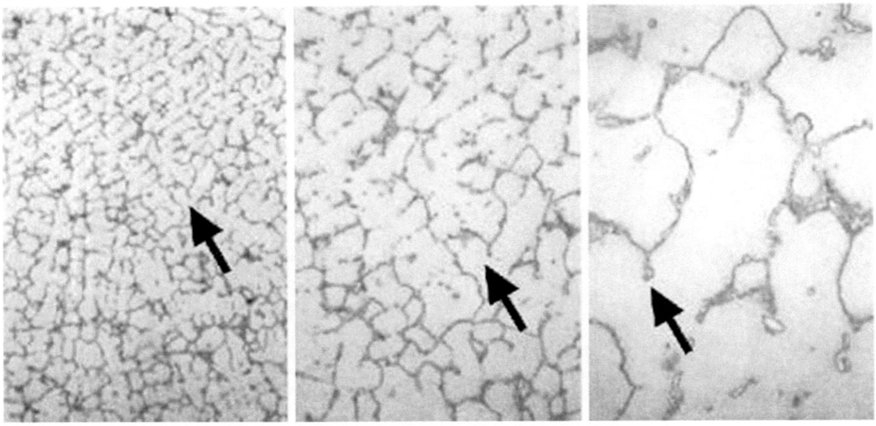

Fig. 3.5 **a**, **b**, **c** Microstructure characteristic of a nickel chromium alloy

(a) (b)

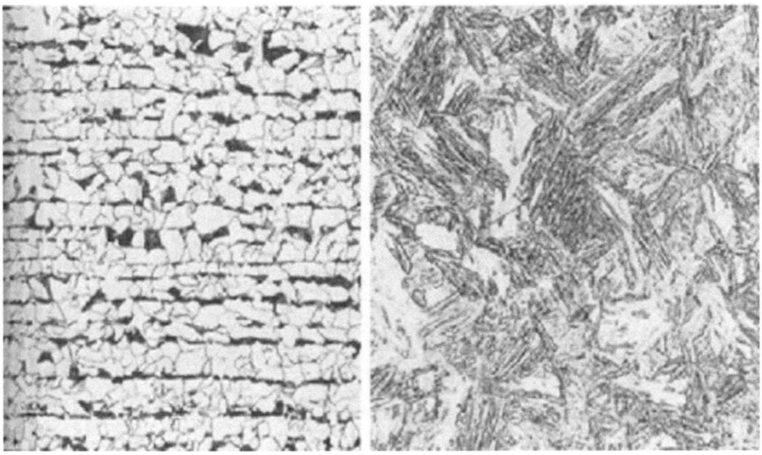

Fig. 3.6 Steels structures (200X) **a** Ferrite pearlite, **b** Martensite

Figure 3.5a, b, c shows the microstructure characteristic of a nickel chromium alloy (25Cr/35Ni/0.4C/1Nb). From left to right: 65X, 130X, and 350X. The arrows indicate the same detail (including grain boundary). Figure 3.6a, b shows two microstructures produced by different heat treatments of structural carbon and low alloy steels:

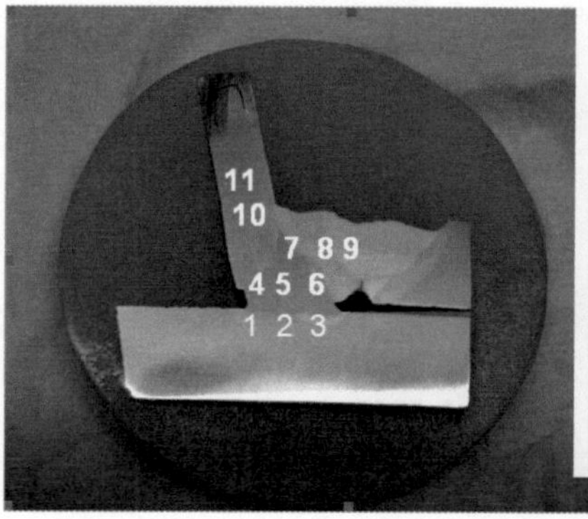

Med. No.	HV 300g	HRC(1)
1	195	<20
2	195	<20
3	203	<20
4	341	34
5	345	34
6	356	36
7	286	28
8	258	23
9	256	23
10	223	<20
11	214	<20

Fig. 3.7 Microhardness mapping

a. Typical ferritic-pearlitic base metal of slowly cooled steels. Banding is common in laminated sheets and pipes, and is due to the deformation of the grains during cold rolling of the plate.
b. Typical martensitic quenched steels. Martensite is hard and strong, but brittle, if not properly treated with a tempering heat treatment.

At intermediate cooling rates, with controlled alloying elements and thermal–mechanical treatments, it is possible to obtain intermediate microstructures. Of these, bainite, when obtained as a second phase in a fine-grained ferritic matrix, enables high strength and ductility.

Martensite can be found even if it is not desired, for example, in heat-affected zones of welds (HAZ). When a failure is suspected to be related to a problem of this type, it is common to perform micro-hardness mappings on the surface of the sample. Microhardness (such as Vicker hardness, HV) refers to hardness tests (Chap. 2) where the indenter is very small. Figure 3.7 shows the result of such mapping. Note in this case that hardness in point 6, in HAZ, is nearly twice that in "normal" base material and weld metal points.

Another way in which mild steel becomes brittle is due to exhaustion of ductility by large plastic deformation, sometimes very localized. The imprint of the Vickers hardness is very small, and allows mapping in very specific areas. In the example of Fig. 3.8 (50X) a section of a pipe suffered gouges and dents when struck accidentally by the teeth of a backhoe. We see surface areas of up to 350 HV, whereas the unaffected base material is less than 280 HV.

An important application of metallographic examination techniques, particularly in the analysis of failures related to wear, corrosion, and fatigue, is the verification of surface treatments. In Fig. 3.9 (a, b, c, approx. X50) there are three examples:

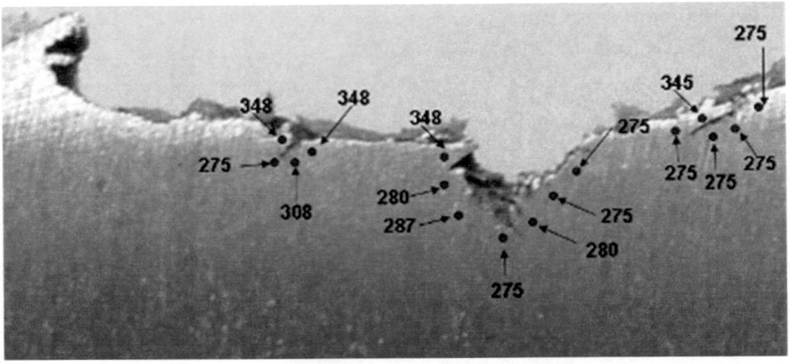

Fig. 3.8 Section of pipe with dents

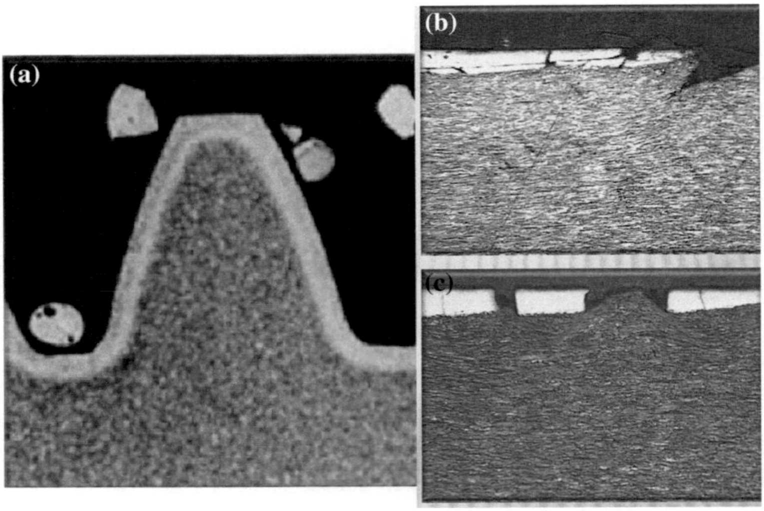

Fig. 3.9 **a** Gear tooth steel (0.16 % C, 0.6 % Mn, 1.65 % Cr, 3.65 % Ni) with a carburizing and surface hardening treatment, **b** Cracking of the hardened martensitic structure in the surface of an engine crankshaft, **c** Surface defects in a galvanized steel wire (zinc surface coating by immersion)

(a) a gear tooth steel (0.16 % C, 0.6 % Mn, 1.65 % Cr, 3.65 % Ni) with a carburizing and surface hardening treatment,
(b) cracking of the hardened martensitic structure in the surface of an engine crankshaft,
(c) surface defects in a galvanized steel wire (zinc surface coating by immersion).

The most salient feature of metallographic analysis in the context of a failure analysis is the characterization of cracks and other defects within the material. An example follows.

Fig. 3.A1 Circumferential fracture at girth weld in a pipeline

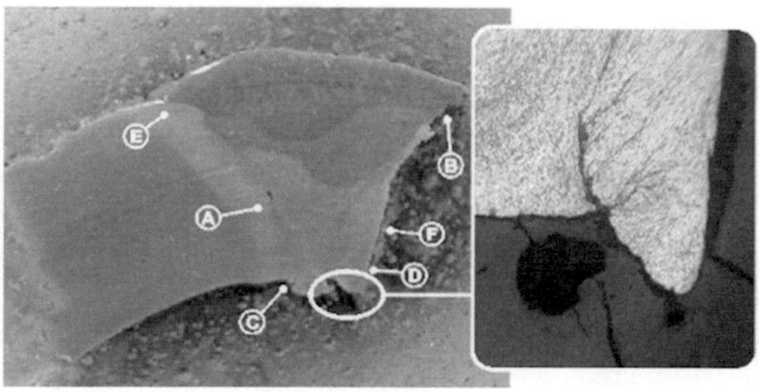

Fig. 3.A2 Metallographic sample of a fractured section after pipe buckle

3.4.1 Example 3.A Metallographic Analysis of a Welded Pipe

Some results of a metallographic analysis are shown in this example. This analysis was performed during the study of the causes that led to a blowout followed by fire in a pipeline in a mountainous area. The blowout was caused by the propagation of a circumferential fracture in the region of a field-welded circumferential (girth) joint, see Fig. 3.A1. The initiation site of the fracture was determined by fractographic analysis, as a localized reduction in thickness of up to 20 % in base material adjacent to the heat-affected zone (HAZ) of the girth weld. Manufacture or construction defects were detected, that could be considered preferential sites of crack initiation (see defects A, B, C in cross-section of Fig. 3.A2). The insert shows the large deformations undergone by the weld HAZ before rupture, indicated by regions D, E, F in Fig. 3.A2. The analyzed fracture was preceded by large plastic deformations (see also diameter variations in the Figure). The blowout was a consequence of large tensile and bending loads in the longitudinal direction, probably due to earth movements that forced the pipe into exhaustion of ductility.

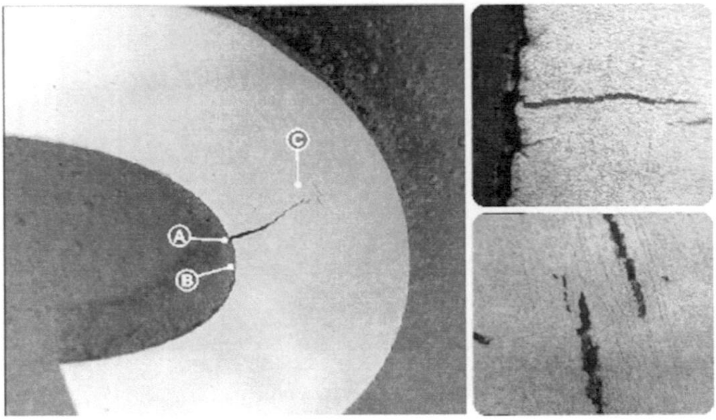

Fig. 3.A3 Cross-section of a buckled pipe, note secondary cracks

Figure 3.A3 (a, b, c) shows the polished and 2 % Nital-etched cross-section of a zone collapsed by plastic buckling in a compressed area of the pipeline. The Figure shows the morphology of microcracks in the buckle, and their interaction with the microstructural characteristics of the material. Note on one hand the severity of the buckle, the local radius of longitudinal curvature is comparable to the thickness of the originally straight tube wall (Fig. 3.A3a). These microcracks are seen at higher magnification in the inserts; note their interaction with the microstructural banding of the ferritic-pearlitic steel.

No cracks can propagate under compressive loads. What happened here is that, immediately upon the occurrence of localized buckling of the section, elastic restitution led to a reopening of the buckle. The material at sites A, B was so deformed that its ductility was exhausted, so that the presence of even small tensile forces were sufficient to propagate the crack AC. The other cracks, in the middle of the thickness, which are perpendicular to crack AC, are secondary cracks produced after large plastic deformation. They were produced by the stresses generated in the thickness direction, due to lateral deformation (Poisson′s ratio will be discussed in a later chapter).

3.5 Microscopic Examination of Fracture Surfaces

One of the main issues with the observation of fracture surfaces is, as seen in Chap. 2, that surfaces are often contaminated. Different cleaning methods are sorted by severity in Fig 3.1. Atmospheric oxidization is a particular problem for fracture surfaces generated during a failure.

Cleaning rusted surfaces, as seen in Fig 3.1, require a chemical cleansing solution. There are commercial products, such as Alconox, but specific formulas can also be prepared. It is convenient to prepare the solution in a container immersed in a water bath at an ultrasonic washer. This is helpful when cleaning small samples. In a container with distilled water at a temperature of 60–90 °C, add citric acid in small amounts while mixing vigorously to dissolve. Once all the citric acid is dissolved, check the pH and temperature of the solution. Add small amounts of ammonia water to raise the pH to 8, mixing at all times. Do not breathe vapors.

The pH number is a measure of acidity and alkalinity; it ranges from close to 0 for strong acid media, up to 14 for strong alkaline media. A pH around 7 indicates neutral solutions. When pH = 6, the solution will act as a "buffer", i.e., it takes time and ammonia water to raise the pH but we must continue to do so in small quantities; at some point the pH of the solution may grow rapidly and exceed 8. It is important to control the temperature of the solution. The mixture has almost no cleaning power below 60 °C. If it reaches 90 °C, the solution will take a caramel brown color and should be discarded. To prepare the passivating solution, pour a tablespoon of sodium nitrite in a container with distilled water at room temperature and stir thoroughly until completely diluted.

Cleaning procedure: immerse the sample in the cleaning solution at a temperature of between 60 and 90 °C. Since the sample is at room temperature, the temperature of the solution decreases slightly depending on the mass of the specimen. Check the temperature of the solution. Shake vigorously. Using ultrasonic washer facilitates the detachment of rust and dirt from the surface of the specimen. For small-sized pieces, 1 min is sufficient for acceptable cleaning. If necessary, re-immerse the sample in the solution, then wipe it again and immerse the sample in the passivating solution. Keep the sample into the solution between 30 s to 10 min. Frequently inspect the state of the specimen. If dark spots begin to appear, quickly remove from water and clean with alcohol or acetone before drying.

The two most powerful tools in a fractographic laboratory are the stereo microscope, for low magnifications, and the SEM, for high magnifications (discussed at the end of the chapter). The stereo microscope is designed for low magnification observation of a sample using incident light illumination. It uses two separate optical paths with two objectives and two eyepieces to provide slightly different viewing angles to the left and right eyes. In this way it produces a three-dimensional visualization of the sample being examined. Stereomicroscopy overlaps macrophotography for recording and examining solid samples with complex surface topography, where a three-dimensional view is essential for analyzing the detail. Stereomicroscopy is widely used when determining fractographic characteristics and in identifying initiation sites.

The two main modes of final failure in pressurized components, machinery and other metallic equipment that can be identified with fractographic analyses are:

Brittle Fracture: as seen in Chap. 2, this is the sudden failure when the stress (residual or applied) reaches a certain value. On a macroscopic scale, brittle fracture occurs without appreciable plastic deformation and due to rapid crack

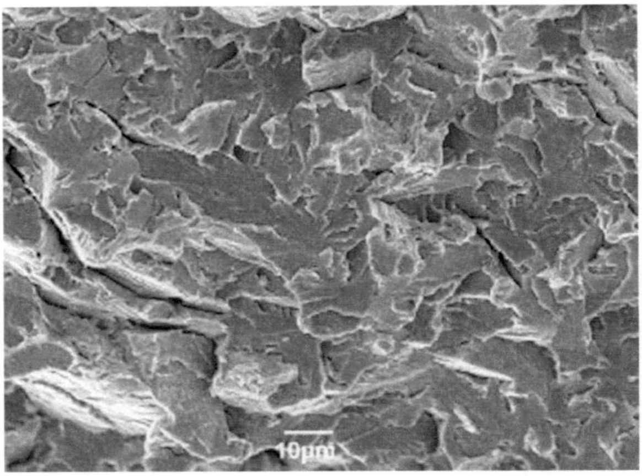

Fig. 3.10 Brittle cleavage fracture at large magnification

propagation. On a microscopic scale, fracture usually occurs along specific crystallographic planes called fracture planes, which are perpendicular to the applied stress. Most brittle fractures are transgranular, that is, propagate through the grains. But if grain boundaries are weak, it is possible to have intergranular fracture propagation. At the microscopic level, brittle fractures in structural steels involve a process called cleavage, where the grains are broken and bright flat surfaces are formed. Figure 3.10 (X1000) shows such brittle cleavage fracture at large magnification.

Ductile Fracture and Excessive Plastic Deformation: this failure mode is very well understood and is the basis of the classic design of structural components. On a macroscopic scale, we have seen (Fig. 2.7) typical 45 degree slant ductile fractures in structural steel plates. The effect of multilayered fracture surface occurs in modern medium to high strength steel pipes (for example, API 5L X70), fabricated with controlled thermal–mechanical treatment. All ductile fractures in structural steels at a microscopic level involve a process called microvoid coalescence. Figure 3.11a 1000X, b 10000X shows this type of process at large magnifications, the small grains in each void are original inclusions in the material.

3.5.1 Example 3.B Fractographic Analysis of a Thick-Walled Tube

This example shows the power of fractographic techniques for determining the physical cause of an unusual failure. Figure 3.B1 shows the surface of a curved thick-walled steel tube, from a reactor used in a petrochemical plant that presented repetitive failures. Pipes are subjected to an internal pressure 2,500 bar. Causes of failure and damage mechanisms were analyzed in ten tubes. In all cases branched

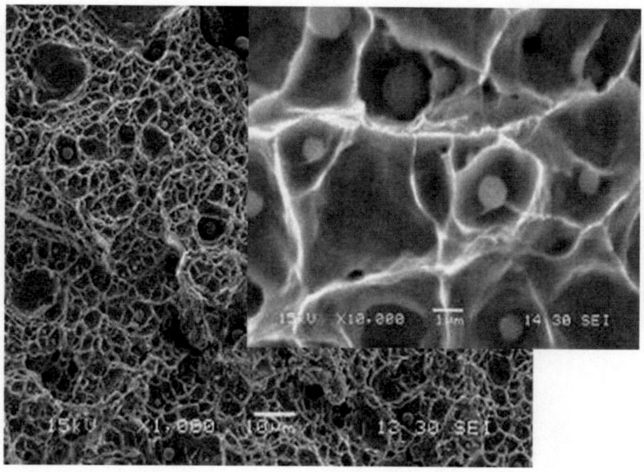

Fig. 3.11 Micro void coalescence in a ductile fracture

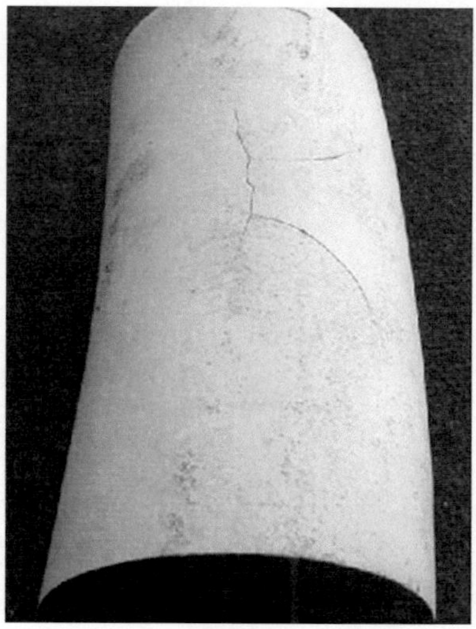

Fig. 3.B1 Curved thick-walled steel tube

cracks were observed, Fig. 3.B2 (a, b). These studies showed a fast or brittle fracture, i.e., without large plastic deformations, that can in some cases propagate as an intergranular fracture.

Figure 3.B3 (a X14, b X200) shows the initial defect on the inner surface of one of the tubes, from which cracks grew in service. All cracks in these tubes are largely circumferential and propagation is intergranular and branched. Grain boundary embrittlement was suspected to be due to some kind of degradation mechanism. This rare form of cracking required specific analysis: residual stress measurement and simulation of mechanical and thermal embrittlement.

SEM micrographs of Fig. 3.B4 (a X1000, b X4000) show an extremely rare case of intergranular fracture with microvoid coalescence, which identified a

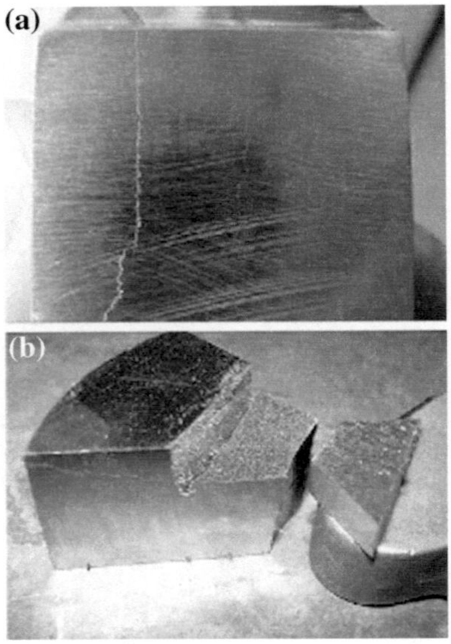

Fig. 3.B2 (**a**, **b**) Branched cracks

Fig. 3.B3 Initial defect at inner surface of thick-walled tube

particular embrittlement mechanism of the tube material. Cracks propagated under the action of tensile stress due to the combination of service loads (thermal and mechanical expansion, internal pressure and moments arising from pipe support) with large axial residual stresses. The root cause of all failures is a grain boundary degradation mechanism that occurred during service; this mechanism could be related to the manufacturing process. In absence of cracks, this embrittlement mechanism could be reversed by further thermal treatment of the tubes.

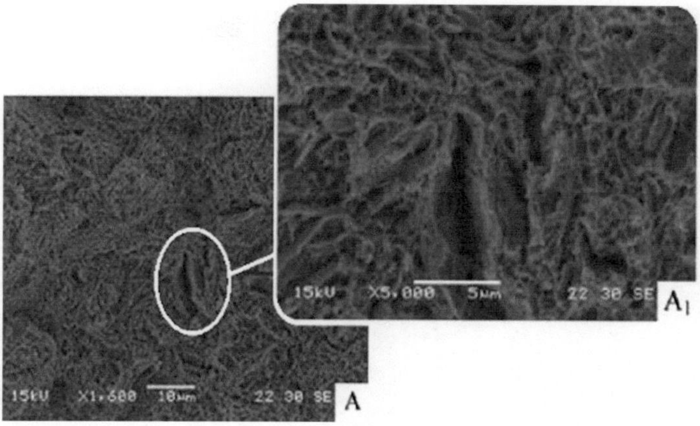

Fig. 3.B4 Intergranular fracture with microvoid coalescence

3.6 Identification of Subcritical Growth Surfaces

Forms of subcritical propagation from a defect until the occurrence of the final failure also lead to fracture surfaces with particular characteristics. In addition to the macroscopic characteristics defined Chap. 2, such as beach, ratchet, and river marks; microscopic characteristics can also be defined. Fatigue cracking is a mechanical form of degradation that occurs when the component sustains cyclic stresses in a period of time.

The most notable feature of transgranular fatigue propagation is the formation of striations. Unless fracture surfaces are degraded after failure by impacts or corrosion, these surfaces clearly indicate when crack growth was due to fatigue. As already mentioned, beach marks in the smooth fracture surface indicate successive positions of the crack front. This enables individualizing initiation sites of crack growth. In some materials, it is possible to identify the propagation mechanism at a microscopic scale. Fatigue crack growth in stainless steels, aluminum, and structural steel welds involve microscopic features called striations, which are very small parallel marks. Figure 3.12 shows this type of process at large magnifications. The crack propagates from left to right; roughness is produced by equally spaced, parallel striations. At lower magnification, the surface is flat and opaque (not shiny or polished).

Subcritical propagation by stress corrosion cracking (SCC) or Creep, for example, is characterized by intergranular, generally branched crack propagation. Figure 3.13 shows an example of SCC. At a magnification of 1000X grain boundaries are seen as a corrugated surface with a corrosive deposit.

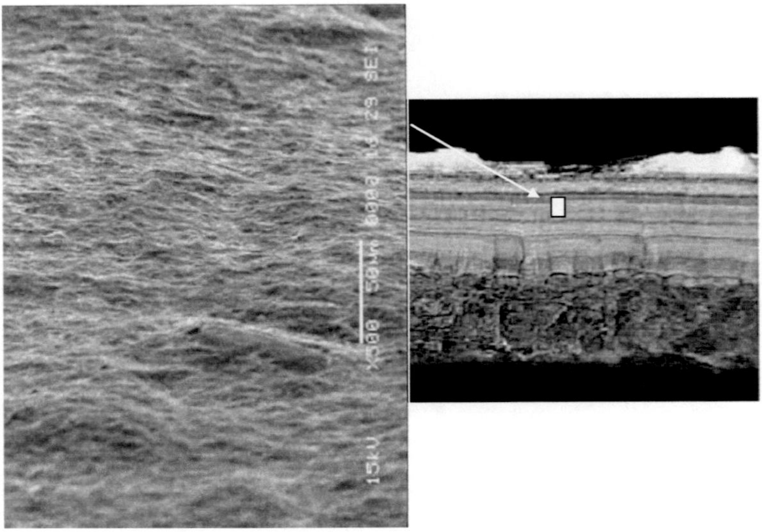

Fig. 3.12 Fatigue crack surface: macroscopic beach marks and microscopic striations

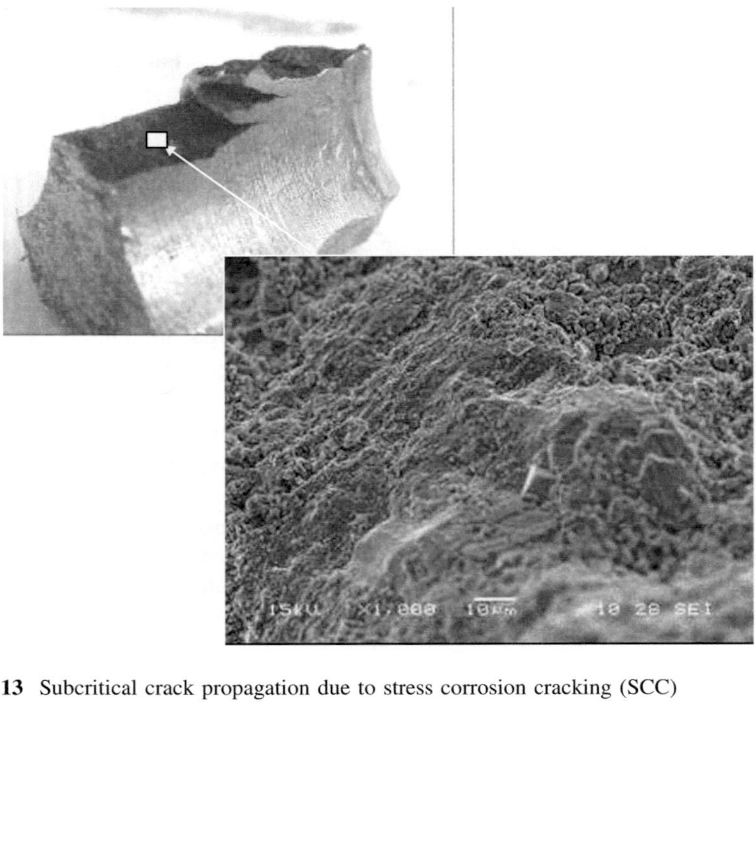

Fig. 3.13 Subcritical crack propagation due to stress corrosion cracking (SCC)

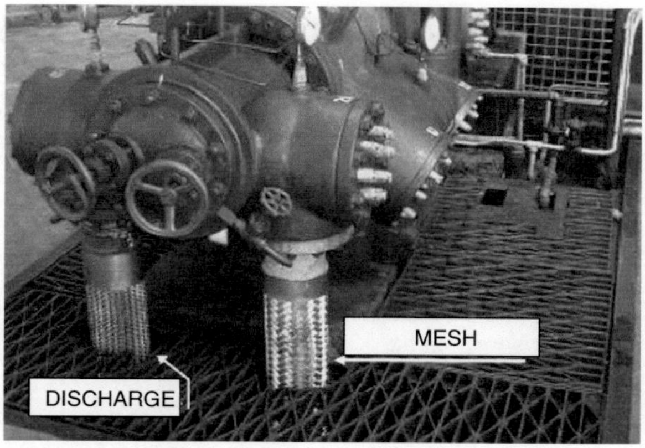

Fig. 3.C1 Position of flexible hose at compressor discharge

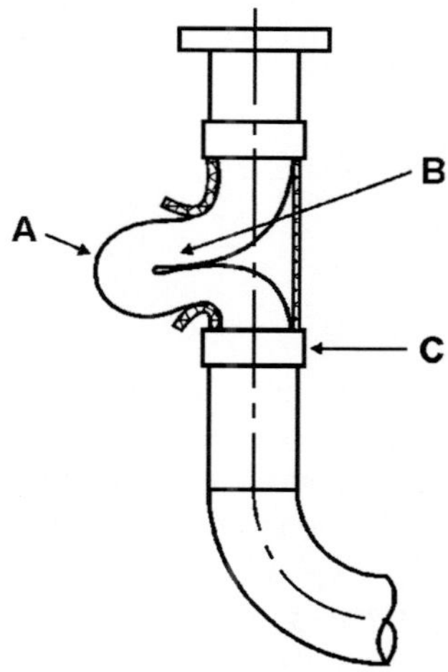

Fig. 3.C2 Failure mechanism by lateral instability of the corrugated tube. Zone **A** External side of the curve; Zone **B** Internal side; Zone **C** Final break and separation from the corrugated tube

3.6.1 Example 3.C Failure of a Flexible Connection

In this example we examine the failure of a flexible connection consisting of a 6″ diameter corrugated stainless steel pipe within a stainless steel wire mesh, see Fig. 3.C1. The connection was placed at the discharge of a propane compressor. To determine the reason for failure, it was sought to verify if the hose was incorrectly selected or if there were deficiencies in placement and maintenance.

Visual analysis allowed determining that in the final stages of failure, the corrugated tube protruded the mesh like a hernia in the external side of this curve, thus producing the corrugated to stretch (Fig. 3.C2). Final failure of the corrugated was by tension; the corrugated yielded after the loss of strength of the mesh, which was gradually breaking due to wear and fatigue of wires.

Fig. 3.C3 Mesh wire with wear marks

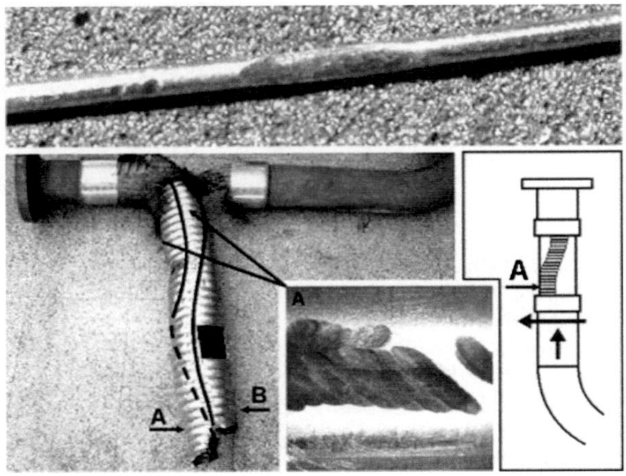

Fig. 3.C4 Failed flexible hose. Wear patterns of corrugated tube are indicated, detail shows wear on the outside of the lobes (20X)

Fractographic analysis allowed detection of the way progressive mesh failure went on. Figure 3.C3 shows that the failure was caused by overloading; after a marked localized reduction of section of the wires due to wear. Wear was caused by friction between the wires and against the corrugated tube during operation, Fig. 3.C4. Fractures in the wires were the result of fatigue crack propagation in previously worn sections, see for example Fig. 3.C5.

No previous defects were found in the wires. Dimensional and mechanical tests confirm compliance with specifications. Analyzing manufacturer and operation data, it was found that operating pressure was above the allowable pressure for the hose, mainly due to the cyclic pressure to which the connection was subjected, downstream the piston compressor.

The failure mode also highlights a problem in placing the flexible joint. The flexible was found to be working slightly compressed, held in position by the flanged ends. This enabled the formation of local contact between the corrugated and the mesh, only in part of its perimeter, Fig. 3.C4. This allowed relative motion and consequent wear. As a solution to the problem, verification procedures of dimensions, position placement, and setting procedure were recommended. Increasing the frequency of inspection and checking vibration level of the flexible and linked elements was also recommended.

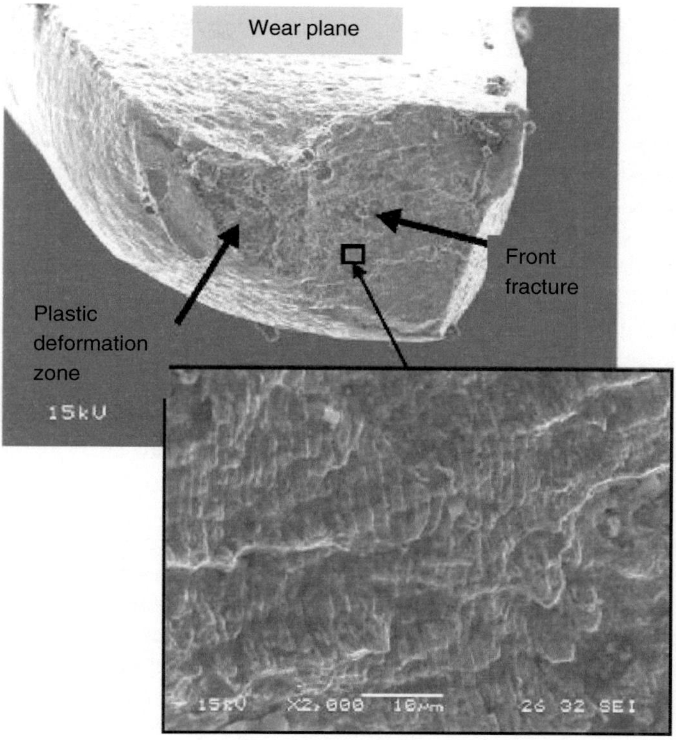

Fig. 3.C5 Micrograph (X100) of a mesh wire; characteristic parts are identified. Detail: fatigue striations (X2000)

3.7 High Magnification Tools, Electron Microscopy

Amplifying the power of an optical microscope is limited by the wavelength of visible light (4,000 Angstroms). Furthermore, depth of field is lost with increasing magnification (as seen in the case of photo cameras in Chap. 2). The electron microscope uses electrons to illuminate an object. Since electrons have a much smaller wavelength than light (about 0.5 Angstroms) they can show much smaller structures. All electron microscopes have several basic elements. An electron gun emits electrons that strike the specimen, creating an enlarged image. Lenses are used to create magnetic fields that guide and focus the electron beam. The vacuum system is an important part of the electron microscope, since air molecules can deflect electrons.

All fracture surfaces at 10X or higher magnifications shown in previous sections were taken with a Scanning Electron Microscope (SEM). SEM sweeps or "scans" a fine electron beam centered over the surface of a bulk sample and builds an image. Brightness at any point represents the intensity of the measured signal when the corresponding point on the sample is illuminated by the beam. The SEM is generally regarded as a technique for surface imaging. It is widely accepted on the basis of its speed, flexibility, simplicity, and minimal sample preparation.

Energy-Dispersive X-Ray Spectroscopy (EDS or EDX) is a technique used to determine the electronic structure of materials by X-ray excitation. EDS has a wide range of applications, especially in determining the crystalline structures of solid samples. An X-ray is electromagnetic radiation with a very high energy. X-rays are capable of penetrating crystal structures: wavelength is of the order of interatomic distances, so that rays diffract, producing diffraction patterns that are characteristic for the crystal. Data obtained are used to determine the crystal structure of each material, and thus its chemical composition.

Characteristic X-rays generated by electron beams provide a means for determining the elemental composition of the sample. X-ray images (maps) are generally noisy, and have a poor spatial resolution, so that analyses are semi-quantitative. They still provide vital information in failure analysis applications such as identification of phases in the microstructure, contamination, corrosive deposits, etc.

In Transmission Electronic Microscopy (TEM) the sample is illuminated with a beam of high energy electrons; focused electrons are transmitted through a very thin sample. The image is projected on a fluorescent screen or electronic image. TEM images are two-dimensional projections of three-dimensional (although very thin) samples. For relatively homogeneous structures, the thin TEM sample can be considered a two-dimensional cross-section, and the projected image can capture the necessary information. However, if features sought for are smaller than the thickness of the sample, the sample may contain important information in the third dimension (beam direction) that cannot be captured in a single projection.

TEM is primarily used for high resolution imaging and crystallography but is not generally used in failure analysis of mechanical components.

3.7.1 Example 3.D SEM Analysis

Here, we can see a failure investigation where the most relevant information was obtained by SEM and EDS analyses. This is a leak that occurred during the preoperational high pressure test of a 12¾″ diameter steel pipeline. Figure 3.D1 shows visual examination of the defect, from the outer surface of the tube. It was not possible to detect the leak from the inner side. Corrosive deposits were found on this inner surface. Chemical composition determinations were made by EDX to identify any possible influence of chemical conditions during the test or previous storage of pipes. Results are shown in Fig. 3.D2. Some of the deposit shows increased density and adhesion; all deposits were found to be iron oxides normally formed at atmospheric conditions.

Observations were performed by optical microscopy and SEM to determine the state of the material and the defects that could contribute to the occurrence of the

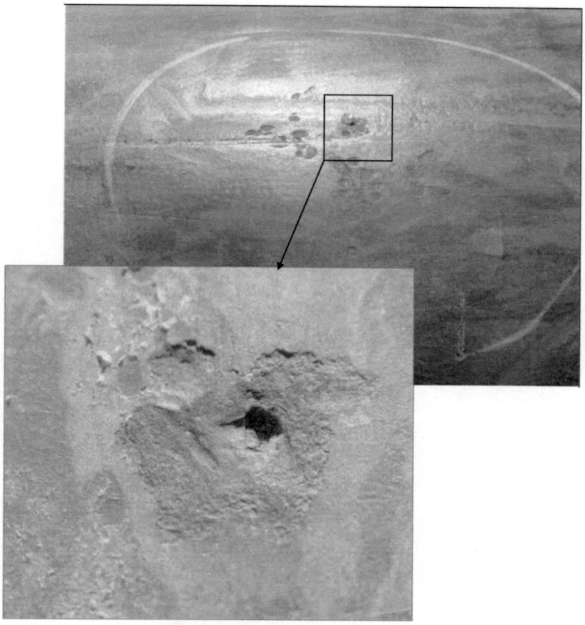

Fig. 3.D1 External examination of the defect

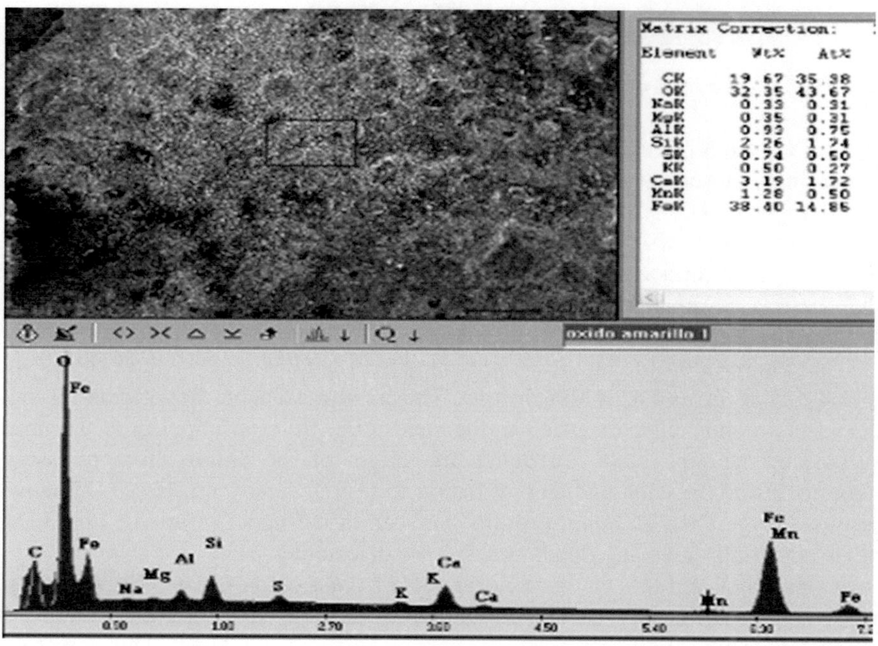

Fig. 3.D2 EDS spectroscopy spectrum

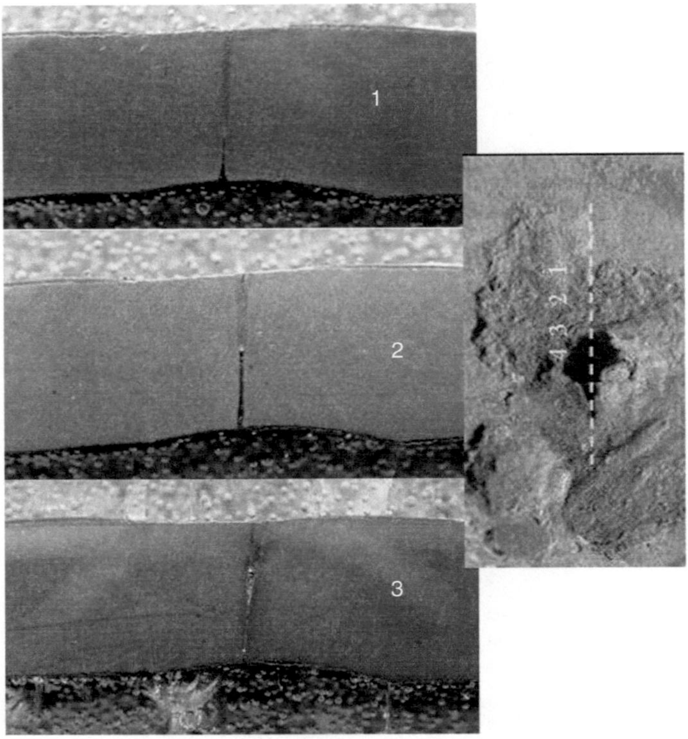

Fig. 3.D3 Cross section of failed ERW weld

failure. Figure 3.D3 composition shows metallographic examination of sections increasingly closer to the middle of the leakage. The dark vertical band is welding metal. Figure 3.D3. 1 shows a planar defect of lack of fusion on the inner surface. Figure 3.D3. 3 shows that this defect joins another, a "cone" of larger diameter, from the outer surface. In this section, the defect is a combination of both. Note that there is no remaining ligament at mid-thickness, and the outer cone reaches the internal lack of fusion.

The composition of Fig. 3.D4 corresponds to a SEM analysis and shows the leakage cone in its larger dimensions. The most reasonable hypothesis for this defect is a nonmetallic inclusion in the weld joint, from an impurity at the inner surface of the pipe plate. To define the origin of the failure more precisely, deposits inside the cone and lack of fusion were EDX analyzed; Fig. 3.D5 shows results. Oxygen and aluminum results are seen in the upper inserts of Fig. 3.D4; other potentially relevant elements were not detectable.

It was concluded that the failure originated from a small, isolated discontinuity: a nonmetallic inclusion with high aluminum content, probably alumina retained in the metal during solidification. Failure origin is then a weld defect. This defect did not lead to a blowout due to its small size, the absence of other discontinuities and

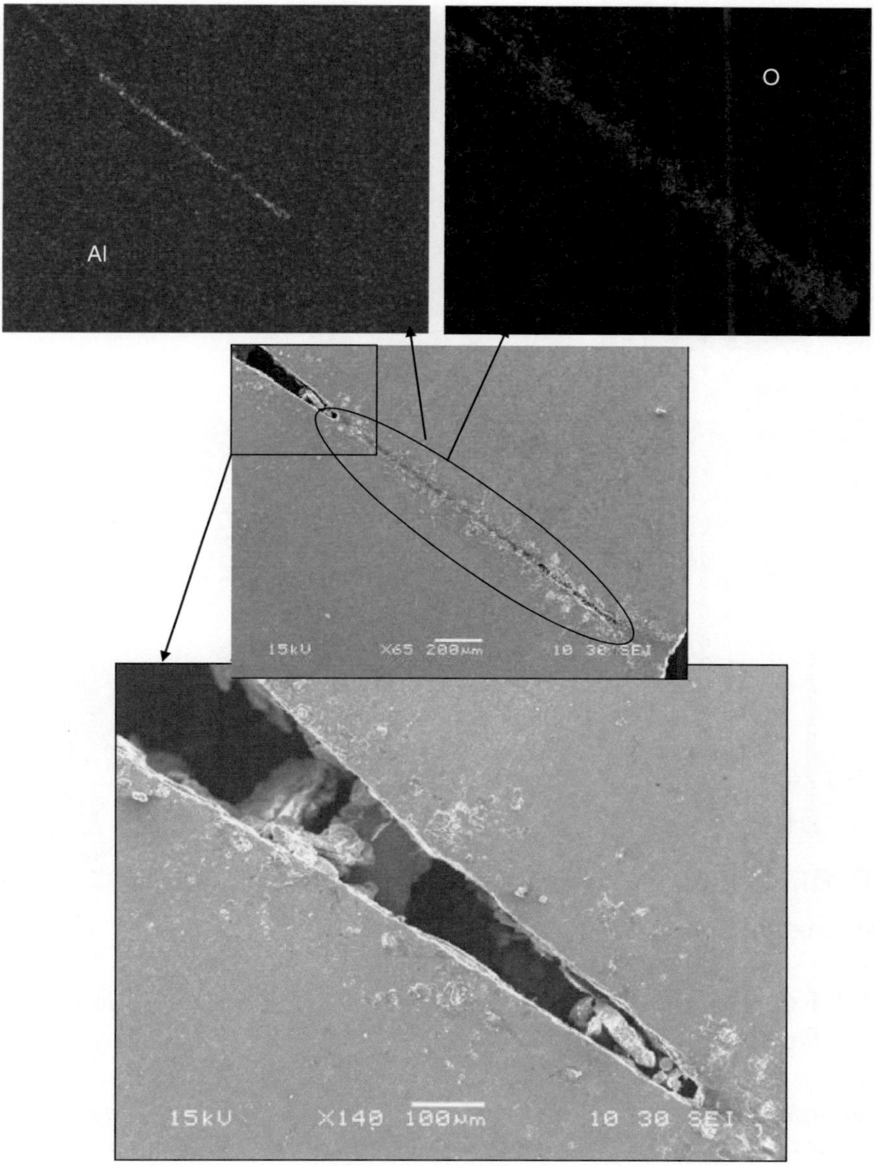

Fig. 3.D4 SEM of ERW defect

good weld ductility (toughness, as will be defined later). This discontinuity may not have been detected during the hydrostatic test of the tube due to the low flow rate of a possible leak, but should have been detected during nondestructive testing.

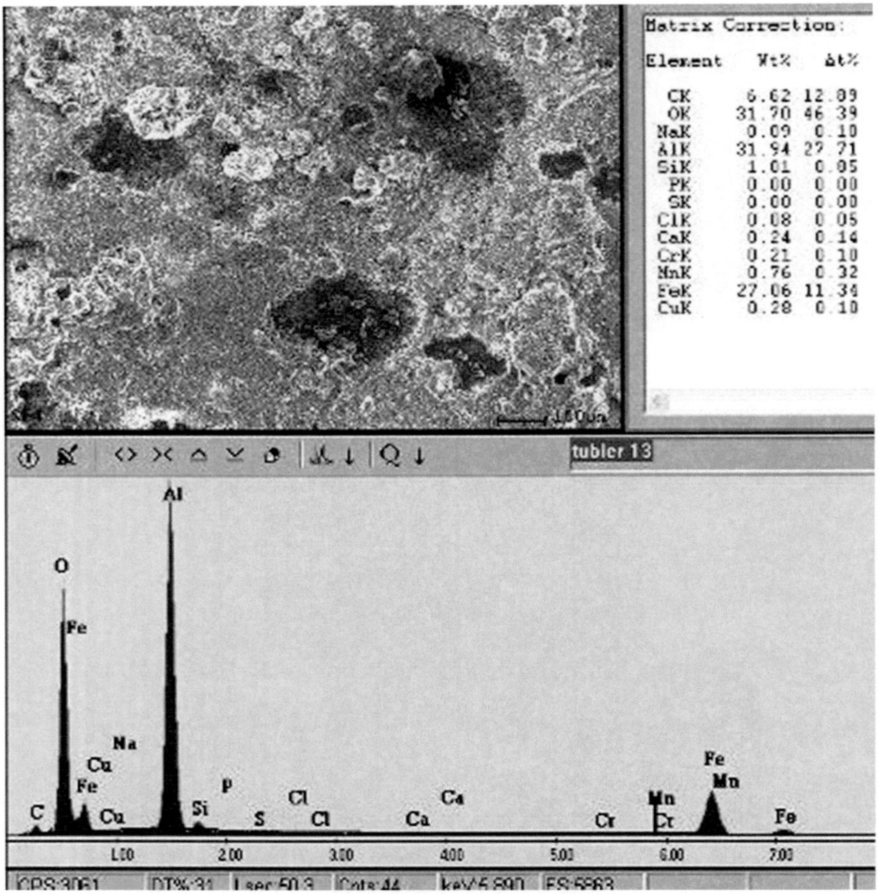

Fig. 3.D5 ED results

3.8 Extra-High Definition Fractographic Analyses, Nano Devices

Advanced techniques, such as extreme resolution SEM (XHR SEM), transmission electron microscopy, and scanning—transmission electron microcopies (S/TEM) allow obtaining sub-nanometer resolution, but differ widely in the information they provide. The selection of the combination of techniques that best meets analytical needs of the problem:

- SEM systems to scan surfaces,
- focused ion beam (FIB) systems and dual beam (FIB/SEM) for imaging sub-surfaces, and
- image analysis systems at atomic scale (S/TEM).

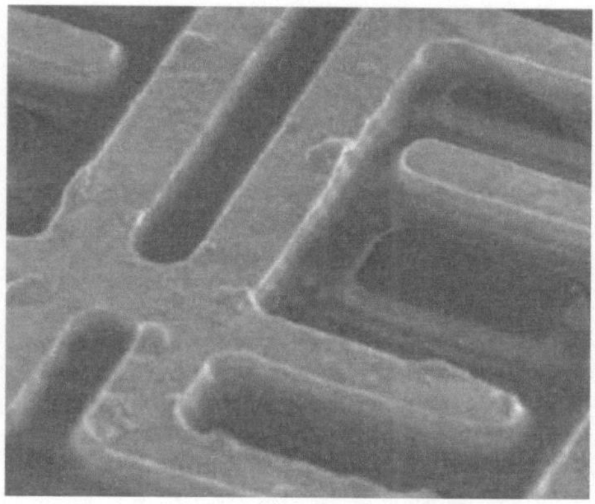

Fig. 3.14 Scanning electron microscopy of low voltage (XHR SEM) with sub-nanometer resolution

Failure analysis of electronic and mechatronic devices is driving the need for high resolution images, due to two main trends. The first and most obvious has been the relentless downsizing of electronic devices to meet the demand for more power and lower costs. When the dimensions of the devices were measured in hundreds of nanometers (nm), the capacity of conventional SEM was more than enough. Now, with few nm devices in development, those same few nanometers constitute a significant fraction of the critical dimension.

The second trend has been changes in the causes of defects, from random defects caused by contamination originating from the environment or processing equipment, to systemic defects arising from the variability of the process itself. Although smaller devices are vulnerable to smaller particles, and there are many more small particles than large in the environment, machinery and equipment manufacturers are very good at reducing contamination as a source of defects. Process variability is typically manifested as variations in critical dimensions, which multiplies the need for accuracy of image-based metrology (Fig. 3.14).

For many years, the SEM has played a major role in the review of defects and dimensional control (CD-SEM), and in failure analysis of parts and packages. SEM is considered as a technique for surface imaging; the practical definition of "surface," however, depends on the application. Beams of high-energy electrons penetrate the sample to a depth determined by their energy and composition of the sample. As electrons penetrate, disperse, and propagate along a region known as the interaction volume, different pictures are originated. Ultimately, the resolution of the SEM is limited by the size of this interaction volume. To analyze nano-components, this volume can be reduced by reducing beam energy. However, conventional SEM cannot focus a sufficiently small low energy electron beam. A low energy beam offers other advantages in semiconductor applications; it reduces the risk of damaging the delicate circuit elements and can reduce or eliminate spurious charges that interfere with the images, and increases the specificity of the image for critical surface phenomena.

The extremely high resolution SEM (XHR-SEM) is a new category of SEM that offers sub-nanometer resolution over the entire range of beam energies from 1to

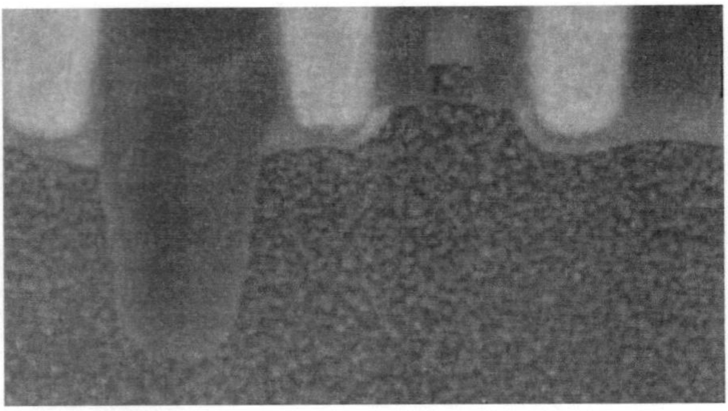

Fig. 3.15 FIB and SEM analysis provides images of structures and subsurface defects

30 keV. It preserves the traditional ease of use, flexibility in sample handling, and minimum requirements for sample preparation of SEM. It also provides the ability to investigate complex three-dimensional surfaces with sub-nanometer resolution, and provides images of specific surfaces that are not available in any other imaging technique available today (Fig. 3.15).

The focused ion beam microscope (FIB) is similar to the SEM, except that the scanning beam is formed from ions (typically of gallium), which has a much larger mass than the electrons. As a result, the FIB can do precisely controlled cuts in the sample. Dual beam systems allow high resolution SEM imaging of FIB cut surfaces.

The need to see the structures and defects below the surface has increased to the extent that innovations in design and processing, such as interconnected multiple layers, have increased dimensional complexity of integrated circuits. FIB systems can perform site-specific cross-sections for representation and measurement. Dual beam instruments can provide high resolution SEM imaging of the cross-section surface. The "cut and see" techniques can reconstruct a three-dimensional detailed model of a structure or a defect from a series of images obtained in the progress of the cross-section through the sample.

With its ability to cut holes accurately located and deposit conductors and insulators in tightly controlled patterns, FIB can also serve to renew a functional integrated circuit. Editing this circuit allows manufacturers to reduce the cost of confirming the validity of proposed changes in circuit design. In the context of sub-nanoscale failure analysis, the most important contribution of the FIB is perhaps its ability to prepare ultra thin samples (less than 100 nm) at a given site for TEM and STEM analyses.

Recent advances in aberration correction have driven TEM images to the sub-Angstrom resolution range, the scale of atoms; while image interpretation has been simplified. Information in the images can now be directly interpreted, eliminating the need for complex reconstruction of images. Image correctors also simplify interpretation of image information near discontinuities in periodic structures, such as crystal defects, grain boundaries, and interfaces. Advances in sample preparation have improved the ease, speed, and reliability of TEM analysis; advances in

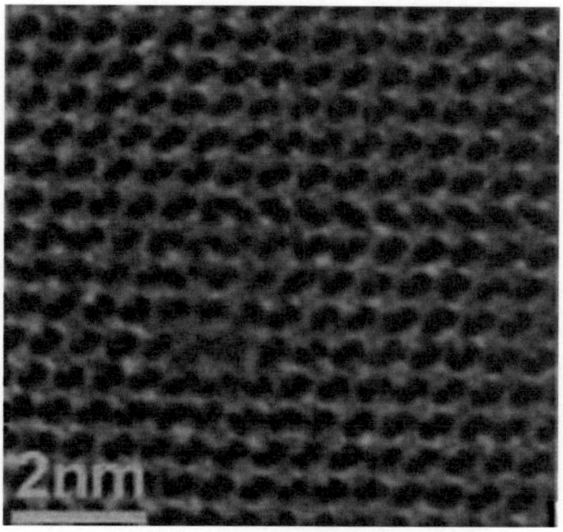

Fig. 3.16 TEM image of the Ga A crystal structure

TEM automation and control systems have also simplified many of the routines for alignment and adjustment of the instruments.

Structures may contain important information in the third dimension (beam direction) that may not be captured in a single projection. Electron tomography reveals this information by combining several images acquired at different viewing angles in a detailed three-dimensional model with almost-nanometer resolution. The energy filtered TEM (EFTEM) forms an image from the electrons with a specific energy loss, and provides the analytical capacity of the full spatial resolution TEM images (Fig. 3.16).

The scanning transmission electron microscope (STEM) scans a fine electron beam focused on a very thin sample and constructs an image from the measured signals at each of the points in the scanned pattern. The construction process of the image is similar to the SEM, but image signals and contrast mechanisms can be very different because the sample is very thin. The STEM is recognized for its ability to provide analytical results with very high spatial resolution. As in the EM, measured signals at any instant of time carry information that is specific to the location of the beam. Unlike the SEM, very thin samples reduce the size of the interaction volume. The spatial resolution for imaging is about the beam diameter. With aberration correction, the STEM has shown a spatial resolution down to 0.5 Angstroms.

References

1. M. Quinteiro, *Apuntes del postgrado en materiales* (Universidad Nacional de San Martín, Argentina, 2008)
2. ASTM E3-11 Standard Guide for Preparation of Metallographic Samples. American Soc. for Testing of Materials, U.S.A., 2011
3. ASTM E407-07 Standard Practice for Microetching Metals and Alloys. American Soc. for Testing of Materials, U.S.A., 2007

4. *ASTM E112/10 Standard Guide for Grain Size Measurement*. American Soc. for Testing of Materials, U.S., 2010
5. P.G. Fazzini, A. Márquez, J.L. Otegui, P. Barcia, Cause and effect assessment after a complex failure of a large ethylene compressor. Eng. Fail. Anal. **13**, 1358–1369 (2006). ISSN 1350-6307
6. P.G. Fazzini, J.C. Belmonte, M.D. Chapetti, J.L. Otegui, Fatigue assessment of a double submerged arc welded gas pipeline. Int. J. Fatigue **29**, 1115–1124 (2006)
7. R. Krueger, T. Morrison, *High Resolution Analysis of Nano-components* (FEI Company, Hillsboro). www.fei.com
8. J.L. Otegui, P.G. Fazzini, P. Barcia, M.J. Schmidt, *Common Causes of Historical Failures in High Pressure Tubes* (Society of Plastics Industry, Houston, 2005), pp. 8–9. Technology Transfer Workshop VI, Nov 8–9
9. J. Massone, J.L. Otegui, H. Ortiz, P.G, Fazzini, Intergranular susceptibility in failures of high pressure tubes. Paper FA 176, ICEFA 2, 2nd. International Conference on Engineering Failure Analysis, Toronto, Canada, 13–5 Sept 2006
10. Report GIE 7103-11-02, Análisis de Fallas de Tres Espárragos, Motocompresores Clark, TGN SA, Argentina, 2002
11. *ASM handbook*. Vol. 11, Failure Analysis and Prevention. ASM International, 2002. (ISBN: 978-0-87170-704-8)
12. Report GIE 2601-04-08, Análisis Pericial de Falla en Prueba Hidrostática de Gasoducto, Petroandina, Argentina, 2008
13. P. Nothnagle et al "Introduction to Stereomicroscopy." http://www.microscopyu.com/articles/stereomicroscopy/stereointro.html

Chapter 4
Mechanisms of Damage and Failure

4.1 Introduction

Components subjected to mechanical loads are designed for a certain period of safe operation, during which failure risk is acceptable. This establishes a design service lifetime for the component. Engineering practices and manufacturing codes consider, during the design stage, which in service damage mechanisms will be acting upon certain components, which will in turn accumulate damage. Virtually all materials react to their environment, most often, these interactions are undesirable and result in economic loss. For example, corrosion represents an annual economic loss of 5 % of gross domestic product in most advanced industrial societies. In America, that means more than 150 billion dollars each year.

Over time, damage mechanisms during operation induce the growth of defects or discontinuities in the material. Sometimes inadequate maintenance practices can produce undesirable effects by activating latent damage mechanisms. The damage mechanisms acting on mechanical and structural components are identified in this chapter, see Table 4.1.

Industrial equipment deteriorates as a result of service conditions. Under normal operation this deterioration is usually, but not always, gradual. It generally occurs as a loss of thickness and occasionally as a chemical or structural change in the material, resulting in weakening or degradation of its mechanical properties. When the equipment is subjected to abnormal operating conditions, such as sudden changes in pressure or temperature, deterioration can be much faster.

A Damage Mechanism is the process by which damage is introduced to the component during service. It has been seen that a failure involves the immediate outage from service, and is usually the result of an accumulation of damage that eventually reaches a certain critical value. All failures generate interruption in operation, economic losses or even lost human lives. The mechanisms that generate immediate or catastrophic failure of a structural component are:

J. L. Otegui, *Failure Analysis*, DOI: 10.1007/978-3-319-03910-7_4,

Table 4.1 Damage mechanisms in mechanical and structural components

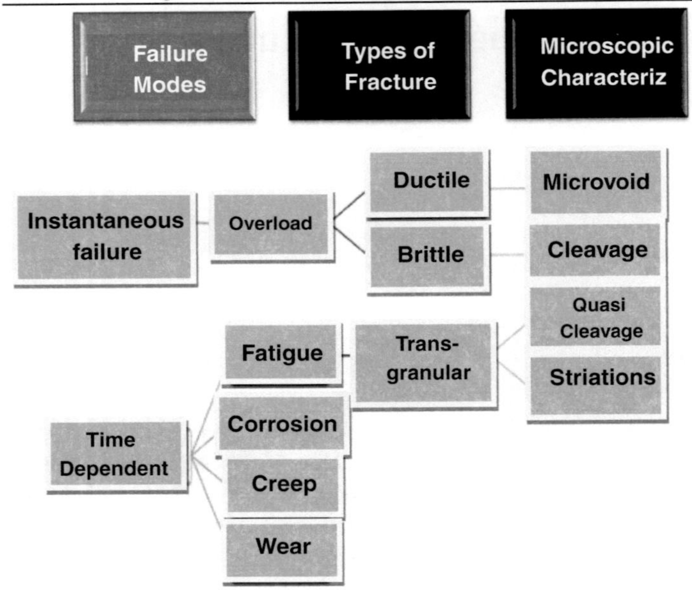

- excessive elastic deformations induced by load or temperature,
- loss of containment (leakage),
- excessive plastic deformation,
- ductile fracture,
- brittle fracture,
- fires,
- elastic or dynamic instability (buckling).

In-service damage mechanisms that can eventually lead to failure include the following:

- Fatigue: high cycle and low cycle crack propagation, thermal shock, associated with impact or wear, etc.
- Corrosion: generalized, localized, direct chemical attack, galvanic, crevice, pitting, intergranular, selective, corrosion-erosion, stress corrosion, microbiological, under insulation, etc.
- creep and thermal relaxation,
- corrosion-fatigue,
- creep-fatigue,
- hydrogen damage,
- wear: adhesive, abrasive, corrosive, sliding, rolling, cavitation, friction,
- surface attacks by gases or liquids,
- other specific and special cases.

Table 4.2 Failure distribution according to mechanism type

Type of failure	Percentage (%)
Corrosion	34
Stress corrosion cracking (SCC)	22
Fatigue	14
Welding defects	8
Erosion	6
Brittle fracture	3.5
Mechanical failure	3.5
Creep	2.5
Overheat	2
Overpressure	2
Others	2.5

Table 4.2 shows the distribution of failures caused by these different damage mechanisms in industrial plants in the UK in the 1990s.

4.2 Failure Mechanisms

The following sections discuss the modes for final failure that can be identified in pipes and pressure vessels, machinery and mechanical equipment. We make a distinction here. Some mechanisms (leaks, fires, etc.) can occur under operating conditions that did not change significantly, simply as a result of gradual degradation. Other mechanisms such as ductile or brittle fracture usually occur after the appearance of stresses higher than normal so far at some point during service. These variations can be small and be within acceptable operating parameters.

Leak: fluid escaping from a pressurized vessel affects service, that only in cases of high toxicity or flammability, represents a safety risk. It is more likely in components with tough materials and with low applied stress (low pressure, large thickness, etc.).

Brittle Fracture: a sudden fracture under applied or residual stresses, without appreciable deformation and due to fast crack propagation. Figure 4.1 shows typical failures caused by brittle fracture. Its propagation characteristics along crystallographic planes were discussed in Chap. 3. The following factors favor the occurrence of brittle fracture:

- low temperatures,
- cracks generated during manufacture or in service,
- large previous plastic deformation,
- low ductility of the material (or toughness, see Chap. 5), common in old materials,
- stress concentration due to discontinuities,
- manufacturing or service residual stresses.

Fig. 4.1 Brittle fracture

Excessive Elastic Deformation: a piece that must keep dimensions within certain limits undergoes elastic deformations that exceed the allowable limits. This could for example lead to interference problems such as clogging, or excessive deflections as in bridges.

Elastic Instability: the most common form of this failure mode is buckling. This phenomenon occurs when a slender structural member such as a column, container or plate is subjected to sufficiently high compression or shear loads.

Ductile Fracture and Excessive Plastic Deformation: as discussed in Chap. 3, this failure mode is best understood and is the basis for design of structural components. Indeed, the main purpose in design is to establish the dimensions necessary to ensure elastic behavior of the component. In theory, this means that no point in a section reaches the yield condition, i.e. plastic deformation. Ductile fracture starts when plastic deformation is large enough to exhaust the ductility of the material at some point in the structure. Ductile (or tough) is opposite to brittle, indicating large amounts of consumed energy and plastic deformation.

As seen in Chap. 2, ductile fractures in metals propagate at an angle (45°) with respect to the direction of the maximum applied stresses. This is due to the physics of the process. Figure 4.2a shows the cross section of a ductile fracture which runs parallel to a longitudinal weld in a gas pipe. Note shear lips and lateral shrinkage, both indicative of plasticity (Chap. 3). Sometimes ductile fracture (also called ductile tearing) can occur in more than one stage. This occurs when crack propagation makes applied stress to decrease, while material toughness increases due to plastic deformation. Both phenomena are rather hard to explain, we will be back to them in due course.

Plastic Instability: under certain conditions, a particular section of a material that has sustained large plasticity becomes unstable and leads quickly to plastic collapse. A known example of this phenomenon is the constriction that precedes break in a tensile test of a ductile material (Fig. 4.2b).

Fig. 4.2 **a** Cross section of a fracture by ductile tearing. **b** Plastic instability

4.3 Mechanisms of In-service Damage

Deterioration or degradation of mechanical and structural components is possible when their surfaces are in contact with a wide range of organic and inorganic compounds, fresh or polluted water, steam or simply the atmosphere. The mode of deterioration can be chemical, electrochemical, mechanical, or a critical combination of them. Also, deterioration may be accelerated by temperature, stress, vibration, high fluid flow rates or flow irregularities, among other factors.

The damage accumulation rate is the rate at which the amount of damage is accumulated during a certain time. This depends mainly on the nature of the mechanism, as well as the conditions to which the structure is exposed. The most common mechanisms will be summarized, although the list is necessarily partial. There is a great deal of damage mechanisms. We begin with the most common in static equipment: corrosion and fatigue, and continue by specificity, ending up with damage mechanisms related to machine components and rotating equipment.

4.4 Corrosion

Corrosion has always been a major problem in chemical processing, mining and oil and gas industries, in naval structures and even in civil constructions. As industries have grown and adopted modern processes, corrosion problems have become more numerous and complex. Table 4.2 shows that 56 % of failure cases are related to corrosive environments. Corrosion is defined as the deterioration of a material due to chemical or electrochemical reactions with the surrounding medium. Corrosion is a natural (thermodynamically possible) and complex process that can occur in various forms and is affected by numerous chemical, physical, electrochemical, mechanical and metallurgical variables.

Corrosion rate is influenced by material dependent variables (chemical composition, metallurgical properties, etc.). Because corrosion is a surface mechanism it is also influenced by variables dependent on the formation of surface films on metal (electrical resistance, thickness, composition, presence of defects, etc.). Therefore, one of the main features of a corrosion mechanism is the vast variety of situations in which this process can occur, depending on the specific combination of material and corrosive medium.

In general, any corrosive process involves loss of material, but the shape of the generated cavity varies greatly with the type of corrosion, and helps in many cases to identify the mechanism. Figure 4.3 shows schematically how the section is affected by some of the most common mechanisms of corrosion, which are described below.

Simply, corrosion is the way nature reverses a manmade material with a high energy state (steel or other metal) to its original condition as natural mineral (ore), as it was found in nature. Iron atoms in steel combine with moisture or acids to form other compounds, such as oxides, sulfates, carbonates, etc.

Water is present in almost all media in different forms and concentrations, and most media contain considerable amounts of impurities and dissolved gases. For example, acid gases such as carbon dioxide (carbon dioxide—CO_2) and hydrogen sulfide (Hydrogen sulfide—H_2S) are mixed with most hydrocarbons, and are highly soluble in water—which tends to lower its pH. The corrosivity of water is a function of the amount of gases contained in the solution. All water with low pH is considered to be corrosive to steel. Lower pH values represent higher acidity or corrosiveness.

Corrosion in steel begins very aggressively but often reduces its action as soon as a protective surface film or corrosion deposit is formed on the metal surface. For example, CO_2 generated iron carbonate scale, a corrosion byproduct, covers the metal and retards the penetration of corrosion, thus decreasing corrosion rates. However, if the film is continuously slit or removed by abrasion, aggressive corrosion remains in the area where the deposit has been removed, which causes deep corrosion pits or cracks.

Galvanic Corrosion occurs when two dissimilar metals are electrically connected in the same corrosive medium (Fig. 4.4a). The driving force for the attack

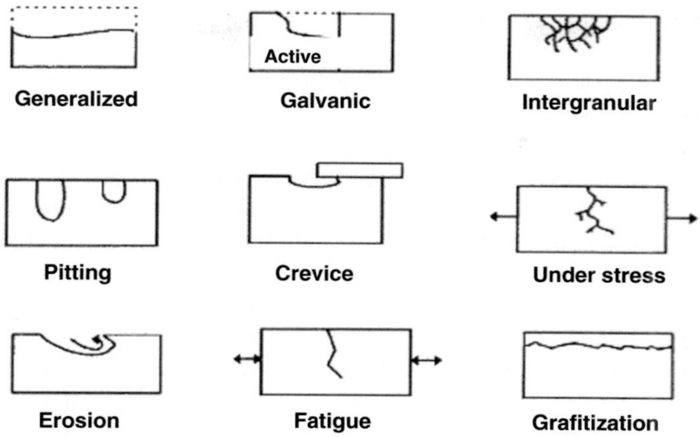

Fig. 4.3 Forms of corrosion degradation (C.P. Dillon 1982)

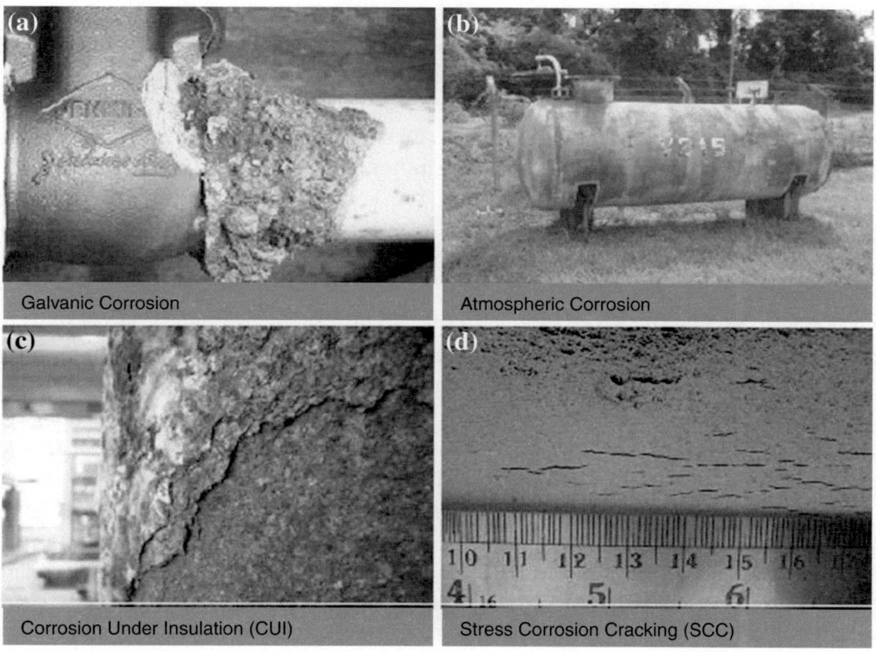

Fig. 4.4 Corrosion mechanisms

is the difference in electrochemical potential between the two metals. As a result, the more noble (cathodic) metal is protected and the less noble (anodic) will corrode faster. The severity of the attack by galvanic action is given by the potential difference between dissimilar metals, electrical conductivity of the medium and the ratio of exposed surfaces. The best prevention is through good design: no dissimilar metals should contact. Other possible solutions include the addition of inhibitors to the corrosive medium and the elimination of electrical contact, for example with the use of dielectric joints.

Atmospheric Corrosion is perhaps the most likely form of generalized metal loss in metallic components (Fig. 4.4b). The loss of thickness is uniform, this being the most "benign" corrosion, as it is predictable. The most important variables in the process are: the type of atmosphere (industrial, marine, rural), relative humidity, temperature and the presence of salts, sulfides and dirt. The most effective action against this type of corrosion is good surface preparation and application of suitable coatings.

Corrosion Under Insulation (CUI) is a major problem for pressure equipment in refineries and petrochemical plants (Fig. 4.4c). The effect of the insulation is primarily mechanical, since it provides an annular space in which water or moisture can be maintained in contact with metal for very long periods of time, with continuous access to air (oxygen). The second effect is that its composition may contain aggressive ions (Cl^-, Br^-, I^-) that provides a corrosive medium for stress corrosion cracking (SCC).

The most prone conditions for corrosion under insulation are frequent hot/cold temperature cycles, lower parts of horizontal vessels, where protective barriers are damaged or broken, venting pipes, and end of insulation on flanges, carbon steel vessels operating between −4 and 121 °C that may suffer condensation, equipment subject to water or steam splashes, intermittent temperatures above 100 °C, etc. Mitigation measures include protecting the surface with appropriate coatings/paints, periodic inspections and, in some cases, removal of the insulation in areas where heat conservation is not that important.

Pitting is one of the most dangerous forms of corrosion (Fig. 4.5a). It occurs in very small surface areas, but as the attack is very fast it can cause perforation of the wall. The attack occurs by the action of certain ions (Cl^-, SO_4^-) which have the property of locally breaking the protective film, exposing bare metal to the corrosive medium. Therefore, materials forming protective films are more susceptible (stainless steels, nickel alloys, etc.). Figure 4.6 shows some typical shapes and distributions of pitting, and their classification according to ASTM G46.

Carbon Dioxide Corrosion occurs when CO_2 is dissolved in water to form carbonic acid (H_2CO_3), thus promoting widespread corrosion and/or pitting on carbon steel. Critical factors of such a mechanism are CO_2 partial pressure, pH and temperature. Pitting corrosion of CO_2 (with round bottom are deep, with steep walls and sharp edges). Pits are usually interconnected in long lines, but sometimes can be individual and isolated. Pits in steel are usually filled with iron carbonate, a gray deposit of low adhesion, which is a byproduct of corrosion.

Sulfide Corrosion is due to acid H_2S gas. Pits are deep with a round bottom and beveled edges; usually small, which occasionally spread over the entire surface of the component. A second corrosion product due to H_2S is the iron sulfide scale. Surfaces and pits are covered with a very sticky black inlay of iron sulfide that is highly insoluble and cathodic to steel; it tends to accelerate penetration rates

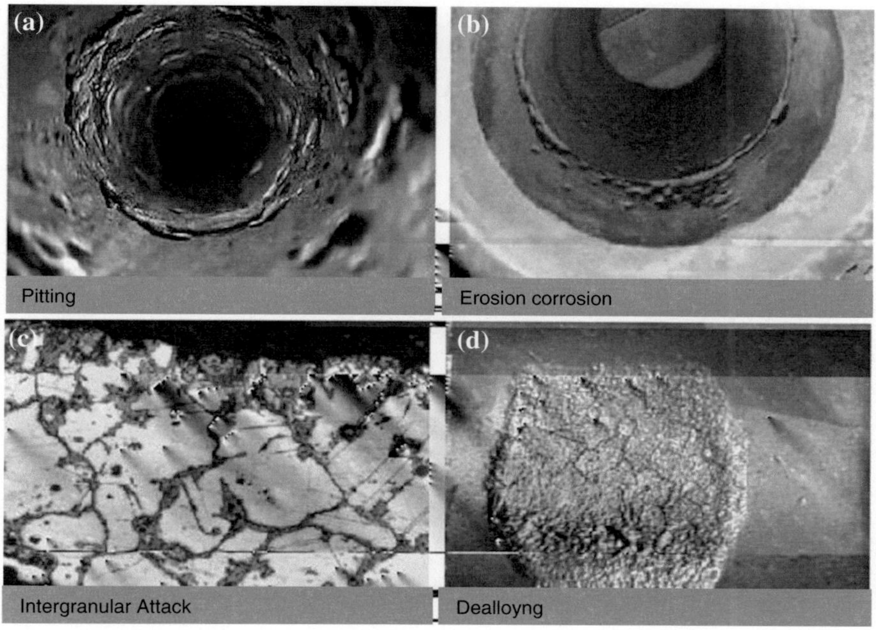

Fig. 4.5 Corrosion mechanisms

of corrosion. A third associated mechanism of corrosion is hydrogen embrittlement, which can cause brittle fractures from the pits due to hydrogen absorption and loss of ductility in structural steel. Although H_2S is a relatively weak acid (when compared to CO_2), trace amounts of H_2S can cause corrosion if some water is also present, thus justifying chemical inhibition programs.

Corrosion by Amines refers to general and/or localized corrosion that mainly occurs in carbon steels in amine treatment processes. Corrosion is not caused by the amine itself, but is the result of dissolved acid gases (CO_2 and H_2S), degradation products; hot stable amine salts and other contaminants.

Crevice Corrosion occurs in the presence of small confined spaces (gaps), where the corrosive environment develops differentiated concentrations.

Erosion–Corrosion: Erosion is the removal of material from the metal surface by the action of numerous individual impacts of solid particles from a fluid (Fig. 4.5b). When erosion is increased by corrosion, it is called erosion–corrosion. The most important factors influencing the mechanism are speed, size and shape of the particles, fluid velocity and the active corrosion mechanism. Erosion–corrosion is characterized by the formation of marks in the shape of a "horseshoe" on the surface in the direction of flow. The most prone areas to this mechanism are those with high turbulence. Mitigation forms are the use of corrosion-resistant alloys and/or altering the process to reduce corrosivity.

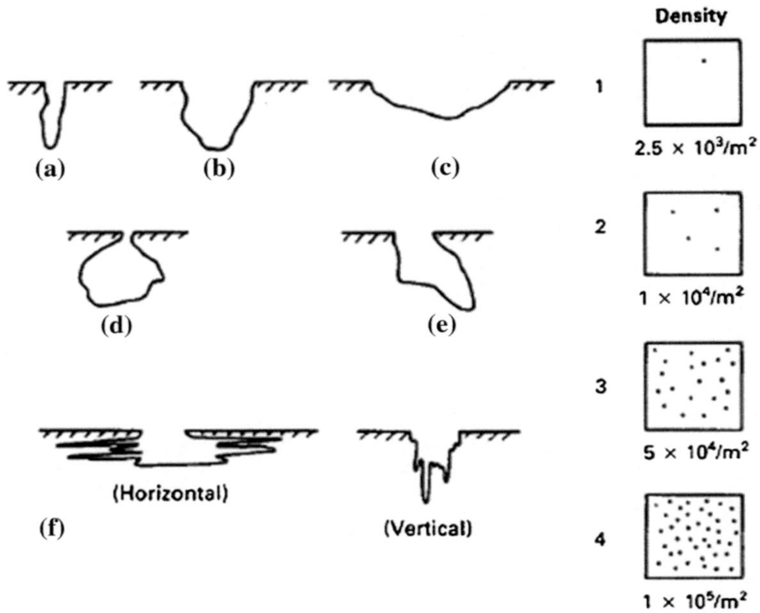

Fig. 4.6 Typical pit shapes and pitting densities by corrosion [3].

Intergranular Attack depends strongly on the microstructure of the material. The attack is concentrated in grain boundaries because they generally have a different composition than the rest of the material (Fig. 4.5c). Stainless steels and nickel alloys are most susceptible. The way to prevent this is to use low-alloy steels, or stabilized with carbide forming elements other than chromium, or by using post weld heat treatments.

Dealloying is a selective corrosion (removal) mechanism in which one or more alloy elements are preferentially attacked, leaving a porous structure of lower density (Fig. 4.5d). Zinc dealloying is typical of brasses with more than 15 % zinc. Removing the zinc leads to a porous, weak copper.

Sensitization: certain manufacturing processes modify local surface composition, allowing the medium to preferentially corrode a portion thereof. Essentially a phenomenon of galvanic corrosion, different compositions function as of materials with different electrochemical potential. One sensitization mechanism is intergranular attack of stainless steel welds. Intermetallic $Cr_{23}C_6$ precipitates in grain boundaries of 304/18-8, 316 type austenitic steels, which are exposed to temperatures between 415 and 815 °C. This carbide is rich in Cr and Cr is taken from adjacent areas (Fig. 4.7a), which then corrode preferentially, in a very thin strip around grain boundaries (Fig. 4.7b). Normally this is avoided by using low carbon steel or stabilizing with non-Cr carbide forming elements (Nb, Ti).

Microbiological Induced Corrosion (MIC) is a corrosion mechanism defined as "those forms of corrosion which are initiated or influenced in their rate or

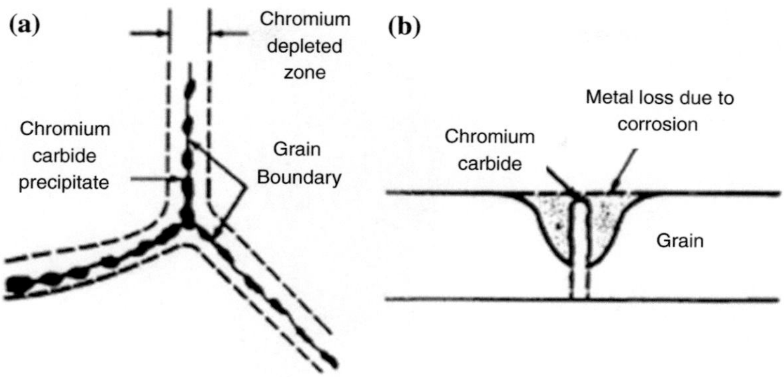

Fig. 4.7 Sensitization

extension by the presence or activity of microorganisms." The microorganisms that contribute to microbiological corrosion cover a wide spectrum, but have a common factor in that they are part of the natural cycle of sulfur in nature, so that this element plays a very important role in the metabolism of these bacteria. It is for this that the bacteria can be classified into two groups: sulfate oxidizing or sulfate reducers.

Sulfate oxidizing bacteria are involved in the corrosion process by oxidizing sulfur compounds to form sulfates, with the concurrent production of strong acids (sulfuric acid for instance). These bacteria require the presence of oxygen (typically obtained from air) for the development of their metabolism, so they are also known as causing aerobic microbiological corrosion. Sulfate-reducing bacteria (SRB), on the other hand, complete the sulfur cycle by reducing sulfates to sulfides. This process is completed in absence or with very low concentrations of oxygen, so that these appear as pitting MIC.

Figure 4.8 shows several examples of MIC. A number of microscopic life forms are present in many media, particularly in untreated oil. Some unicellular organisms are able to live in all kinds of conditions and multiply with incredible speed, commonly referred to as bacteria or "bugs". Bacteria are classified according to their need for oxygen: aerobic (requiring O_2), anaerobic (do not require O_2) and optional (any form of requirement). Some bacteria produce H_2S, some produce organic acids or enzymes, water soluble iron oxides, or any combination thereof. MIC is very aggressive and corrodes components rapidly.

Figure 4.9b is an example of Acid Producing Bacteria (APB). Corrosion pits due to APB have the same basic features as those from CO_2 corrosion. They look cavernous with sharp edges and the base is usually striated or grainy. The pit does not contain deposits. Figure 4.9a is an example of corrosion by SRB. SRB producing H_2S probably cause more problems in oil well equipment than any other type of bacteria. Corrosion due to the SRB has the same basic characteristics of the H_2S acid gas corrosion. They are often associated with multiple cracks in the bottom of the pit and tunnels around the edges, clusters of pits and/or unusual

Fig. 4.8 MIC: examples in oil field sucker rods

anomalies as bright spots on the surface of oil field sucker rods. Pits and cracks are the result of hydrogen sulfide, a byproduct of the metabolism of the bacteria, which corrodes and weakens the steel surface that is under the colony.

SRB are also causing major problems in pipes in refineries and petrochemical plants. They are mainly found in the soil and attack buried components. Other components prone to this attack are water-cooling systems (heat exchangers) and storage tanks.

Stray Current Corrosion occurs due to induced stray or parasite electrical currents. Pits are usually unique and isolated. They can be caused by grounding, nearby cathodic protection systems, etc. Electrical arcs leave deep pits, irregularly shaped with smooth sides and edges, and a small cone at the base of the sting.

4.4.1 Example 4.A Corrosion in Exchanger Tubes

Figure 4.A1 shows a schematic diagram of the heat exchanger. Horizontal tubes failed in several sections, due to multiple pitting corrosion leakage, see Fig. 4.A2. Natural Gas (NG) flows inside the tubes, downstream a gas compression plant, and is saturated with water. Chromatographic chemical composition of the NG flowing inside the tubes and deposits in the inner surface of separators located upstream of the failed heat exchanger are available.

Figure 4.A3a, b shows metallographic samples from longitudinal and transverse sections of the tubes. The thinner areas coincide with the bottom of the tubes. Corrosion pits initiate from the inside surface, and are symmetrical. This allows the identification of a corrosion mechanism without an erosive component; pits are oriented in the direction of fluid flow, see Fig. 4.A4. Table 4.A1 shows chemical composition of the tubes; they correspond well with standard specifications.

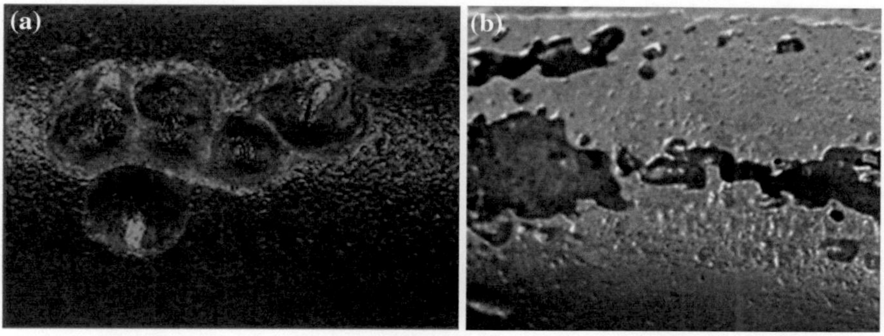

Fig. 4.9 **a** Sulfate oxidizing bacteria. **b** Acid producing bacteria

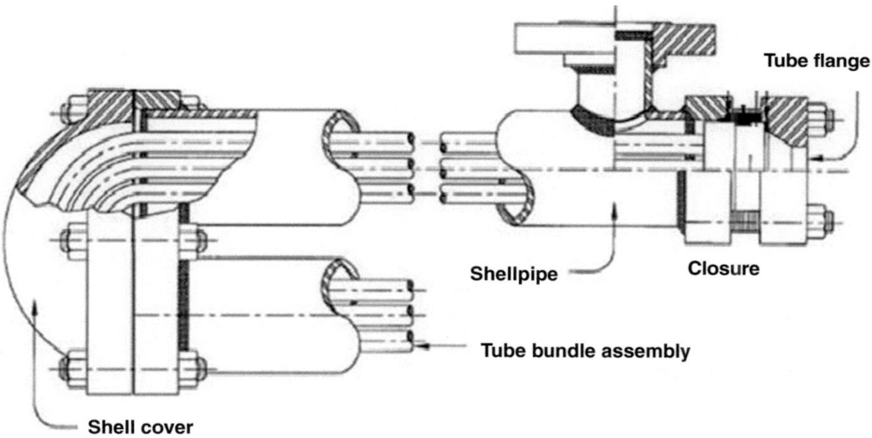

Fig. 4.A1 Schematic diagram of the heat exchanger

Fig. 4.A2 Horizontal tubes failed in several sections, due to multiple pitting corrosion leakage

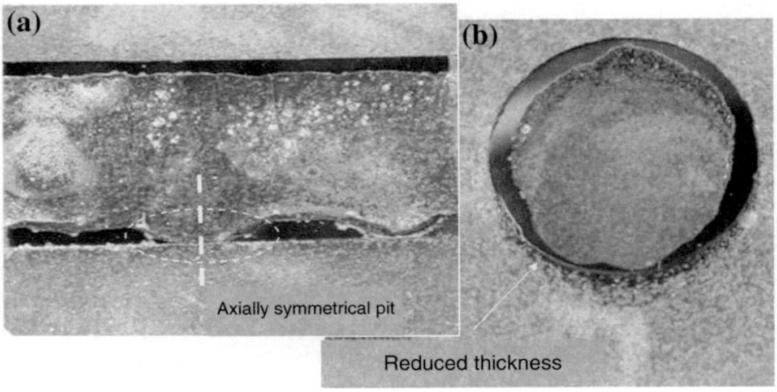

Fig. 4.A3 Metallographic samples **a** Longitudinal **b** Transverse

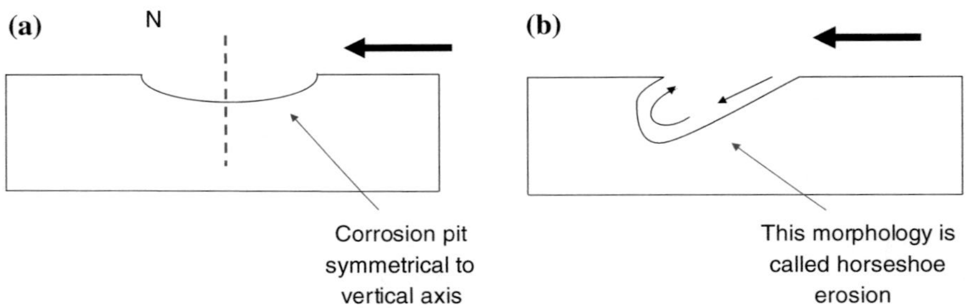

Fig. 4.A4 Shematic pit shapes for **a** Corrosion **b** Erosion

Table 4.A1 Chemical composition of tube material

Element	Material Exchange [%]	ASME Specification SA-214 [%]
Carbon (C)	0.02–0.03	0.18
Manganese (Mn)	0.45	0.27–0.63
Molybdenum (Mo)	0.22	–
Copper (Cu)	0.13	–
Phosphorus (P)	0.01	0.035 (Max.)
Sulfur (S)	0.008	0.035 (Max.)
Chrome (Cr)	0.053	–
Nickel (Ni)	0.042	–

Table 4.A2 shows chemical composition of deposits inside the tubes. Upon confirmation of the existence of sulfur, a sample was reacted with hydrochloric acid thereof. Vapors of sulfide acid formed (with typical odor to rotten eggs), which confirmed the presence of sulfides in the corrosion deposits. This study showed that the composition of the transported gas and operational characteristics

Table 4.A2 Chemical composition of internal and external deposits

Corrosive deposit	Interior		Exterior	
Element %	General	Specific	General	Specific
Carbon (C)	1.73	0.42	0.91	10.4
Oxygen (O)	36.39	34.23	34.05	33.65
Aluminum (Al)	0.35	0.39	0.35	0.74
Silicon (Si)	0.49	0.39	0.62	0.67
Sulfur (S)	1.96	2.13	1.60	1.68
Calcium (Ca)	–	–	0.54	0.54
Manganese (Mn)	0.63	0.58	0.48	0.47
Iron (Fe)	58.45	61.96	61.46	61.21

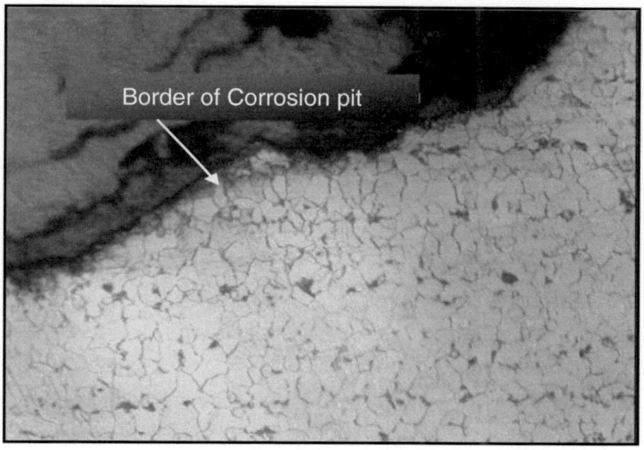

Fig. 4.A5 (X200) Metallographic inspection of corrosive attack

of the component determined the medium for the generation of internal corrosion pits, which resulted in perforation of the tube walls.

NG transported inside the heat exchanger tubes is saturated with water and has concentrations of hydrogen sulfide. Metallographic inspection (Fig. 4.A5 200X) helped to confirm that the corrosive attack did not follow any preferential metallographic phase; no evidence of any problem with the metallurgy of the tubes was found.

The location of the pits also allowed relating the problem to winter times. In cold weather the heat exchanger is not used; condensed water is formed inside the exchanger, reaching the lower parts of the tubes and generating preferential pitting along the lowest point in the perimeter of the tube. Retained water saturated with hydrogen sulfide and generated two corrosive processes acting simultaneously:

A—acid attack, and

B—corrosion by differential aeration.

The acid attack is governed by the equation:

$$H_2S + Fe \rightarrow H_2 + FeS$$

Corrosion by differential aeration is generated by varying the concentration of oxygen in the areas with retained fluid. The top of the tube is exposed to high concentrations of oxygen and acts as the cathode in the cell, while the portion beneath the retained liquid is deficient in oxygen and acts as an anode.

Although fluid retention and the range of service temperatures provided the ideal medium for growth of bacterial colonies, there was no evidence of MIC in the morphology of the pits. To conclusively rule out this mechanism damage, it would be advisable to analyze bacteria cultures in samples of the liquid carried by the gas.

4.5 Propagation of Cracks, Fatigue

Pre-existing defects or cracks within a material directly reduce or eliminate the time required for nucleation or initiation of cracks. When the size of a crack is small, its growth rate is slow; growth rate increases with crack size (mechanics of crack growth is studied in Chap. 6). Fatigue cracking is a mechanical form of degradation that occurs when the component is exposed to cyclic stresses in a period of time, and often results in an unexpected failure. Cyclic stresses can result from mechanical loads, thermal cycling or vibration, and are typically below the yield stress of the material. A discontinuity that acts as a stress raiser (for example at a welded joint) may initiate a fatigue crack under cyclic loading. This defect would propagate slowly during service to reach a critical size; from which the crack grows fast and leads to almost instantaneous collapse of the structure.

As we have seen fracture surfaces are clearly indicative of whether crack growth is due to fatigue, unless they are degraded after failure by scratching or corrosion. Beach marks on smooth fracture surfaces indicate successive positions of the crack front; this enables individualizing initiation sites. We also saw in Chap. 3 that in some materials it is possible to identify the mechanism of propagation at a microscopic scale. Fatigue cracking in stainless steels, aluminum and some structural steels involve a microscopic process called striation (very small parallel marks).

4.5.1 Example 4.B Fatigue Failure in a Pipeline

An uncommon failure in a pipeline was caused by a fracture in the double submerged arc (DSAW), longitudinal seam weld, see Fig. 4.B1. Cracks propagated from a very long internal or embedded weld defect due to lack of fusion; defect length is three feet, defect depth is 3 mm. What makes this study particularly interesting is that previous In Line Inspection (ILI) of the line was unable to detect the defect; no evidence of third party damage was found in the nearly 50 years of operation of the pipeline. Fortunately, the lack of atmospheric moisture in the area

Fig. 4.B1 Pipeline fracture

and readiness of the operator to protect the initiation site allowed recovering the fracture surfaces in very good condition. Figure 4.B2 shows a portion of the fracture near the failure initiation. Indicated are:

- 1, 2: areas of unstable propagation, formed during the blowout,
- 3, 4: in-service propagation surfaces, smooth surface with beach marks,
- 5: surface of the machined plate edge, indicating lack of fusion.

Infrequent for a gas pipeline, it is observed that the cracks had propagated by fatigue during operation. Normally pressure variations in natural gas pipelines are insufficient to cause fatigue. Although few large pressure cycles have been recorded over the last 5 years of service, apparently these years do not represent the entire service life of the pipeline.

A fatigue propagation model of the lack of fusion defect was carried out. Figure 4.B3 shows a polished and etched section of failure initiation. The trace of the fracture is shown vertical to the right of the photo. Figure 4.B4 identifies areas as sketched in Fig. 4.B3. The lack of fusion is due to a severe misalignment of both weld beads (outer and inner), although it is worst from the inside (top of the Figure).

The fracture path includes the lack of fusion and fatigue cracks. Fractures' surfaces near both tube surfaces are slant. These are not shear lips, since they are due to fatigue. Crack inclination is perpendicular to the direction taken by hoop stresses in these ligaments, see Fig. 4.B4.

Figure 4.B5 (a) shows a composite photograph of half of the initial defect, which is almost symmetrical; the inner tube surface is indicated at the top. Note that the horizontal scale is 100 times smaller than the vertical (the defect is very long and shallow). Figure 4.B5(b, c, d) shows the sequence of events during the lifetime of the pipeline, arrows indicate in each case the direction of propagation of crack front. First, the crack opened the inner surface of the tube wall, and then (possibly very shortly before the blowout) the crack opened up the outer surface. At that time the

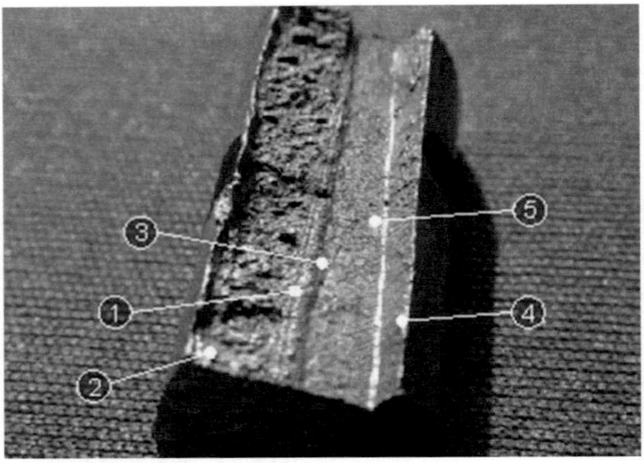

Fig. 4.B2 Detail of fracture

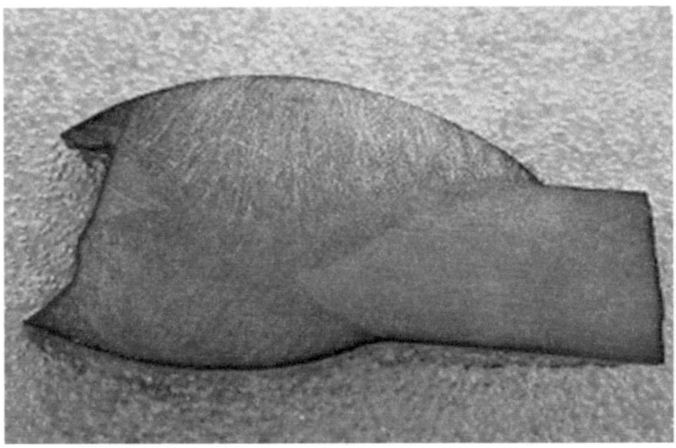

Fig. 4.B3 Fracture initiation zone

criticality of the crack increased significantly (as discussed in Chap. 9), leading to final fracture. No time available, therefore, to detect a gas leak.

Figure 4.B6 (a, b, X10.000) shows at high magnification crack propagation due to fatigue. The crack propagates from left to right (note the equally spaced parallel striations); spacing between striations is between 0.1 and 0.3 microns. With this experimental data the number and approximate magnitude of pressure cycles leading to crack growth could be determined.

Experimental tests were conducted to characterize fatigue propagation of fatigue cracks in weld metal. It was noted that relatively few large amplitude cycles (zero to maximum pressure) were sufficient to propagate the crack. With these data, Fracture Mechanics models (Chap. 6) were used to relate striation spacing, as measured in the fracture surfaces of the failed pipeline, with pressure cycles that should have provoked them. It was found that in the early life of the line there should have been many more large pressure cycles than previous assumptions.

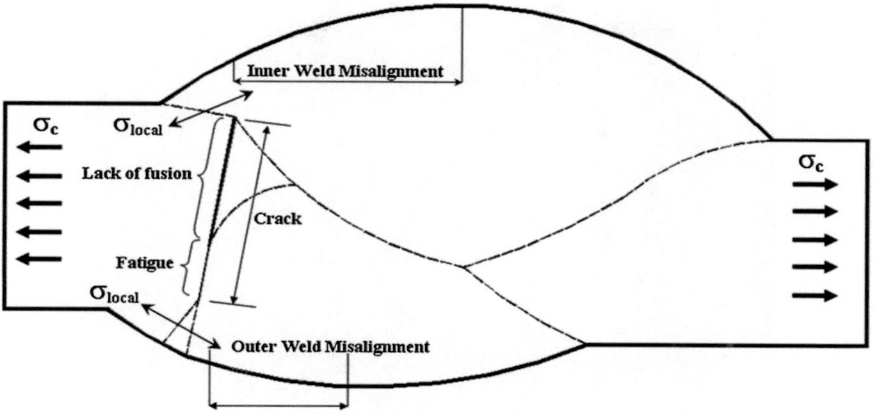

Fig. 4.B4 Detail of stress state in misaligned weld

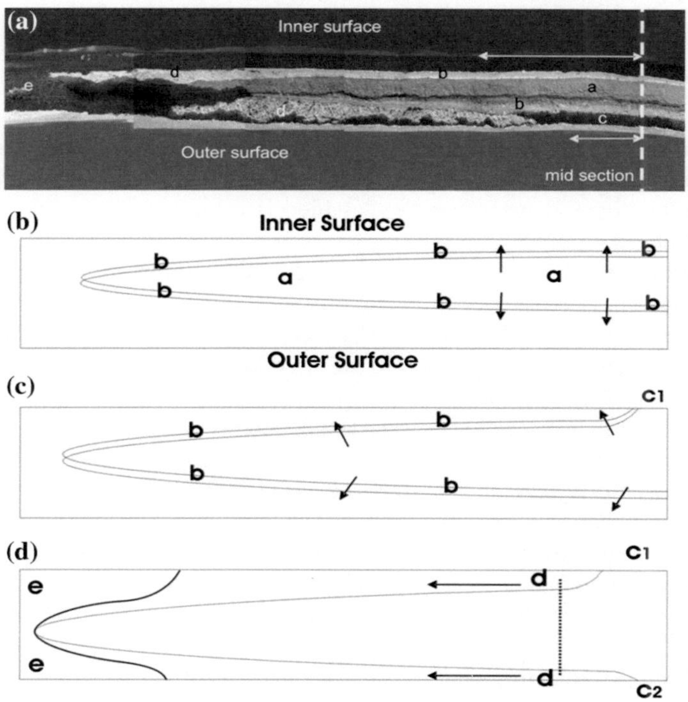

Fig. 4.B5 **a**, **b**. **c**, **d** Model of successive crack growth patterns

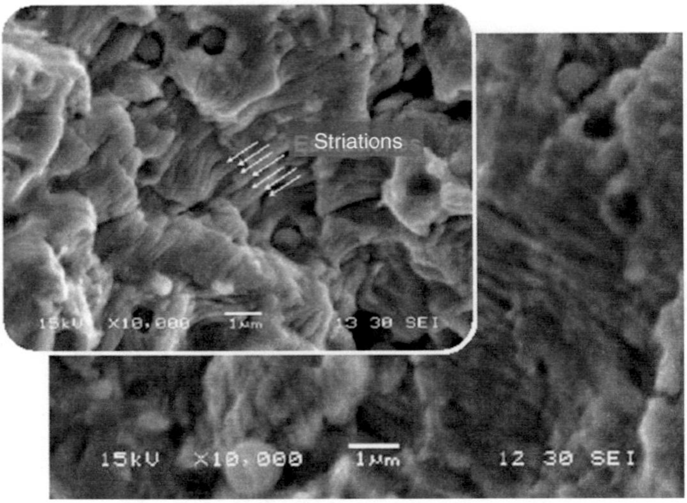

Fig. 4.B6 High magnification SEM images of fatigue crack

4.6 Hydrogen Damage

Hydrogen damage is a term that covers a variety of mechanisms for degradation of mechanical properties in steel, due to adsorption and diffusion of atomic hydrogen. In service hydrogen damage can be classified into three mechanisms, which are described below.

Blistering: this phenomenon is manifested by the appearance of blisters on the material surface. A cross section of the blister would often show the presence of cracks, generally stepped-like, near the surface. A feature of this mechanism is that cracks have a significant component in the direction of wall thickness (Fig. 4.10a). Blistering is probably the least dangerous of the three cases because it is easily detectable by visual inspection before danger of catastrophic failure.

This type of damage is particularly severe in equipment containing hydrogen sulfide (H_2S), but is generally limited to devices with low yield strength, i.e. low strength steels. We saw that these steels generally have a ferrite-pearlite microstructure, usually with banding parallel to the rolling plane. These bands typically have a strong segregation of nonmetallic inclusions (mostly MnS). The matrix/ inclusion interface is the preferred location for initiation of cracks that cause blistering.

Hydrogen Attack: this type of steel embrittlement occurs when exposed to hydrogen at elevated pressures and temperatures (above 200 °C). In this case the steel undergoes a marked reduction in strength and ductility; due to decarburization. Atomic hydrogen can diffuse through the steel and react with iron carbides (cementite), to generate methane:

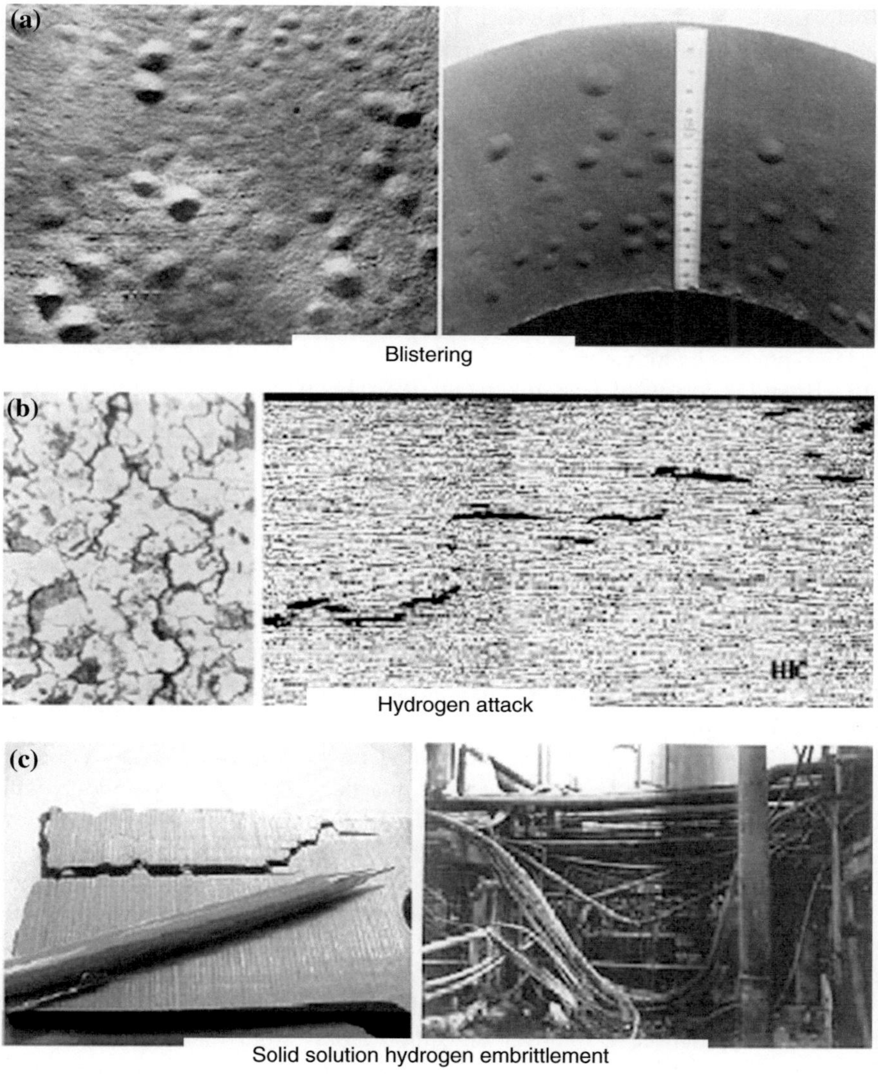

Fig. 4.10 **a**, **b**, **c** Types of damage due to hydrogen

$$\Rightarrow Fe_3C + CH_4 + 3Fe_4H$$

As the molecule of methane (CH_4) cannot diffuse through the steel due to its size, great pressure accumulates and eventually leads to intergranular cracking (Fig. 4.10b). This mechanism leads to an irreversible loss of ductility in the steel. The factors that influence this type of attack are temperature and stress.

Temperature is important because it promotes the diffusion of hydrogen, and applied stress adds up to the stress generated by methane pressure.

Thus, hydrogen attack is basically a mechanism for decarburization, wherein the diffusion of hydrogen within the metal leads to an internal reaction with the formation of cracks that eventually cause failure of the component.

Solid Solution Hydrogen Embrittlement: this mechanism of damage is probably the most dangerous and least known. Determining factors are: metallurgical characteristics of the material, level of tensile stress (applied or residual), and hydrogen concentration and temperature (Fig. 4.10c).

Quenched (martensite) microstructures are highly susceptible, especially welding and heat affected zones. A widespread rule of thumb indicates that in order to prevent hydrogen embrittlement hardness should not exceed 22 Rc or 200 HB (Brinell). Post-weld heat treatment (PWHT), if effective, tends to reduce residual stress and hardness of HAZ. Hydrogen concentration capable of inducing this phenomenon is generally very low (about 1 ml/100 g of steel). Figure 4.11 shows H cracks parallel to the banded microstructure. Hydrogen cracking tends to occur in grain boundaries, inclusions and other weak parts of the microstructure.

4.7 Damage by Prolonged Exposure to High Temperature

Proper selection of materials for high temperature service is an essential factor in Creep-resistant design. In general, these metal alloys contain elements such as Cr, Ni, and Co in various proportions depending on specific characteristics sought. Creep phenomena can lead to excessive plastic deformation or end up with breakage in a structural element (Fig. 4.12).

Creep damage usually involves noticeable microstructural changes, especially in grain boundaries. The example of Fig. 4.13 shows the final stages of Creep aging of chrome-nickel alloys at high temperatures: (A) stage I at up to 600 °C, (B) stage II at 600–700 °C, (C) stage III at 700–800 °C, (D) stage IV at 800–900 °C, (E) stage V at 900–1000 °C, (F) stage VI over 1000 °C.

Graphitization or Decarburization: these damage mechanisms are characterized by a progressive removal of carbon atoms from the matrix in steels, forming carbides and leaving only a matrix of iron. Decarburization can occur during exposure to high temperatures during heat treatment, or exposure to direct heat or high temperature service in a gas (non-carburizing) atmosphere. Figure 4.14 shows from left to right:

(a) ASTM A36, rolled steel plate, 9.5 mm thick. The structure consists of equiaxed ferrite (white) and pearlite (black). Nital etched, 250X;
(b) ASTM rolled steel plate, 25–36 mm thick. Pearlite and ferrite still present, along with some small inclusions;
(c) graphitizing of a ASTM A201 Gr.A steel plate after 5 years of service at temperatures between 595 and 650 °C. Nital etched, 100X.

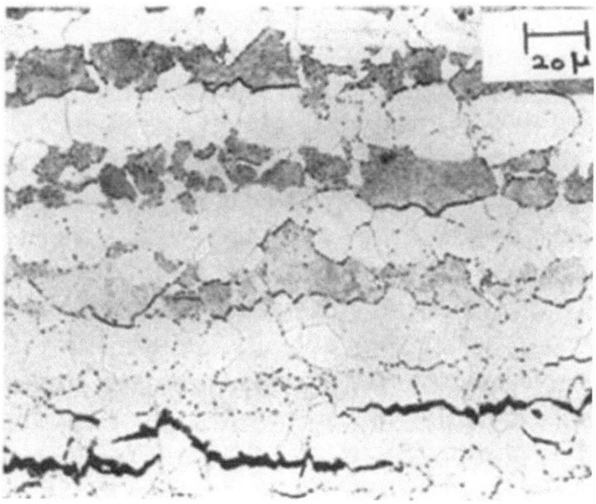

Fig. 4.11 Microstructural sensitivity in hydrogen cracking (X500)

Carburization: when the surface of the steel at high temperature is in contact with a carbon-rich atmosphere, the damage mechanism involves the progressive enrichment of carbon into the microstructure. Carburization modifies the microstructure, increasing the amount of pearlite or even promoting martensitic transformation. The net result is embrittlement.

4.7.1 Example 4.C Fire in a Hydrocarbon Furnace

Metallurgical degradation of structural steels at elevated temperature and the action of aggressive media, both inside and outside the tubes, is the most common cause of leakage and explosions in oil furnaces. In this example we examine a failure that caused fire in a fuel JP1 heating open fire furnace. Failure originated from a tube in the lower rows of the furnace, above one of the burners. The crack that led to the leak is about the size of the tube diameter and is quite open, showing a "fish mouth"; radial deformations indicate large plastic strains before failure. Tube nominal thickness is 6.85 mm; smallest thickness at failure is 2.6 mm. Thickness reduction is most noticeable near the edges of the main crack. This crack propagates in an inclined plane with respect to the thickness, another indication of failure by ductile tearing.

Low magnification fractography shows a series of longitudinal and branched secondary cracks on both sides of the fracture; along with thick, brittle and dark deposits on the inner tube surface. Brittle cracks in the deposit also indicate that the metal substrate suffered significant plastic deformation before and during the leak. Thick deposits were not observed on the outer surface. After cleaning, pitting defects were also found on the inner surface, up to 1 mm deep. Some pits are

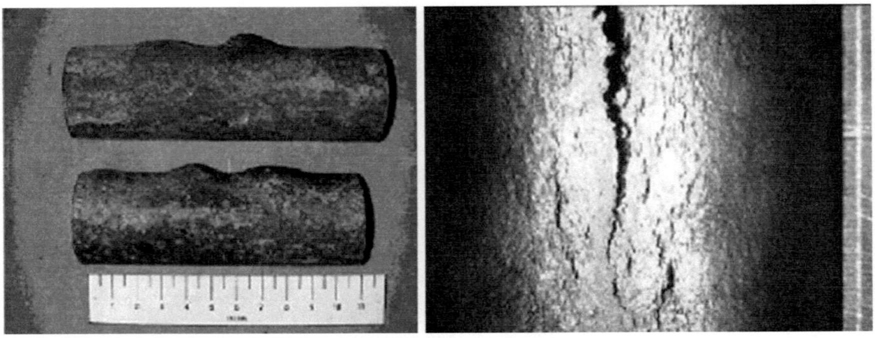

Fig. 4.12 Creep

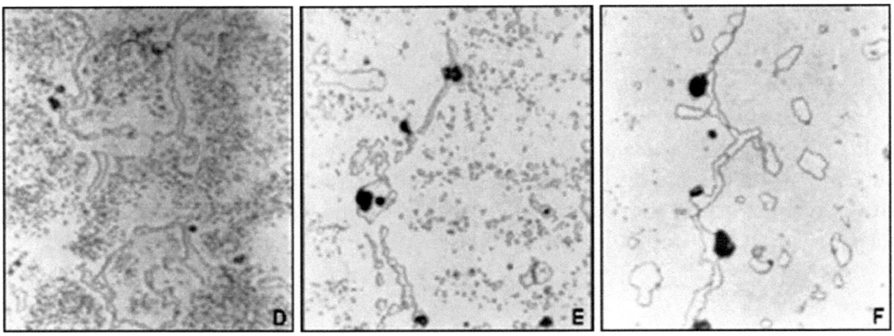

Fig. 4.13 Creep aging Microstructure

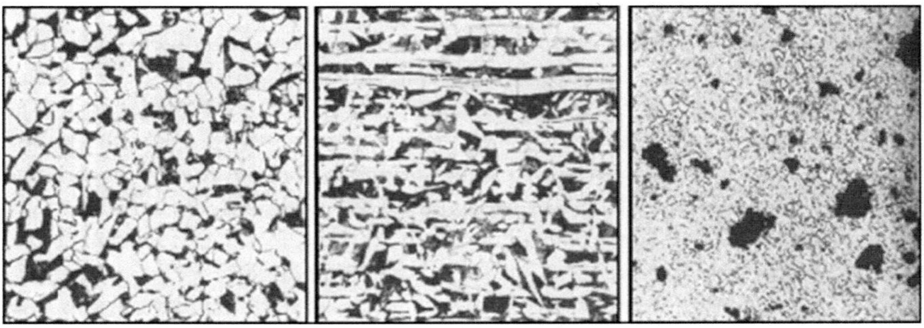

Fig. 4.14 Degradation of a ferritic pearlitic microstructure

equiaxed, others are elongated in the flow direction, see Fig. 4.C1. The depth of these defects does not exceed 1 mm.

Fig. 4.C1 Surface damage

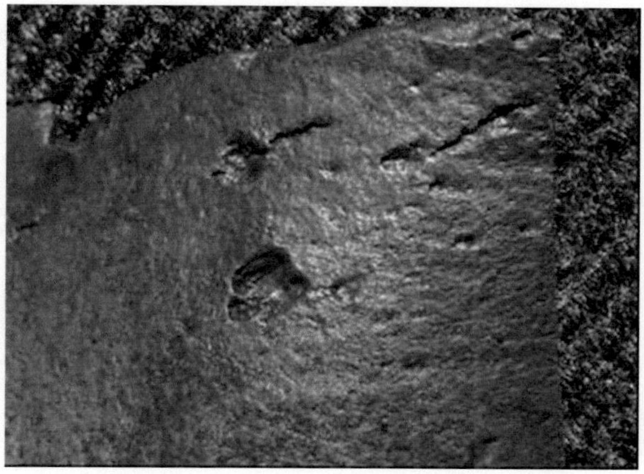

No chlorine but large amounts of sulfur were found in samples of the corrosion deposits; sulfides are normal products of degradation by sulfur circulating in JP1. Mechanical properties of the tube material correspond to standard specifications, i.e. tube metal is not excessively degraded; the steel shows good strength and ductility.

Figure 4.C2 (5X) shows the surface of the sample extracted from the probable site of fracture initiation. The trace of the fracture that caused the leak is seen to the right of the photo. Other cracks initiated at both surfaces are also observed, they are wide open due to large deformations during the opening of the main fracture.

Notice a dark band adjacent to the inner surface (top of photo), and a light band adjacent to the outer surface. These bands are due to carburization of the material adjacent to the inner surface and decarburization at the outer surface. These bands were found in all sections in the side of the tube exposed to fire, but are wider and more severe in sections near the site of the leak. The microstructure on the outer surface shows globulized pearlite. The microstructure near the inner surface, with higher carbon content, is virtually 100 % pearlite. The portions of the tube unexposed to fire do not show a carburized layer, and show the typical microstructure of a ferritic pearlitic structural steel.

The propagation of cracks shown in Fig. 4.C2 is intergranular and branched. There are at least three main cracks, two initiated at the outer surface, and the mid one initiated in the inner surface. A large number of small cracks initiated at the outer surface are also intergranular and branched. Crack lengths vary between 0.05 and 0.2 mm, cracks are entirely within the decarburized area, see detail of Fig. 4.C3 (200X). Figure 4.C4 (500X) shows the carburized band at the bottom of a pit, on the inner surface of the tube; two small defects propagate into the tube thickness from the bottom of the pit. They are full of corrosion products.

A colony of microcracks parallel to the surface of the tube is also seen in the carburized band near the inner surface, see Fig. 4.C5 (500X); the total length of the colony is less than 0.2 mm. The large opening of these microcracks is probably

Fig. 4.C2 (X5) Low magnification cross section

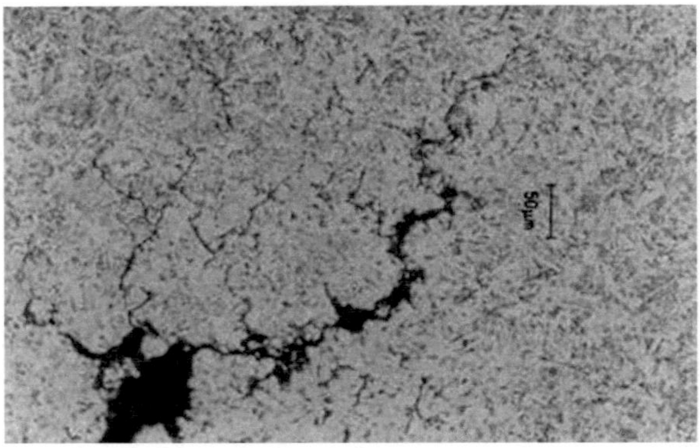

Fig. 4.C3 (X200) detail of cracks near outer surface

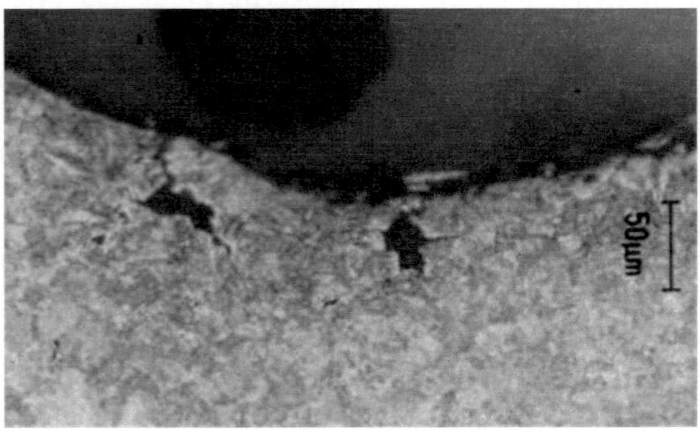

Fig. 4.C4 (X500) detail of cracks near inner surface

not due to plastic deformation but indicative of a possible process of hydrogen attack.

It was concluded that the failure occurred due to the propagation of cracks from the inside of the tube, due to carburization of the steel near the inner surface.

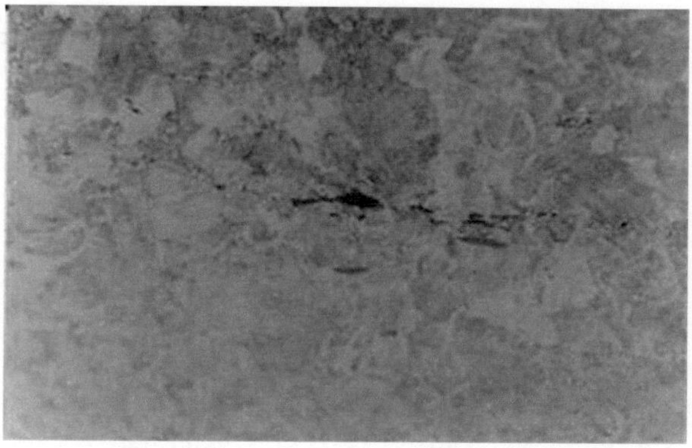

Fig. 4.C5 (X500) Microcracks in carburized microstructure

Mechanical stress concentrations at the bottom of pits and hydrogen damage drove initial cracking. Pitting, corrosion-erosion and cavitation are associated with the internal flow, favored by a high concentration of sulfur. The damage was probably accelerated by increased temperature in the tubes around the burners. Getting a homogeneous temperature distribution in the tubes is a key part of the design and operation of such furnaces.

Creep microstructural damage and micro-cracking detected on the outer surface of the tubes were only an important influence on the final stages of crack propagation, accelerating the occurrence of the leak. Carburization, pitting and hydrogen cracking in the inside surface are not easily detected by the usual methods of nondestructive inspection from outside. Conducting a hydrostatic test at a high enough pressure would ensure adequate residual life in similar situations before onset of a leak. These and other mitigation techniques are discussed in Chap. 9.

4.8 Environment Assisted Cracking

Environment Assisted Cracking (EAC) is a generic term that describes the formation of cracks by several factors combined with the medium surrounding the component. When water (electrolyte) comes into contact with the steel, minerals, ions, and gases present in the water, it may attack or corrode the metal. These chemical or electrochemical reactions can lead to corrosion as a generalized thickness reduction, pits and/or cracking. EAC is the term used to describe component failures in service that occur after a certain amount of subcritical crack growth induced by the corrosive environment (see for example Fig. 4.4d). This phenomenon is associated with the combination of mechanical stresses beyond a certain critical value, a corrosive species specific for each material and, in some systems, metallurgical conditions that lead to growth of cracks (susceptible material) (Table 4.3).

Table 4.3 Susceptibility of various materials commonly used in mechanical components to EAC

Material	Specific media
Carbon steel	Nitrates, carbonates and hydroxides
High strength steels	Aqueous solutions containing H_2S
Stainless steels	Chlorides (Cl^-), organic acids
High nickel alloys	High purity steam
Copper alloys	Ammonia solutions
Aluminum alloys	Solutions containing Cl^-, Br^-, I^-
Titanium alloys	Solutions containing Cl^-, Br^-, I^-, organic solutions

Only specific combinations of material and medium can produce EAC. A medium that leads to EAC in a metal may not produce it in another. There are many mechanisms of EAC, depending mainly on the combination Metal/Medium. Cracks can be either intergranular or transgranular, and are generally highly branched. There are two basic mechanisms: corrosion fatigue and SCC, which in turn can be divided into several mechanisms:

- SCC by amines
- PASCC (Polythionic Acid SCC)
- chloride SCC
- caustic SCC
- ammonia SCC
- carbonates SCC.

Corrosion Fatigue

Corrosion Fatigue occurs when corrosive agents penetrate into fatigue cracks; these agents can accelerate crack growth as chemical conditions within the crack can be more aggressive than in the component's surface. Even though the metal surface can be passivated (with a stable barrier of oxide deposits), even at crack surfaces, on the crack tip every change in load tends to break the brittle deposits and reactivate the process. Thus, corrosion fatigue resulting from the combined action of a stress cycle and a corrosive environment reduces the required number of cycles to failure.

Stress Corrosion Cracking (SCC) of Buried Components

SCC involves corrosion mechanisms that depend both in the aggressive media and the stress state. Tensile stresses cause cracks to open; these stresses can originate from applied loads or be residual stresses from fabrication. SCC occurs under the action of tensile stress, while corrosion fatigue requires cyclic loading.

Since SCC develops slowly, propagating cracks in a colony may exist in the component for many years without causing operational problems. Cracks are generally spaced; cracks of a colony may vary both in length and depth. Over time, some individual cracks in the colony become longer and deeper, coalesce and in some cases lead to failure. If these cracks are sufficiently spaced, they may first grow in thickness until a leak occurs. SCC in pipelines and other buried structures

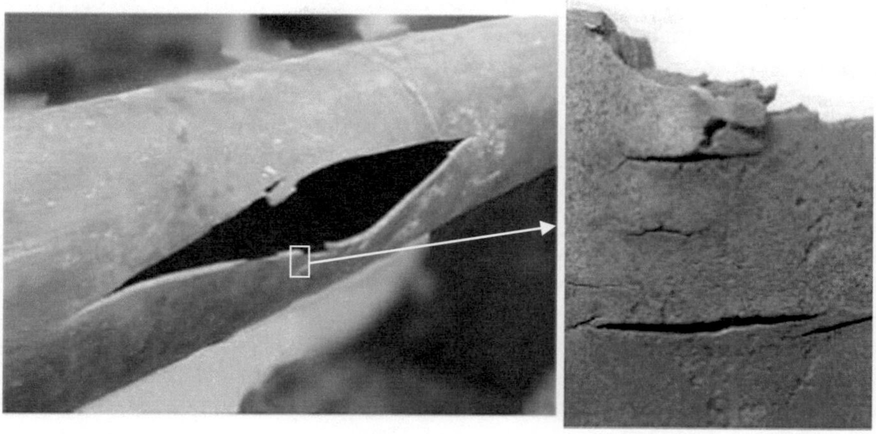

Fig. 4.15 Failure by SCC in a buried pipeline

can be characterized as "high pH SCC" or "neutral SCC"; referring to the pH of the surface of the component in the area of the crack, which may not be the same as the pH of the ground. The most significant feature for the identification of SCC in pipelines is the appearance of colonies of parallel cracks in the outer surface of the tubes. Figure 4.15 shows a SCC failure in a pipeline. These cracks were initially invisible to the naked eye, and then they opened by deformation during final fracture.

Although pipelines and other buried structures are coated to protect them against corrosion, there is always the risk that the metal is exposed to the surrounding environment. To prevent corrosion where coating fails, cathodic protection is generally used. A current is circulated through the soil into the pipes, giving the structure cathodic potential.

The carbonate/bicarbonate (CO_3–HCO_3) solution has been identified as responsible for high pH SCC. This medium may develop as a result of the interaction of hydrile ions, a cathodic protection byproduct, and carbon dioxide (CO_2) into the ground, generally coming from the decomposition of organic matter. Cathodic protection further causes pH of the electrolyte beneath the damaged coating to increase. Carbon dioxide dissolves in the high pH environment, resulting in the formation of a concentrated carbonate/bicarbonate electrolyte.

High pH SCC occurs only in a narrow and limited range of cathodic potential in the presence of carbonate/bicarbonate media with a pH greater than 9. Under these circumstances, the electrochemical reaction forms a thin protective (magnetite) film on the surface of the steel; no loss of metal by corrosion is associated with high pH SCC. If the passivation film is intact, SCC cannot start, since this film acts as a barrier between the metal and the environment. But if the steel is subjected to any deformation, the brittle film breaks, exposing the metal to the medium thus allowing SCC initiation. Eventually the protective film is formed again on the exposed surfaces, restoring the protective barrier. Cracks may start and stop cyclically, depending on the level of stress or strain in the steel. Since it takes time for forming the film, cracks may grow only if the deformation occurs more rapidly than the rate at which the film is formed. Consequently, strain rate is a condition which determines the growth of high pH SCC.

The low (or neutral) pH SCC was first identified in a buried pipeline in 1980 in Canada, in areas where there was water with pH between 5.5 and 7.5 trapped under the coating. The cracking environment is ground water containing dissolved carbon dioxide from decaying organic matter. This mechanism occurs with little current at the surface for a long time, either by a highly resistive ground or by shielding of the cathodic protection system. Unlike high pH SCC, cracks frequently initiate in corrosion pits, within which a medium having a pH low enough to produce atomic hydrogen is generated. Hydrogen enters the steel and locally degrades mechanical properties, so that cracks are initiated or grow by a combination of hydrogen dissolution and embrittlement.

There are many similarities between these high and low pH SCC mechanisms. Colonies of multiple parallel cracks on the surface of the product are typical of both, generally perpendicular to the largest tensile stresses. Among the most obvious differences is high pH SCC sensitivity to temperature, fractographic morphology, pH of the environment and the way in which cracks grow. High pH SCC tends to propagate branched, intergranular cracks, since grain boundaries are more susceptible to dissolution. In near neutral pH SCC cracks are transgranular and wide open. Transgranular crack propagation, although not as common, can occur in high pH SCC, especially when the cracks are deep or are subjected to high or fluctuating stress.

Caustic Embrittlement

Carbon steels operating in environments associated with boilers are susceptible to caustic SCC, particularly those of high strength under tensile stress and an environment rich in NaOH at temperatures of 200–250 °C. Soluble salts can accumulate in surface discontinuities, such as gaps between tube and tube plate, which are fairly highly stressed areas. Propagation velocity can be very high, but it is difficult to determine a priori due to the nature of the mechanism. Caustic (or alkaline) SCC may remain inactive and accelerate under certain circumstances; the design criterion is to avoid crack initiation.

4.8.1 Example 4.D Cracks in a Heat Exchanger Tube Plate

Leaks appeared in joints between 40 mm diameter, 5 mm thick tubes, and the 35 mm thick tube plate, all made of steel, in secondary exchangers at a petrochemical plant (Fig. 4.D1). Design conditions are:

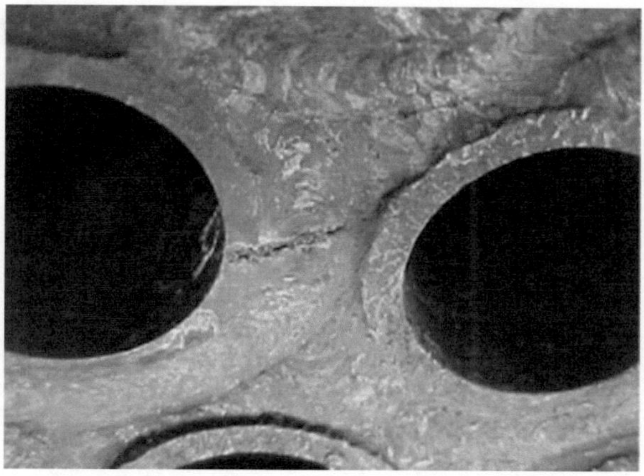

Fig. 4.D1 Welded ends of exchanger tubes

- water shell, 120 bar, 250 °C,
- cracked gas pipe; 7 Bar; 600–400 °C.

Tube to tube plate joints are welded, with previous pipe expansion. Microstructures and chemical compositions of plates and tubes corresponded with manufacturer's records.

Cracks developed in the circumferential seal welds. Crack initiation occurred in the root of the welds, in the bottom of the notch between tube and tube plate. Figure 4.D2a, b shows polished sections of the joints between tubes and tube plate; note the effect of crevice and stress concentration at the root of the welds.

Crack propagation is strongly influenced by the microstructural characteristics and mechanical properties of the HAZ; its features define a microstructural embrittlement mechanism. One possible cause of failure is alkaline SCC from the gap between tube and tubeplate. However, caustic elements were not found within the gap.

Final crack propagation occurred by corrosion fatigue. Figure 4.D3 is a detail of Fig. 4.D2b. Voids or cavities are seen in the gaps between tubes and tube plate and within cracks due to the effect of galvanic corrosion, with copper particles vapor deposited inside the crack (copper contamination). This is an additional effect to the stress concentration at the root of the weld.

Bainitic microstructures in non-failed sections of the tube plate correspond to quenching and tempering heat treatment, as specified by the manufacturer, Fig. 4.D2a. However, failed sections the tube plate have a ferritic -matrix microstructure, with lower strength. Hardness at failed HAZs is higher than allowable by specification; this denotes an ineffective stress relief heat treatment. It is concluded that thermal damage to microstructure, high residual stresses in the tube—tube plate welds and high applied mechanical stresses contributed to crack propagation.

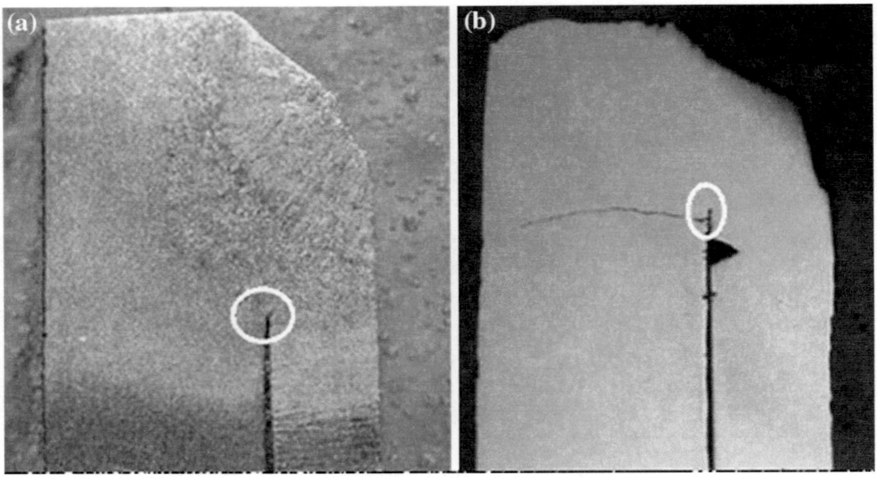

Fig. 4.D2 **a**, **b** Polished sections of the joints between tubes and tube plate

Fig. 4.D3 Crack propagation occurred by corrosion fatigue

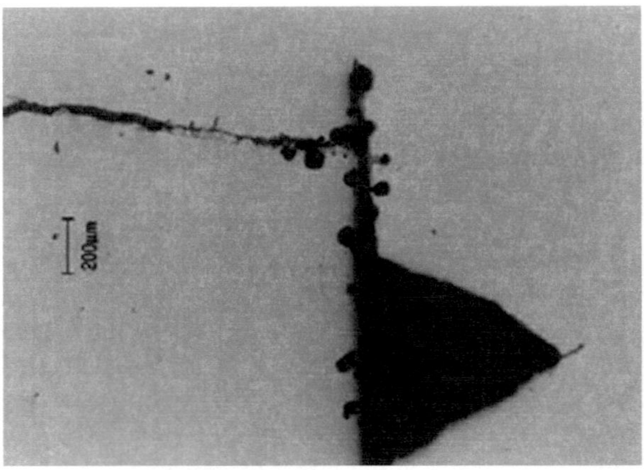

4.9 Discontinuities or Defects Introduced During Manufacture

While manufacturing and pre-operational testing methods substantially reduce the possibility of a significant reduction in reliability of metallic components, discontinuities are frequently involved in failures, especially in old components with many years of operation. These are some of the most frequent:

Laminations

Laminations or exfoliations occur in steel processing, due to bubbles trapped inside the raw steel during solidification. When the steel is rolled to form plates, these

Fig. 4.16 Laminations/Exfoliations

bubbles are "squashed" and trapped within the plate, forming a thin discontinuity within the material. Small laminations are frequent and mostly benign, since they are aligned with applied stresses and therefore they neither reduce thickness nor generate stress concentrations. Laminations from nonmetallic inclusions, most commonly Mn and complex sulfides, are typically found in old technology steels (Fig. 4.16).

Discontinuities in Welds

Welds often have a variety of geometric discontinuities, metallurgical and mechanical, because the very process of localized melting and solidification. The type of defect depends on the welding process; pipes and vessels are mostly welded by electrical resistance welding (ERW) or arc forming methods (manual electrode, submerged arc welding, gas shielded weld, etc.).

A substantial fraction of failures in pressure equipment initiate in weld discontinuities, which become critical defects by some damage mechanism during subsequent service. As an example, Fig. 4.17 (5X) is a polished section of an ERW longitudinal seam weld in a pipeline. The trace of a fracture initiated at stitching defects (localized lack of fusion) is seen at the top (see insert, 100X). Another typical flaw in ERW welds are hook or J cracks (Fig. 4.18, 5X). When the plate that makes up the tube has previous laminations at the edges to be joined, during metal ejection by the forging action of the ERW process these laminations are configured in a "J".

The following mechanisms are those that most frequently contribute to failures in arc welding (depicted in Fig. 4.19).

Misalignment: when the axes of two components are not in line, or thickness of one side is different from the other, misalignment occurs between the joining parts. Misalignments are common in old double submerged arc welds (DSAW).

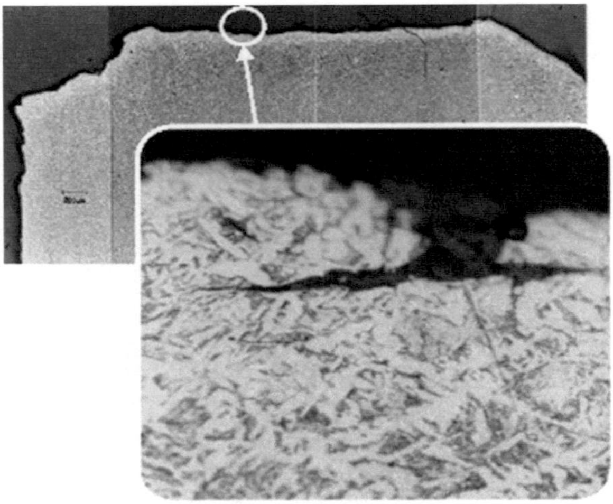

Fig. 4.17 ERW welding microstructure

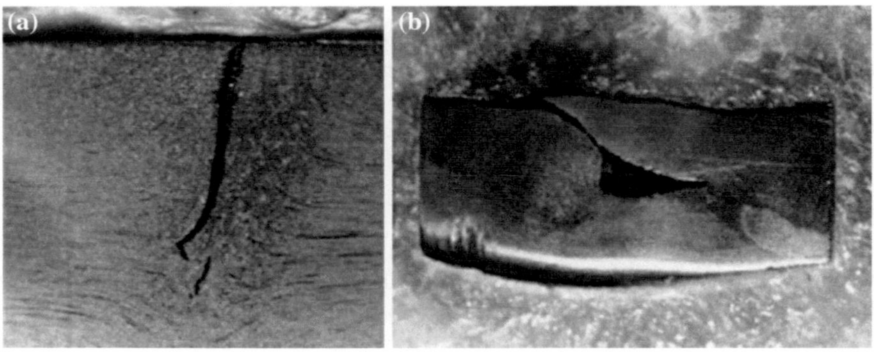

Fig. 4.18 **a**, **b** Examples of hook cracks in ERW welds

Alignment is done by system calibration; plate ends to be joined may become misaligned, misalignment is often combined with lack of fusion

Slag Inclusions: nonmetallic oxides are found sometimes in the form of elongated and globular inclusions in weld metal. During deposition and subsequent solidification of the weld metal, many chemical reactions occur between molten metal and slag. Some of the products of these reactions are nonmetallic compounds, soluble in the molten metal only to a certain degree. Their lower density makes them to seek the outer surface of the molten metal; when these products find restrictions to leave the metal, they get trapped.

Porosity: globular bubbles or holes are frequently found in weld beads. Gases in the pores are released during cooling of weld metal as a result of chemical reactions taking place inside the welding and as a result of reduced solubility as temperature lows down. Pores form when solidification is too fast for these

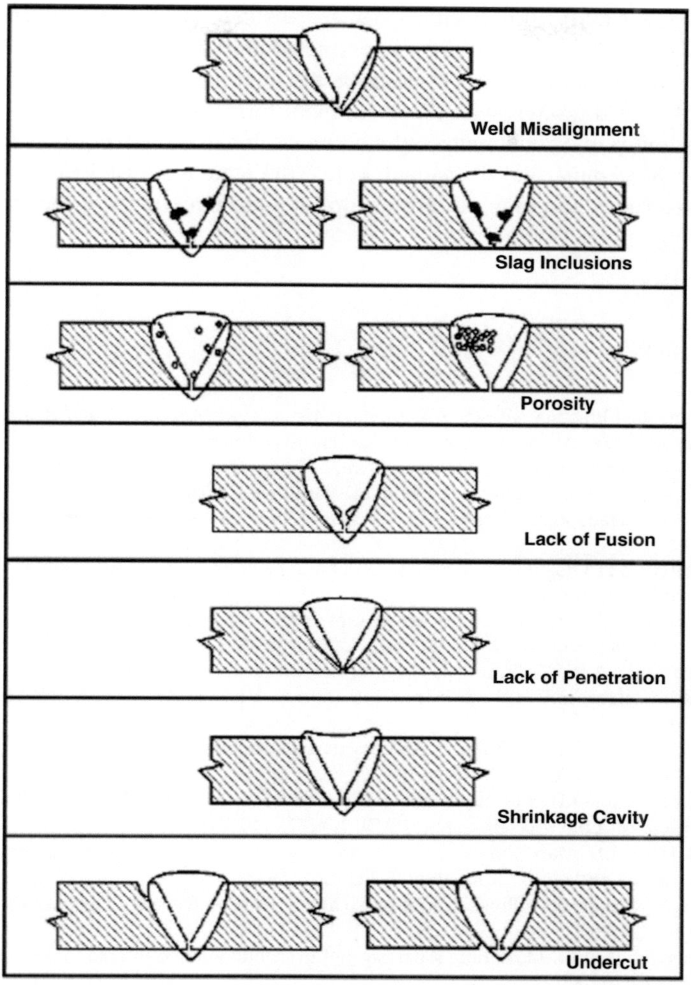

Fig. 4.19 Typical defects in arc welds

bubbles to float out the weld pool. These defects are readily detectable by NDT and seldom result in major failures.

Lack of Fusion: it occurs when weld heat does not melt the base material. This is a potentially dangerous, planar discontinuity between base and weld materials.

Lack of Penetration: this term is used to describe the situation in which the weld and base metals are not fused integrally at the root of the weld. It can occur when the root of the weld groove does not reach melting temperature, or when weld metal does not reach the root of a fillet weld. Lack of penetration usually leads to severe stress concentrations and may lead to failure, particularly if the weld root is subject to tensile stress.

Shrinkage Cavity: this occurs when the weld is large and subjected to fast cooling, so that the center portion is contracted and shrinks inward.

Undercut: this term is used to describe the melting and incomplete refilling of the wall of a weld groove, with the subsequent formation of a marked depression. Sharpness and size of undercuts and amount of thickness reduction in base metal depend on procedure, and in manual welds, the technique used by the operator. Certain electrodes, too high current or too long arc may increase the tendency to undercutting.

References

1. Recommended Practice API RP 571, *Damage Mechanisms Affecting Fixed Equipment in the Refining Industry* (American Petroleum Institute, Washington, D.C, 2003)
2. Guide for Inspection of Refinery Equipment, *Chapter II: Conditions Causing Deterioration or Failures* (American Petroleum Institute, Washington,. D.C, 1973)
3. ASM Metals Handbook, *Volume 13: Corrosion* (American Society for Metals, Metals Park, 1987)
4. C. Manfredi, J.L. Otegui, Failures by SCC in buried pipelines. Eng. Fail. Anal. **9**(5), 495–509 (2002). ISSN 1350-6307
5. P.G. Fazzini, J.L Otegui, Failure analysis of tube-to-tubesheet welds in cracked gas heat exchangers. Eng. Fail. Anal. **11**, 903–913 (2004). ISSN 1350-6307
6. P.G. Fazzini, A.P. Cisilino, J.L Otegui, Experimental validation of the influence of lamination defects in E.R.W. pipelines. Int. J. Press. Vessels Pip. **82**, 896–904 (2005). ISSN 0308-0161
7. Informe Técnico GIE 2103-05/05 Análisis de Falla de Horno de JP1 Afectado por Fuego. (TGS, Argentina, 2005)
8. B.S. Delanty, J.O'Beirne, Low-pH stress corrosion cracking. In: Proceedings of 6th Symposium on Line Pipe Research. PRCI. L30175 (1991)
9. B.S. Delanty, J.E. Marr, Stress corrosion cracking severity rating model. In: Proceedings of International Conference on Pipeline Reliability (1992)
10. Cañerías y Recipientes a Presión, José Luis Otegui, Esteban Rubertis. Eudem ISBN 978-987-1371-18-1 (2008)
11. ASM handbook. Vol. 11, Failure Analysis and Prevention. ASM International, 2002. (ISBN: 978-0-87170-704-8)
12. P.G. Fazzini, J.C. Belmonte, M.D. Chapetti, J.L. Otegui, Fatigue assessment of a double submerged arc welded gas pipeline. Int. J. Fatigue. ISSN 0142-1123 (2006)
13. J. Massone, J.L. Otegui, H. Ortiz, P.G. Fazzini, Intergranular susceptibility in failures of high pressure tubes. Paper FA 176, ICEFA 2, 2nd. International Conference on Engineering Failure Analysis. Toronto, 13–15 Sept 2006
14. H.P. Bloch, F. Geitner, *Machinery Failure Analysis and Troubleshooting*, 3rd edn. (Elsevier Publishing Company, London, 1998). ISBN 0-88415-662-1
15. M.G. Fontana, *Corrosion Engineering* (McGraw Hill, New York, 1986)
16. C.P. Dillon, *Forms of Corrosion: Recognition and Prevention* (NACE International, Houston, 1982)
17. P.R. Roberge, M.A.A. Tullmin, K. Trethewey, Knowledge discovery from case histories of corrosion problems. CORROSION 97, Paper 319. NACE International, Houston, 1997
18. L.M. Wyatt, D.S. Bagley, M.A. Moore et al., *An Atlas of Corrosion and Related Failures* (Materials Technology Institute, St. Louis, 1987)

Chapter 5
Damage Resistance Tests of Materials

Experimental tests discussed in this chapter refer to those chemical and mechanical analyses of materials involved in a failed element. In this context, we call material to each part of the component suspected to possess particular properties that may have influenced the failure mechanism. Features to search for include, base, weld and HAZ materials, surface treatments, corrosion deposits, microstructural changes by temperature or radiation, and in general any noncompliance possibly occurring during manufacture or in-service.

To define the characteristics and qualities of materials, samples are taken and standard tests are performed as appropriate. It is important to define what standard is used to define the properties. In general, these are determined according to standards developed by the American standards for testing of materials (ASTM) and other national standards. In some cases other standards may also apply, for example, pipe strength is usually defined according to the criteria by API 5L from the American Petroleum Institute, which differs slightly from the definitions of ASTM. We will see these features in the corresponding sections.

Chemical and mechanical characterizations should be performed in specialized laboratories with relevant qualification of operators and calibration of experimental equipment. The details of the tests required in any failure analysis depend on the particular conditions and origin of the failure. Typically, test protocols for base, HAZ and weld materials after a mechanical failure include:

a. Charpy impact tests and correlations with toughness, K_{IC}.
b. Tensile testing, true strain—true stress curves.
c. Hardness mapping, as appropriate, in areas of failure initiation.
d. Global spark interferometry chemical analysis.
e. Microprobe chemical analysis (EDS) of metals and deposits.

The results of these tests are first compared with specifications and applicable regulations and manufacturing licenses. Secondly, they are fundamental for the implementation of mechanical models that allow understanding the mechanism of damage and eventual failure (Chap. 6).

J. L. Otegui, *Failure Analysis*, DOI: 10.1007/978-3-319-03910-7_5,

We have seen that the conditions for ductile tearing or brittle fracture usually require, at some point during service, higher than historical stresses. In some cases, ultimate failure comes after degradation mechanisms alter material properties by modifying its microstructure (Creep, decarburization, carburization, and various forms of embrittlement) or generate discontinuities that reduce the strength of a section to mechanical loads (fatigue, corrosion, etc.).

In the next chapter, we will see the rules and methods for performing analytical mechanical models to verify stresses in the failed component, depending on the conditions of mechanical loading and environmental effects. By comparing these models with the actual properties of the materials in the areas of interest, the conditions that led to final failure, modes of damage propagation, and their relation to the combined effect of material preconditions and loading are defined.

5.1 Tensile Testing

The uniaxial tensile test is well known and widely used for determining mechanical properties of a structural material (Fig. 5.1).

The information is usually obtained from an applied force F and a measured elongation Δl in a calibrated part of the sample of length l and initial section A_0. Whereas Δl is homogeneously distributed along the measurement length L, stresses and strains are constant at any section:

$$\sigma = \frac{F}{A} \tag{5.1}$$

$$\varepsilon = \Delta 1/1 \tag{5.2}$$

The relation between (engineering) longitudinal stress σ and strain ε results in a diagram such as in Fig. 5.2.

The initial (elastic) slope is called Young's modulus (E, or elastic stiffness) of the material. Elongation is accompanied by a shrinkage in the transverse directions (defined by Poisson's ratio, v), which leads to a decreasing in cross-sectional area that is instantly taking values A_{in} ($A_{in} < A_o$). This leads to an "instantaneous" value of the stress σ_{in}:

$$\sigma_{in} = \frac{F}{A_{in}} > \sigma \tag{5.3}$$

This is represented by the dashed line in Fig. 5.2. There can be significant discrepancies between σ and σ_{in} in the elastoplastic range. Until the engineering stress–strain diagram reaches a maximum load (point G in Fig. 5.2), elastoplastic deformation may be considered constant throughout the calibrated section L. Until that point, the engineering strain can be calculated in terms of final and initial sample lengths as follows:

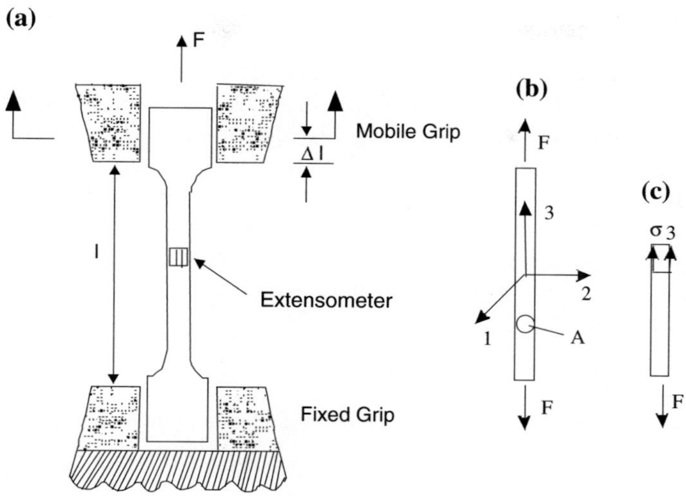

Fig. 5.1 General dispositions for tensile testing. **a** Load rig, **b** Calibrated region of specimen, **c** Constant axial stress in specimen

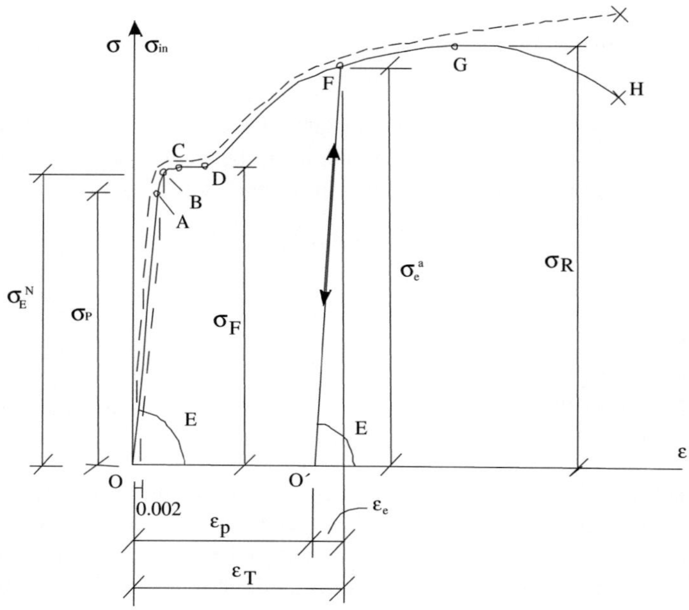

Fig. 5.2 Stress–strain diagram for the tensile test of a steel bar

$$(L_{\mathrm{fin}} - L_{\mathrm{ini}})/L_{\mathrm{ini}} = L_{\mathrm{fin}}/L_{\mathrm{ini}} - 1 \tag{5.4}$$

This is an average value based on the final displacement. For points to the right of maximum load, the engineering stress drops due to a localized reduction in section, a phenomenon called necking. To determine true strain, sections are monitored during the test. This can be done with sensors placed on the surface of the specimen, since where necking occurs is unknown prior to the test, normally several measurements are made during the test. Considering constant volume during deformation, the percentage increase in length is equal to the percentage reduction in area, so the engineering strain is defined in terms of final and initial sample sections as:

$$\varepsilon_{\mathrm{L}} = \Delta l/l = (L_{\mathrm{fin}}/L_{\mathrm{ini}})/L_{\mathrm{ini}} = (A_{\mathrm{ini}} - A_{\mathrm{fin}})/A_{\mathrm{inic}} \tag{5.5}$$

The true strain is defined with respect to the instantaneous differential variation of displacement, i.e.,:

$$d\varepsilon = dL/L$$

so that, integrating, we obtain the true strain at a given time of the test:

$$\varepsilon = \ln(L_{\mathrm{fin}}/L_0)$$

thus,

$$\varepsilon = \ln(\varepsilon_{\mathrm{L}} + 1) \tag{5.6}$$

A common way of defining the elastoplastic behavior is from the definition of strain as a hardening coefficient. The relationship between stress and strain in the plastic region can be defined in the form of an equation of the type:

$$\sigma = \sigma_0 \varepsilon^{\mathrm{m}} \tag{5.7}$$

where

σ = true stress
σ_0 = coefficient of strain hardening
ε = true plastic strain
m = strain hardening exponent.

Yield strength is defined as the ability of a material to withstand stresses without plastic deformation (Fig. 5.2). For some material the departure from linearity in the stress–strain curve is quite gradual, so that yield strength is not easy to identify. That is why a standard yield stress is defined at 0.2 % total strain. The ultimate tensile strength (UTS) is a measure of the ability of a material to withstand stress before failure. Sometimes an intermediate strength value is defined, the flow strength σ_f, which is used in certain failure models (also to be discussed in next chapter). With this test it is also possible to measure other material properties:

- Ductility: elongation and reduction of area to rupture.
- Toughness: energy required to cause failure, equal to the area under the stress–strain curve.

There are other more accurate methods to determine toughness; we will see them later in this chapter. An important aspect in determining the strength of a material is the test temperature compared to operation. At higher temperatures, all materials tend to decrease yield strength. In applications for high temperature steels alloying elements such as Cr and Mo are commonly used, which reduce drop in strength with temperature. The following example shows a case of a component that failed due to this phenomenon.

5.1.1 Example 5.A A Stainless Steel Tube at High Temperature

This study was conducted to determine the causes that led to abnormal longitudinal elongation in the vertical pipe (stand pipe) of a catalytic reactor, in operation in a petrochemical plant. Pipe length is 16 m and is suspended from the top. The original pipe was replaced due to operating problems; the new pipe was in service for 2 years before the equipment had to be stopped due to excessive elongation of this pipe. During this period, the equipment recorded three stops; tube skin temperature is not monitored. There is no history of this type of failure in this component, neither in this nor in the previous pipe, which had been in service for 14 years.

The pipe is made of AISI 304H stainless steel and subject to an outer operating temperature of 740 °C. Diameter is 20″ and thickness 15.5 mm. It has an inner concrete coating, with a metal mesh for support; coating thickness is between 25 and 27 mm. The tube contains a catalyst which consistency is similar to fine sand. The pipe is suspended vertically and supports its weight, the concrete cover, and the catalyst, resulting in an applied axial tensile stress of 4 MPa to the pipe wall. An excessive accumulation of plastic deformation of the stand pipe forced a repair that consisted of cutting a 12-cm-long slice in the bottom of the pipe; deformation was therefore 0.8 %, four times greater than the elastic limit (strain at yield strength).

Mechanical tests were performed on longitudinal and circumferential tensile samples. Yield strength is 350 MPa, UTS is 700 MPa; these are well above requirements for an A-213-GR TP 304H material at room temperature (205 and 517 MPa). According to ASME BPVC the maximum working temperature for this material is 815 °C; at this temperature the maximum allowable stress is 9 MPa. However, the German standards set maximum service temperature for the same material at 570 °C.

Figure 5.A1 shows the variation with temperature of yield stress and allowable stress according to European and American standards; mechanical strength of the steel drops markedly with temperature. The final temperature of each graph can be interpreted as the maximum temperature allowed by each standard for this item. Allowable stress (ASME BPVC) falls from 137 to 15 MPa at 760 °C (down to 10 %). Strength reduction is not linear, and the slope of the curve falls markedly after 600 °C. If for any operational problem service temperature increased by only 10 %, and rose from 740 to 815 °C, the allowable stress of the material would fall to less than 9 MPa. If the temperature rises above 900 °C, the allowable stress of the material falls to less than the required 4 MPa. At this temperature range the behavior for this material is unpredictable, so the European standard does not permit the use of the material beyond this temperature.

The fall with the temperature of strength in the stand pipe material is probably greater than that estimated in Fig. 5.A1. The evidence shows that the allowable stress at the maximum operating temperature of the material is less than the applied stress. Figure 5.A2 shows a typical microstructure in the tube wall. Figure 5.A3 shows room temperature tensile behavior of pipe material; as in all austenitic steels, tensile strength is more than twice the yield strength, and the deviation from linearity in the stress–strain curve (i.e., the accumulation of plastic strain) begins at stresses lower than the Standard (0.2 % strain) yield stress.

Microstructures show dense slip bands within the grains. This indicates that the material has been subjected to severe plastic deformation (Fig. 5.A2). Precipitated carbides are also observed in grain boundaries. Metallographic evidence shows a material subjected to severe plastic deformation at high temperature, confirming that excessive deformation was due to a material with strong susceptibility to creep relaxation at service temperature.

Due to the limited microstructural damage observed, the occurrence of creep fracture would require many more years of service. However, further plastic elongation would not be avoided at current operating conditions. In this case, plastic deformations represent a limitation to the functionality of the equipment. It is possible that strain hardening and appearance of secondary creep would tend to reduce over time the rate of elongation. The occurrence of transient start-ups makes recurrence of primary creep and accelerates deformation. It was recommended that the equipment operates in steady state and avoids, if possible, start–stop cycles.

5.2 Ductile–Brittle Transition Temperature

The case of the Liberty ships in the mid-twentieth century mentioned in the Introduction was the catalyst for the emergence of the discipline called Fracture Mechanics. Even before, the industry involved in the construction of structures and components that worked at low temperature had adopted the concept of toughness. The experimental determination of the "transition temperature" by impact tests

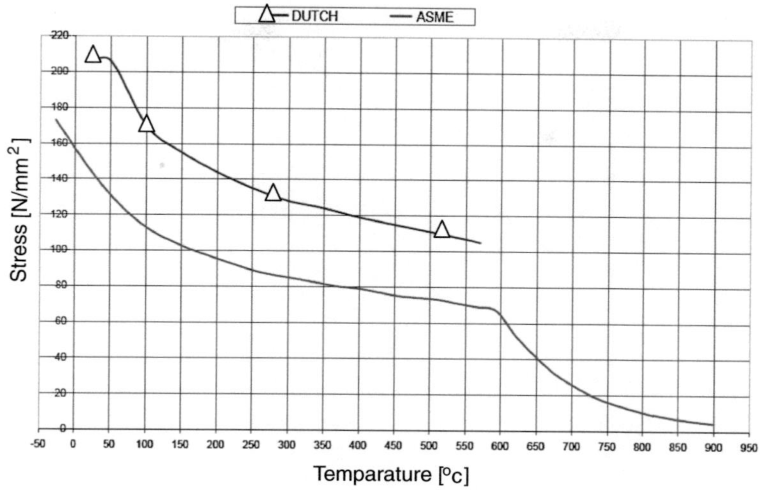

Fig. 5.A1 Dependence of strength upon temperature

Fig. 5.A2 Effect of plastic deformation in microstructure

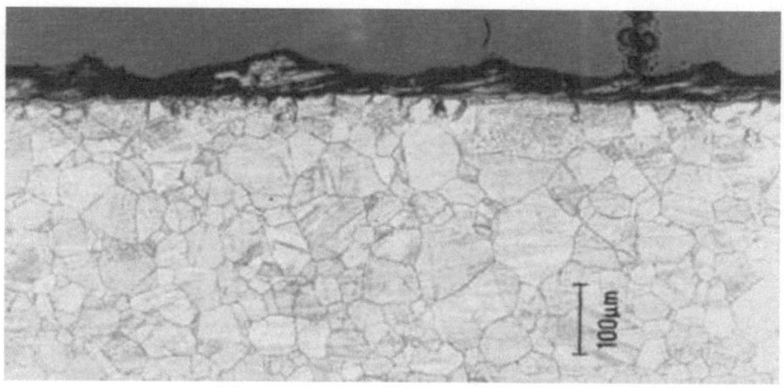

was originally developed by the German scientist Charpy in the late nineteenth century. The tendency of materials to brittle fracture when subjected to high deformation rates is the basis for the Charpy V-notch (CVN) impact test.

Charpy tests are currently used by all manufacturing codes and are regulated in the manufacturing standards and documents for integrity assessment and life extension. Small prisms of material with a central notch of precise dimensions are struck at their midspan by a pendulum or hammer with determined radius and angle of attack (Fig. 5.3). Standard samples are 10.0 × 10.0 mm section, but subsize samples can be used in case of thinner parts. The procedure is described in ASTM E 23. Test results are interpreted in terms of absorbed energy during breakage, as determined by the difference between initial and final angles of the pendulum.

The transition temperature (Tt) is a property of some materials that is related to the tendency to fracture initiation, notably including structural steels and polymers. Each of these materials shows a clear change in its tendency to ductile or brittle

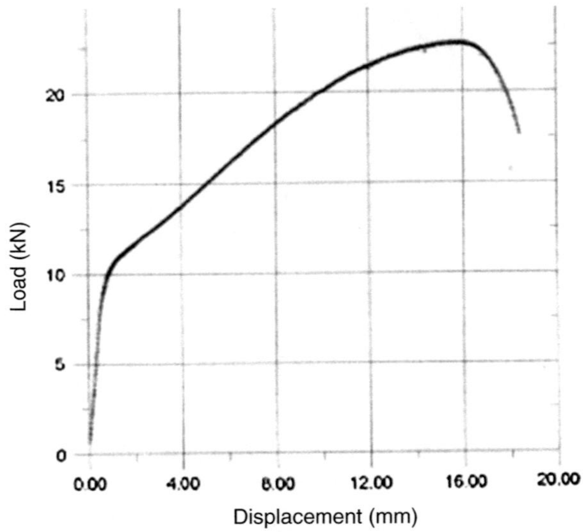

Fig. 5.A3 Tensile test result

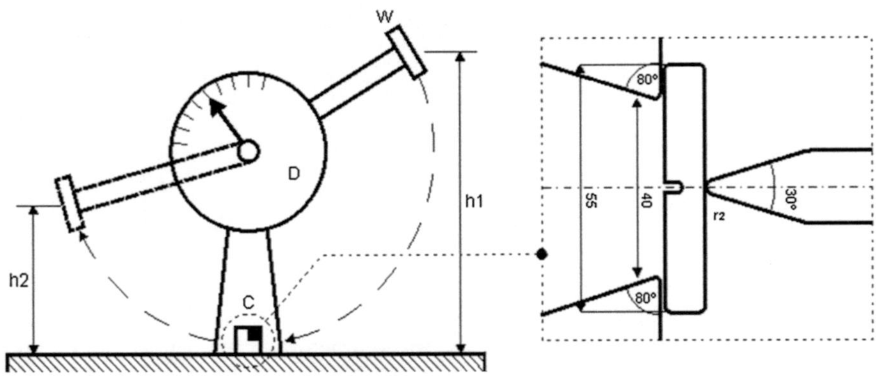

Fig. 5.3 Charpy V-notch impact test

fracture at a certain temperature, which depends on the type of material, its composition, and its previous thermal and mechanical treatment. Below this Tt temperature, the break is brittle, showing little or no deformation. Above Tt the material behaves ductile, or better to say, tough. Ductile behavior is evidence that the material experiences a large amount of deformation, that is, absorbs large amounts of energy, before rupture.

This change in the energy absorbed by the material is related to changes in the physical mechanism by which fractures propagate in the material, which manifest themselves in different aspects of fracture surfaces as seen in Chap. 3. Table 5.1 shows that not only the temperature favors the occurrence of brittle fracture.

Table 5.1 Parameters affecting toughness

	Ductile	Brittle
Temperature	High	Low
Load velocity	Low	High
Geometry	Without concentrator	With concentrator
Size	Thin	Thick
Load type	Torsion	Tension/compression
Material strength	Low	High

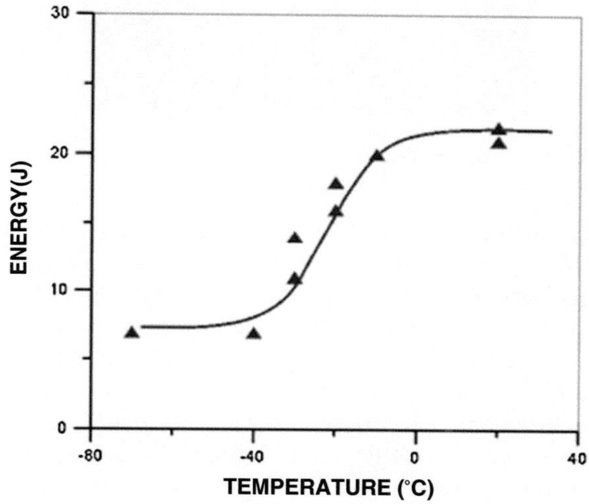

Fig. 5.4 Ductile–brittle transition for a structural steel

We saw the size effect when analyzing shear lips and chevrons during the instantaneous propagation of a fracture (Chap. 2). Figure 5.4 shows a typical brittle–ductile behavior for metals; it corresponds to a steel from a pipeline built in the late 1950s. In this case, transition temperature is roughly –20 °C.

5.3 Experimental Determination of Fracture Toughness

The Charpy test allows qualitatively characterizing toughness and transition temperature of a material. However, it is often required to quantitatively define material toughness in terms of the "critical stress intensity factor" K_{IC}. This measure of toughness is useful because it can be directly compared with the applied stress intensity factor K_{app} (which we will discuss in the next chapter). Determination of K_{IC} requires special tests or conversions from the energy absorbed in Charpy impact tests (CVN).

Experimental measurement of K_{IC} is performed with cracked samples, loaded in tension or bending, see Fig. 5.5. Samples can be three-point bending (3 PB, Fig. 5.5a) or compact tension (CT Fig. 5.5b); these are loaded until unstable crack

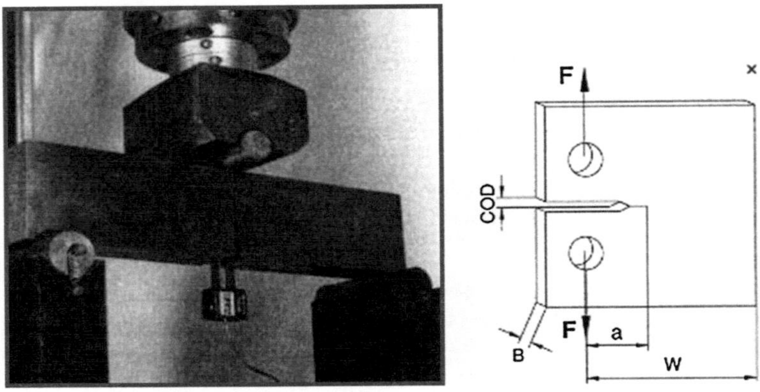

Fig. 5.5 Fracture toughness specimens

propagation. Propagation and loading of the crack tip opening displacement (CTOD) are measured with devices called clip gauges. To obtain valid results, the Standard requires verifying that the sample dimensions are large enough to ensure small-scale plasticity at the crack tip (also discussed in the next chapter).

However, most materials currently used in pipes and pressure vessels are quite tough, so that K_{IC} testing requires samples of enormous dimensions. That is why another approach has been developed, based on a parameter called the critical *J* integral, or J_{IC}. This setting supports the existence of large-scale plasticity in the remaining ligament and thus allows the use of smaller samples. The characteristics of the samples are similar to those in Fig. 5.5. Further details about the procedure can be found in standard ASTM E1820.

cDetermining J_{IC} requires testing several samples with the same initial crack length, loaded to obtain different amounts of crack propagation. Figure 5.6 shows an example of two samples, after complete fracture. The surface of the machined notch is shown at the top; the dark rough area immediately below corresponds to the precrack, fatigue propagated to achieve an original defect with adequate geometry. Crescent-shaped marks correspond to stable fracture propagation during loading. The sample at the right was subjected to a crack propagation Δa greater than that to the left. The clear zone below is the cryogenic fracture made after the test to separate the two halves of the samples for measuring crack propagation during the test.

Values of applied *J* to each sample are proportional to the area under the load–displacement curve for each test. An alternative is the use of a single specimen, in which crack propagation is measured in an indirect way. Variations in sample stiffness are detected by measuring crack mouth opening displacements using a clip gauge (Fig. 5.7).

Having identified the different values of applied *J* and each corresponding crack propagation Δa, these are plotted as shown in Fig. 5.8. Then extrapolation is done to zero crack extension to define an applied *J* for crack initiation, that is, J_{IC} for the material. In practice, this point is defined by the intersection of the J–Δa curve and

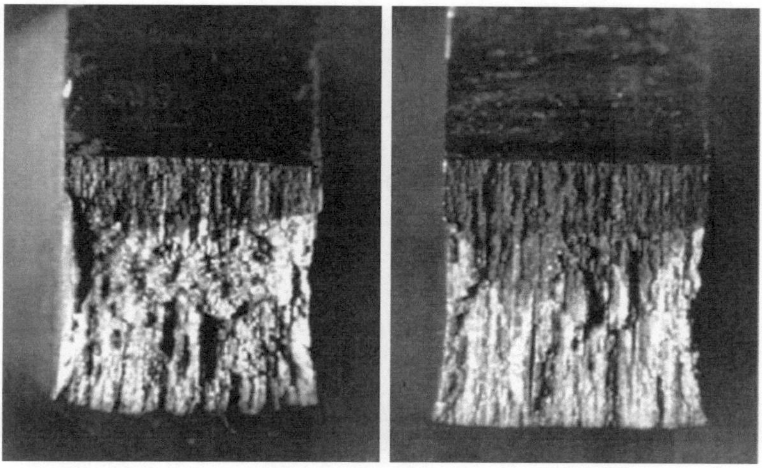

Fig. 5.6 J_{IC} specimens fractured at different crack lengths

Fig. 5.7 Clip gauge arrangement in a fracture mechanics test

the so-called blunting line, which represents an apparent crack extension due to rounding of the crack tip. This rounding is due to the large plastic deformations generated in this area. Once the J_{IC} parameter is defined, a linear elastic relationship relates it with K_{IC}:

$$K^2_{IC} = J_{IC}E \,/\left(1 - \nu^2\right) \tag{5.8}$$

Fig. 5.8 J_{IC} test results

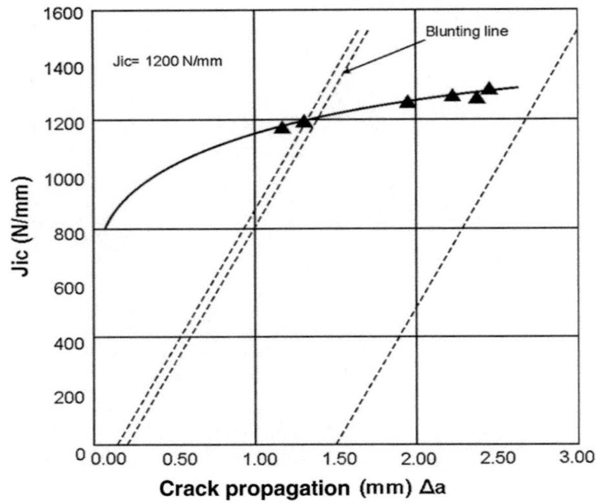

where E is Young's modulus (elastic stiffness) of the material and v is Poisson's ratio.

Fracture toughness K_{IC} for a material can also be estimated from empirical correlations with the value of energy absorbed in the impact test (CVN). These estimates are conservative and provide a lower limit for K_{IC}. Roberts–Newton correlation estimates a lower limit value of K_{IC} for ferritic steels in the ductile–brittle transition regime:

$$K_{IC} = 9.35(CVN)^{0.63} \tag{5.9}$$

where CVN is the energy absorbed in the impact test (Charpy) in ft-lb and K_{IC} is fracture toughness in ksi(in)$^{1/2}$. Another relationship between Charpy energy and K_{IC}, suitable for medium- and high-strength steels above transition, is:

$$\left(K_{IC}/\sigma_{ys}\right)^2 = 5\left(CVN - \sigma_{ys}/20\right)/\sigma_{ys} \tag{5.10}$$

where σ_{ys} is the yield stress or flow stress of the material. The units are the same as in the previous case. In both cases, the conversion to the SI system is relatively simple.

5.4 Spectrometry Analysis and Hardness Tests

Chemical analysis of metals and other materials is almost mandatory in any failure analysis, since it identifies chemical compositions of samples and can be used to contrast with standards and manufacturer's or operator's specifications. When an electric arc or spark is struck against the sample surface, a small portion (few mm^2) of

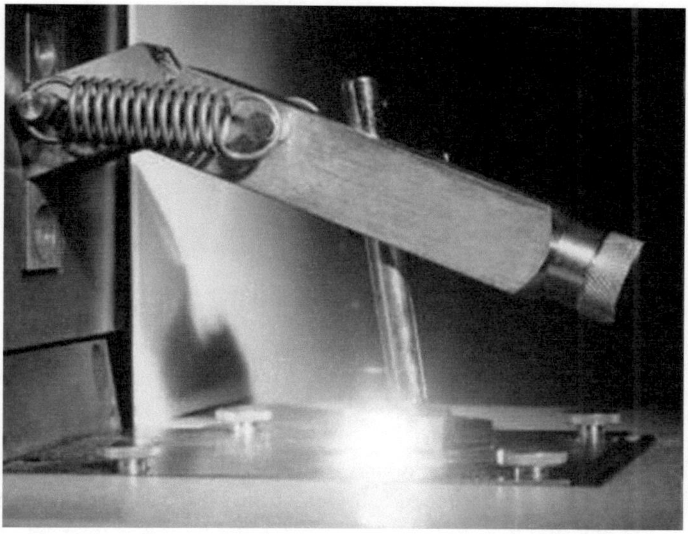

Fig. 5.9 Spark spectrometry chemical analysis

material heats up to a very high temperature, see Fig. 5.9. This excites atoms, which emit a characteristic light or glow. This light is emitted at different wavelengths that are characteristic of that element, and are detected and analyzed by spectroscopic methods. Another method is the microprobe X-ray spectroscopy (EDS) chemical analysis, which is part of the techniques of scanning electron microscopy (SEM, Chap. 3).

Hardness is defined as the resistance of a material to be penetrated by a hard and sharp element, called indenter (also mentioned in Chap. 3). Hardness is defined by a combination of the maximum applied load and the size of the imprint. Hardness tests are very old, and there are several versions. The most common are:

- Brinell [HB] uses a very hard ball of a few millimeters in diameter, which leaves a hemispherical imprint (Fig. 5.10a). Originally driven by a hammer, in modern methods it is driven by a controlled force or by a known kinetic energy. Laboratory (fixed) and field (portable) units are available.
- Rockwell [HRC, HRB]: The indenter is a diamond pyramid, which is applied by a load according to the type of material. The load defines the hardness grade (the most common are *B* and *C*). It is a laboratory test. Once the indentation is done, its size is measured and compared with standard parameters. The larger the footprint, the lower the hardness of the material.
- Vickers [HV] is similar to Rockwell, but load and indenter are much smaller. A small indentation and the precision of its location in the sample makes this test very useful for the characterization of different phases in a microstructure, and definition of surface treatments, etc. (Fig. 5.10b). This is also a laboratory test.

Fig. 5.10 Indentations from **a** Brinnel and **b** Vickers hardness tests

The above list is chronological and also by decreasing size of indentation. Hardness is a measure of resistance to indentation, which is related to strength to mechanical loading and also to friction wear, so that hardness tests are often used to characterize the rheological properties of materials and surface coatings.

In most failure analyses, the interest of determining hardness in carbon steel lies in the correlation between hardness and mechanical strength. When samples from a failed part are too small for machining a sample for a tensile test, or when it is required to estimate strength in small areas potentially affected by thermal cycling, plastic deformation, or damage, it is customary to resort to the indentation test for estimating the strength of the material. According to various literature sources, the relation between UTS and Brinell hardness for structural steels is:

$$\text{UTS} = 0.45\ \text{HB (Kpsi)} = 3.10\ \text{HB (MPa)} \tag{5.11}$$

Several researchers reproduced the relationship UTS = 3.45 HB, based on data from different sources. Figure 5.11 shows correlations with Brinell and Vickers hardness. While there is a relationship between hardness and UTS, the estimation of this relationship has large margins of error and variability. API STD 579—ASME FFS-1 ("Fitness for Purpose") standard presents in Annex F, a table that provides a rough equivalence between hardness and UTS for carbon and low alloy steels in annealing, normalized, and quenched and tempered conditions. At high and low values of hardness and stress the relationship is not linear, data are scattered, and linear correlations found do not fit properly.

Another limitation of the method is that hardness correlates well with UTS, but most standards require yield strength as most relevant material property. There is a recently developed evolution of this method to accurately measure mechanical strength for cases where it is impossible to remove the equipment from service to take samples for destructive tensile tests. It is called automatic ball indentation (ABI); it involves measurement of displacement and force as a cyclic indentation is made on the component surface. The equipment consists of the machine performing the test, a control and data acquisition unit, and processing software. This

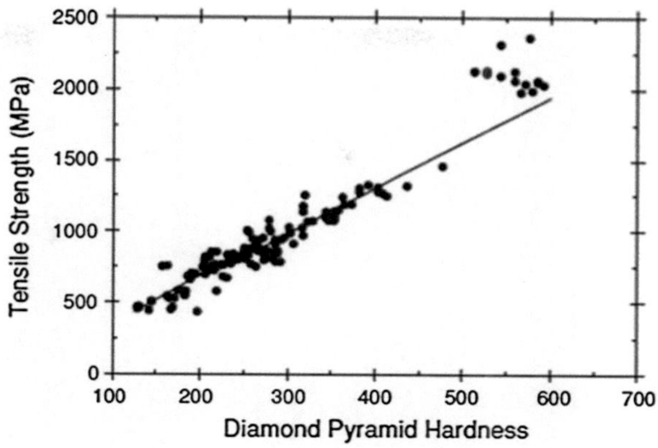

Fig. 5.11 Relationship between hardness and strength (E.J. Pavlina, ASM Intl., 2008)

system greatly improves the stability and repeatability of the measurement over the traditional hardness methods. The explanation of the instrumented indentation method requires some knowledge of mechanical modeling, which is presented in Chap. 6. We will be back with a description of this method in Chap. 9.

5.5 Experimental Measurement of Residual Stresses

Residual stress (usually generated during manufacture or previous service of the component) is characterized by existing in the material in the absence of any applied external loads. Manufacturing residual stresses are largely self-balanced in the section, and have an approximately linear through thickness distribution. Such bending stresses cover small areas. In cases where parts are heavily restrained, i.e., when they cannot move to accommodate differential expansion or contraction, residual stresses can be non-self-balanced, encompassing larger parts of the piece.

The fusion welding process involves adding a weld metal into the base material that, once solidified, cools down from solidification to room temperature. Material's coefficients of thermal expansion make the weld metal to shrink during cooling. If the welded plates have low stiffness, they distort; if they are rigid with respect to the weld bead, material adjacent to the weld bead must stretch to accommodate shrinkage of the bead, whereby tensile stresses appear. These tensile stresses produced during cooling of the weld may get high values, both longitudinal and transverse to the weld bead, but they are very localized, typically to a few millimeters around the weld. Longitudinal welding residual stresses can reach the yield stress of the base plate and transverse residual stresses are about half that. The residual stress field is self-balanced in the thickness, and is destroyed if the section sustains large plastic deformation.

Surface measurements are performed to experimentally check the level of residual stresses actually present in a component. There are several methods, all

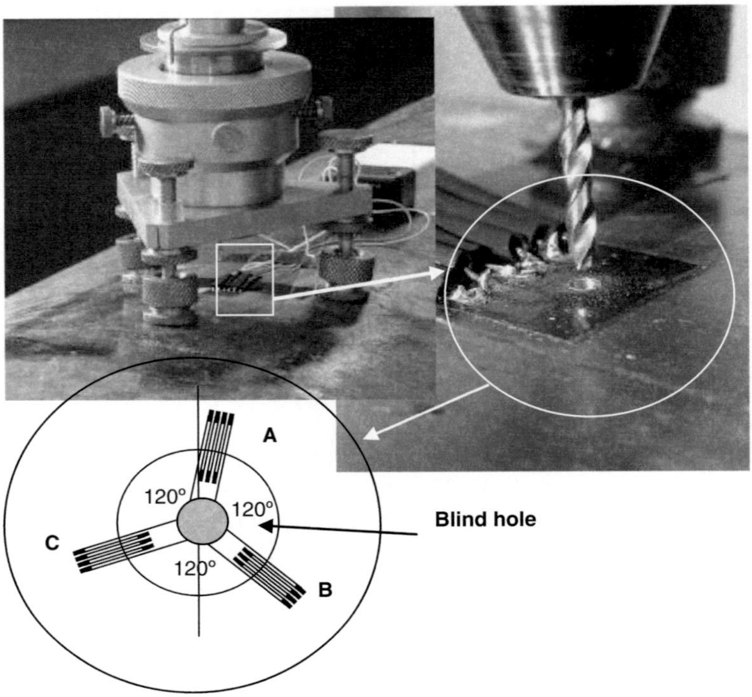

Fig. 5.12 (*a, b, c*) Measurement of residual stresses by the method of the blind hole

based on the gradual removal of material and the measurement of the deformations that it generates on the surface, usually with electrical strain gauges. Most common methods involve machining a slot or a blind hole. The slot method is based in placing strain gauges in the direction of the stresses to be measured, and perform a perpendicular notch. As the groove deepens, strain gauges indicate the relaxation of strain, so that the distribution of strains (and therefore stresses) present in the material thickness can be calculated before the notch is mechanized.

Figure 5.12a shows a device for measurement of residual stresses by the method of the blind hole. A "rosette" strain gauge containing three extensometers (Fig. 5.12c) is bonded on the previously polished piece. A hole is machined at the center of the rosette, and as it goes deeper (Fig. 5.12b), relaxation of strains is measured. Then this relaxation is related to the initial stress state through a series of equations.

If the relationship between stresses and displacements in the piece is known, it is not always necessary to instrument with strain gauges. For example, to quantify the residual stresses in cylindrical containers or tubes, a simple experimental analysis can be conducted. This involves a lengthwise cut of a ring removed from the tube, and accurately measure the displacement between the cut edges (Fig. 5.13). This displacement is proportional to the residual stress distribution in the pipe wall.

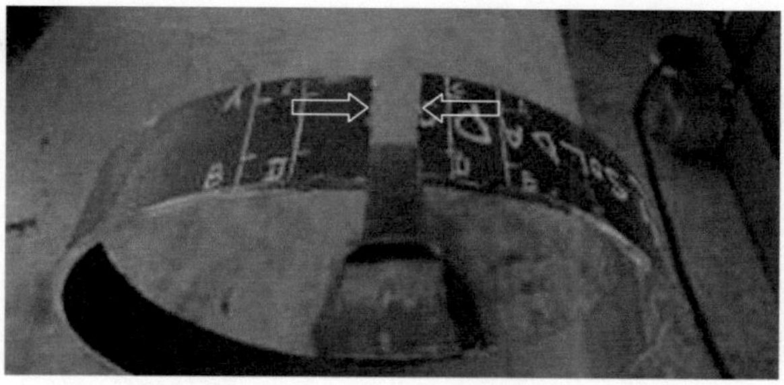

Fig. 5.13 Circumferential residual stress in a line pipe

According to Roark equation (Formulas for Stress and Strain, 4th ed.) maximum residual stress on the outer surface σ_R is:

$$\sigma_{\mathrm{R}} = \mathrm{E}.\delta.\mathrm{T}/\left(\Pi.\ \mathrm{D}^2\right) \tag{5.12}$$

where:

δ = elastic opening after longitudinal cut
E = Steel elastic modulus
t = tube thickness
D = tube diameter.

Manufacturing residual stresses usually have a leading role in the onset of in-service damage mechanisms and in brittle fracture. For example, for the construction of weld tubes the manufacturer places the rolled strip in-between jaws that close the gap between plate ends, and then the weld is performed. When welding by ERW, electrical contact is made; if the weld is DSAW, the arc is performed. These processes produce shrinkage and therefore elastic hoop stresses in the material. Some of these stresses remain after cooling of the weld, with a bending distribution through the thickness. This stress distribution is such that tensile hoop stresses generated on the outer surface add up to service stresses due to internal pressure; this combination may favor initiation of surface cracks by SCC or other damage mechanisms.

5.6 Experimental Determination of the Resistance to In-service Damage

In the following sections, some methodologies for evaluating the resistance of the material to the commonest forms of in-service damage are briefly described:

- Fatigue
- Corrosion
- Creep
- Stress Corrosion Cracking.

For these tests to be representative of the actual material response to each damage mechanism, it is important to ensure that relevant parameters used in the tests are similar to those actually experienced in service.

5.7 Testing of Fatigue Life

The fatigue strength of a material is experimentally defined from polished cylindrical samples subjected to rotating bending. These are simple machines, see Fig. 5.14a, rotating at high speed. A variable weight is hung at one end of the specimen, creating an alternate stress ($S_{max} = -S_{min}$) that is maximum in the calibrated zone (indicated by arrow). Measuring the number of rotations until final breakage, curves of amplitude of cyclic stress S ($S = S_{max} = -S_{min}$) versus number of cycles N are constructed. These are called S–N or Wohler curves, as shown in Fig. 5.14. If S is close to the material strength, S_{UT}, failure would occur within a few cycles. If the stress is below what is called the fatigue limit, S_e, life becomes infinite. In the example shown, each curve corresponds to the fatigue resistance of steel with different levels (wt%) of an alloy element.

These tests are useful for the design (and failure analysis) of smooth parts, typically machine elements subjected to many millions of fatigue cycles. But in the case of static components (pressure vessels, structures, etc.), construction processes generate surface discontinuities (welds, holes, machining marks, thickness changes, etc.) that reduce fatigue resistance when compared with the polished material. Specific S–N curves are determined experimentally for each of these cases, which are similar to that in Fig. 5.14, but at lower values of S. This allows defining fatigue strength of design details, which are classified into specific ranges of severity.

The fatigue limit, as in Fig. 5.14, means infinite life, but is usually defined as a cyclic stress for which the fatigue life is more than a sufficiently large number of cycles. It is common for structural steels that the boundary be located at approximately 2 million cycles. Some materials do not show fatigue limit, as in the case of aluminum. The fatigue limit is directly related to the fatigue threshold as will be defined below. The effect of corrosion (corrosion fatigue, see Chap. 4) generally prevents the formation of a fatigue limit.

This method of analysis is based on the determination of the total fatigue life of a component until it fails, and does not consider the different phases of the process. Mechanical fatigue of materials can be classified into the following discrete but related phenomena:

1. initial cyclic damage,
2. creation of the initial microscopic flaws (micro crack initiation),
3. coalescence of these micro cracks to form a detectable defect (growth of micro cracks),
4. subsequent propagation of the defect (engineering crack growth),
5. final failure or crack instability.

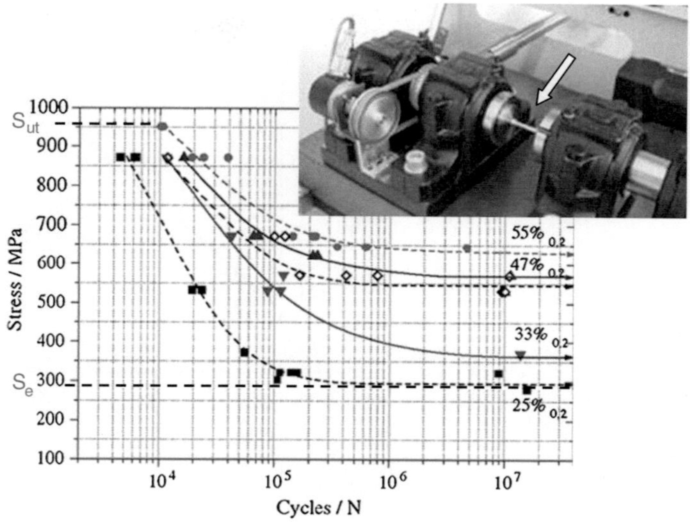

Fig. 5.14 (*a, b*) fatigue life and stress amplitude

In engineering terms, the first three steps are considered as macro crack initiation. Small manufacturing defects, such as surface roughness, weld undercuts, slag inclusions, or small cracks can act as initial cracks so that the "initiation" stage may be short. Fracture mechanics modeling of fatigue (as will be seen in Chap. 6) requires the experimental determination of the functional relationship between the crack growth rate (da/dN, usually much less than a tenth of a millimeter per cycle), and the parameter ΔK:

$$\Delta K = K_{\text{max}} - K_{\text{min}} \tag{5.13}$$

produced by one load cycle. Metals usually show the behavior shown in Fig. 5.15. In the logarithmic plot, the curve is a sigmoid with three distinct regions. The definition and calculation methods of the parameter *K*, widely used in linear elastic fracture mechanics, will be also detailed in Chap. 6.

As shown in the figure, region II of the curve is very close to a straight line. Growth in this region is adequately expressed by the following equation, known as the "Paris law" (named not after the beautiful French capital city, but after a bright scientist, pursuing his Ph.D. at the time of this discovery):

$$\frac{da}{dN} = C \cdot \Delta K^{m} \tag{5.14}$$

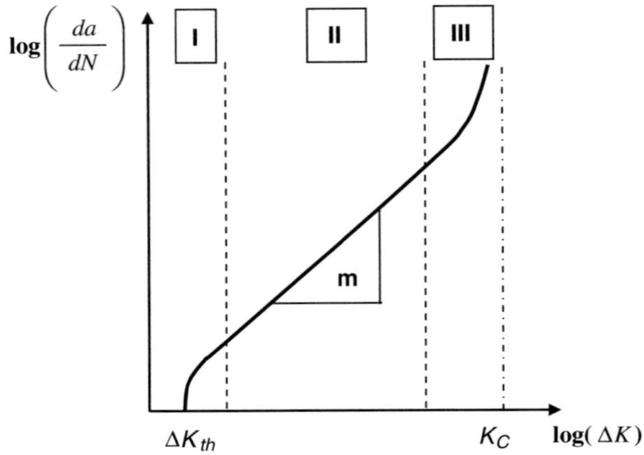

Fig. 5.15 Regions for fatigue crack growth

where C and m are material constants that are determined experimentally. In metals, m usually takes values between 2 and 5. Clearly, the Paris equation does not describe the fatigue behavior in region I, where the curve approaches a threshold value ΔK_{th} below which there is no propagation. Neither does it describe region III, where growth is fast as K_{max} is approaching material toughness K_{IC}, as defined in Sect. 5.3.

Fracture mechanics constants C and m are determined by testing of laboratory samples. The values of applied K_{max} and K_{min} depend on the maximum and minimum applied stresses, sample dimensions, and crack depth, as will be seen in Chap. 6. Test procedures are described in ASTM E647/13e1. A three-point bending sample with a central notch or crack undergoes cyclic tensile stresses in a device similar to that shown in Fig. 5.4. Not only loads are measured, but also crack lengths, as the crack grows during the test. Beach marks are deliberately formed, as defined in Chap. 2, or indirect estimates of crack depth are made.

At times, it is required to fatigue test materials from parts whose shape and dimensions do not allow the extraction of standard samples. In these cases, special samples and procedures can be developed to analyze the mechanisms of interest.

5.8 Techniques for Corrosion Monitoring

All mechanisms of corrosion damage are time dependent. Some mechanisms are relatively easy to detect and monitor, others are difficult, if not impossible. There are plenty of methods and/or inspection techniques to define the progress of corrosion, including:

- Visual examination of exposed surfaces.
- Radiographic Inspection.
- Ultrasonic inspection (UT).

- Eddy current inspection.
- Liquid penetrant and magnetic particle tests.

Inspection methods and predominant corrosion mechanisms have already been discussed in Chaps. 2 and 4. In this section, we will focus on techniques for monitoring corrosion rate. Corrosion rate determines how long the lifetime of a component will be, and therefore determines its safety against final failure. Techniques to monitor corrosion rate allow:

1. Early warning of potential damage that would occur in the structures, if corrosive conditions are kept.
2. To study the correlation of changes in process parameters with their effect on the corrosiveness of media.
3. Diagnosing a particular corrosion problem, identifying the predominant corrosion mechanisms.
4. Evaluate the effectiveness of a corrosion prevention/control applied to the system.
5. Provide information related to conditions of the component and maintenance requirements.

There are a number of techniques for monitoring corrosion rates and mechanisms. The most used in industrial applications are:

- Weight loss corrosion coupons.
- Electrical resistance (ER) coupons.
- Linear polarization resistance (LPR) samples.
- Electrochemical impedance spectroscopy.
- Electrochemical noise.
- Probes for detection of microbial activity (MIC).
- Galvanic probes (ZRA)/potential measurement.
- Hydrogen penetration.
- Erosion sensors.

Following, is a brief description of the fundamentals of corrosion monitoring using weight loss coupons, ER, LPR, electrochemical impedance spectroscopy and electrochemical noise.

Weight Loss Corrosion Coupons: This technique is the simplest and best known of all corrosion monitoring methods. Corrosion coupons are installed to obtain information on the corrosivity of a fluid/soil through visual and gravimetric (weight loss) inspection. This technique is based on exposure for a given time of a coupon made with the same material as the monitored structure, in the same corrosive environment to which the structure is exposed. Measurements of weight loss in the samples after given time periods allow defining corrosion rates.

Electrical Resistance (ER) Coupons: The basis for this technique is that corrosion on the surface of a metallic element results in decreased section with a corresponding increase in electrical resistance. An increase with time of exposure of the resistance can be directly related to loss of metal, and therefore corrosion rates can be obtained. ER samples provide a basic measurement of metal loss, but unlike weight loss coupons, measured values may be taken at any time and at any frequency. As ER coupons are permanently exposed to the corrosion process, online measurements can be taken and response can be much faster than with weight loss coupons.

LPR: The electrochemical technique of linear polarization resistance is a methodology to measure corrosion rates directly and in real- time (corrosion rate snapshots).

Electrochemical Impedance Spectroscopy: This has recently been applied to field monitoring. This technique records the response of the sample to a low voltage alternating current signal. By measuring the response of the sample to different frequency signals, many corrosion features, including corrosion mechanisms, rates, and formation of surface films can be determined.

Electrochemical Noise Monitoring: It consists essentially of monitoring small changes in potential of a sample or monitoring variations in corrosion current between two identical coupons. These variations increase with corrosion rates. This technique is sensitive to localized corrosion such as pitting. If stressed samples are tested, the noise technique is also very sensitive to stress corrosion cracking.

5.9 Estimation of SCC Propagation Rates and Threshold Stress

We saw in Chap. 4 that to have SCC damage three conditions are simultaneously required: an aggressive environment on the surface of the component, susceptibility of the material, and an applied stress that exceeds a threshold value, with a cyclic component. Laboratory and field data indicate that strain has an effect on crack initiation (nucleation of surface cracks) and subsequent propagation rates.

Metallurgy (both chemical composition and microstructure) of the component affects its response to SCC.

SCC crack growth is defined as a process in four stages. The first stage (1) involves the development of conditions for SCC to occur, followed by crack initiation (2). Then these cracks continue to grow and coalesce, while new ones may be initiating as well (3). Finally, the larger cracks coalesce and eventually reach the critical size for failure (4). Step 1 is defined by the time required to damage coatings and developing the electrochemical environment for SCC. It is probably the most difficult period to predict and may vary in orders of magnitude. When developing a propagation model in a component known to have SCC, this stage becomes irrelevant.

The key to predict SCC life of a product with SCC colonies is to properly estimate Step 3. In laboratory tests, the average crack propagation rate is typically

Fig. 5.16 Experimental setup for SCC tests

calculated by measuring the final length of largest cracks and dividing by the total test time. In the final stage of SCC crack growth, when tearing of remaining ligaments develops, the medium is no longer important. In many cases, the tearing step is preceded by a brief period where fatigue is the main mechanism for crack propagation.

Test methods involve the use of variable section (tapered) samples which are tensile loaded and immersed in the media with the appropriate (most severe) electrochemical conditions. The method is based upon the one developed by the American Gas Association, Project 146 Report NG-18. This test allows obtaining the following information for each item:

- Threshold stress σ_{TH}: stress below which no cracks are initiated, and
- crack propagation rate.

The samples are subjected to cyclic tensile stress in a testing machine, such as that seen in Fig. 5.16. This is a load-controlled machine. The sample is positioned vertically between two clamps, which are attached to a load cell. A variable load is applied through a lever arm; weights are attached to a vertical rod, ending in a coil spring with variable controlled rigidity. The experimental equipment is completed with an electrochemical cell for establishing the desired potential, which is maintained via a potentiostat and a reference electrode.

The samples are submerged in the solution with the required pH. To expedite the process, the test temperature is that of highest susceptibility. Typical tests last between 10 and 30 days. In tapered samples, maximum applied stress is at the smallest cross-section, which is set close to the yield strength of the material. Weld samples are tested transverse to the weld, and are usually of constant section, because there is not enough material to taper a weld section. Applying cyclic stress during the test accelerates it, since in these conditions fracture of the surface layer of corrosion products (such as magnetite in steels) at the crack tip is faster, allowing the material anodic dissolution at the crack tip.

After the test, the samples are covered with a black magnetite film (Fig. 5.17). Cross-sections are cut in a longitudinal axis of each specimen, and metallographic samples are prepared, which are then observed in light microscope. The length of each crack and the position of the section are recorded. The smallest section where no micro cracks are detected defines the threshold stress for susceptibility to SCC in the particular test conditions.

5.10 Creep Resistance and Monitoring of In-service Damage

Design lives of components subjected to creep are generally very long (10^5 h). A test this long to define creep strength of a material is impracticable; that is why accelerated creep tests are performed, with a duration of 10^2 or 10^3 h. This acceleration is achieved by increasing either temperature or applied stress above operating parameters. The results are then extrapolated to operating conditions.

Creep tests involve measuring four variables: stress, strain, temperature, and time. Tension is applied by a testing machine; it can be of the constant load or constant stress types (Fig. 5.18a). Typically, the test is performed under a constant load, which only requires the application of a weight to the sample, either directly or indirectly by means of a lever arm, while avoiding load eccentricity. Sample geometries are similar to those used for tensile tests, Fig. 5.19a, usually with extended ends so as to provide a gap for thermocouples to monitor temperature.

Deformations must be measured very accurately; one way is to use two mobile telescopes focus into two previously calibrated marks. Sometimes the samples are not tested in air, but in liquid metal salts or other types of baths, or in vacuum. The test comprises the following steps:

a. gradual heating of the sample, up to 3 h up to the set temperature,
b. soaking of the sample at set temperature for a set time,
c. applying constant tensile load.

In its mechanical part, creep resistance tests are similar to those of SCC. Machines are used in which the samples are subjected to stress up to failure within an oven that maintains the defined test temperature (Fig. 5.18b). Test conditions are defined by standards such as ASTM E139, which includes a brief description of the heating apparatus to be used when performing creep rupture tests in metallic materials. The frame of Fig. 5.18a has a capacity of 6,000 kg and the furnace of Fig. 5.18b has a temperature range from 300 to 1,200 °C. Loading bars and screwed heads of the test machine should be made of a special alloy for working

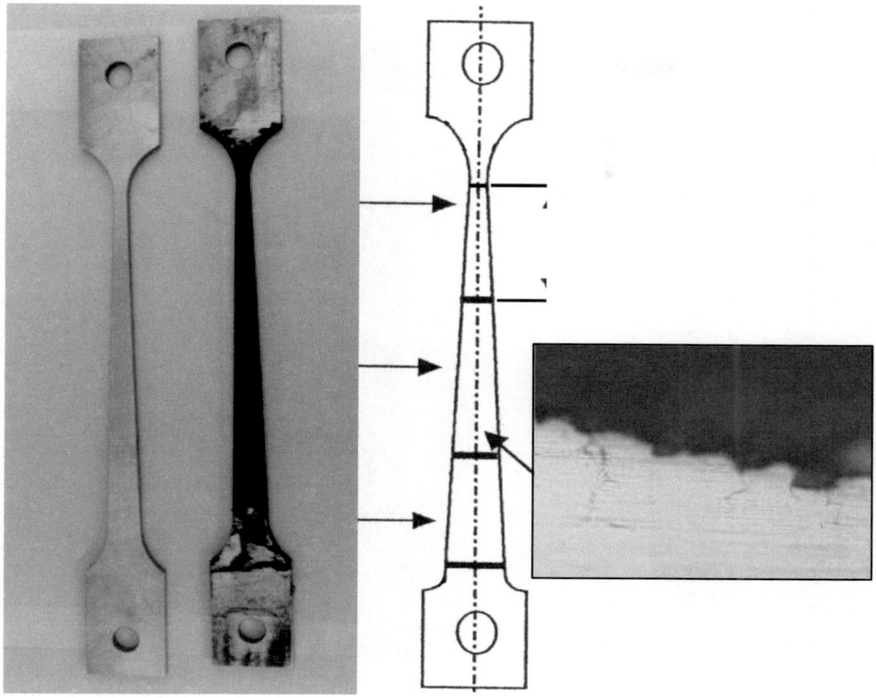

Fig. 5.17 SCC evaluation of a tapered specimen after an SCC test

inside the furnace. The system can also be equipped with a control system for unloading the weight to stop the test after a predetermined time.

Tests are also used to characterize creep crack propagation. For this, a compact tension (CT) fracture mechanics sample is placed in the furnace, as shown in Fig. 5.19b. Fracture mechanics parameters are defined as applied *K* (see Sect. 5.6) and crack propagation rates da/dt. Sophisticated devices are required to estimate the instantaneous length of the crack as it propagates during the test, similar to what was seen for fatigue tests. Removing the sample at any time during the test invalidates results.

Monitoring in-service creep damage in a component at high temperature is based on the technique of metallographic replicas or field metallography. Damage features can be identified by conducting periodic inspections and results interpreted in terms of remaining life. In recent years there have been a number of metallographic parameters in order to quantify different degrees of microstructural damage. Among these parameters are the volume fraction occupied by microcavities, the number of cavities per unit area, Neubauer parameter, and the fraction of damaged grain boundaries (parameter 'A').

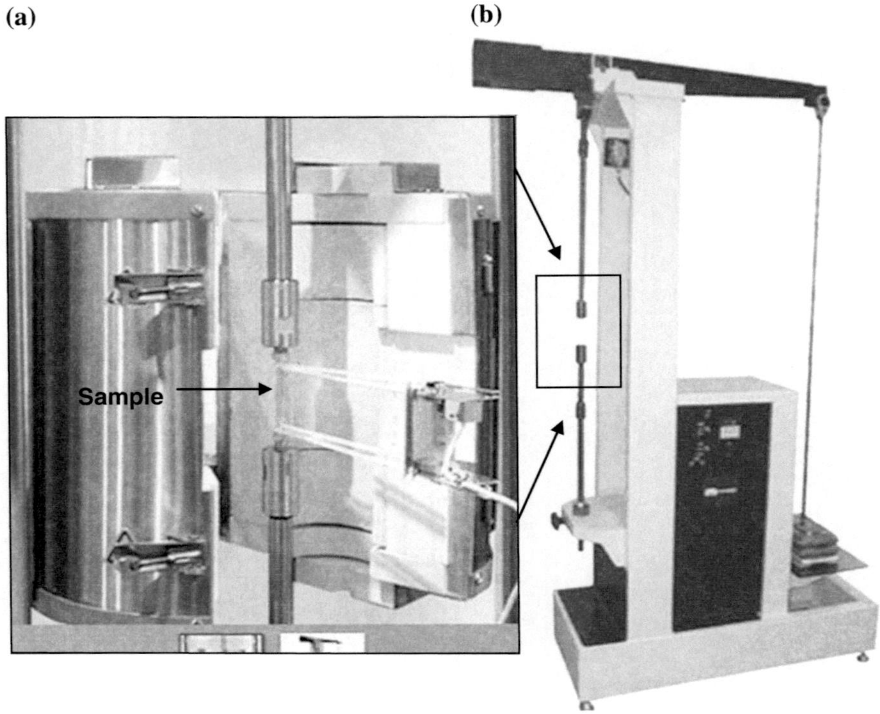

Fig. 5.18 Creep testing a Strain frame b Sample inside furnace

Fig. 5.19 a Uniaxial tension sample b Compact tension (CT) sample

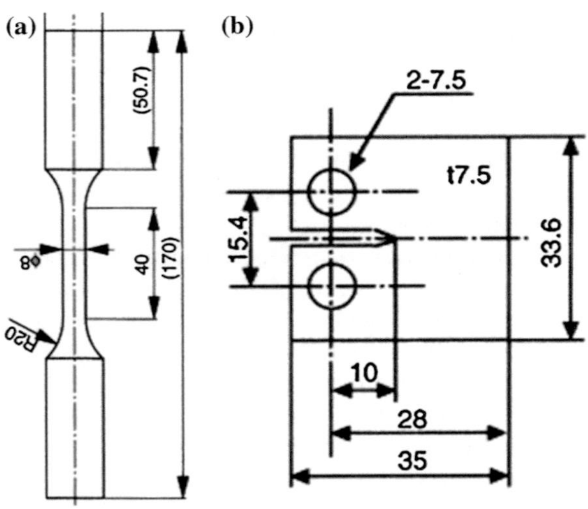

The Neubauer parameter classifies damage in four stages according to the quantity and location of microcavities found:

- Stage A (isolated): only a few isolated microcavities.
- Step B (oriented) microcavities are oriented in grain boundaries.
- Stage C (micro cracks) cavities coalesce to form micro cracks.
- Stage D (macro cracks) measurable cracks are observed in grain boundaries.

In spite of the thoroughly empirical classification, quantifications of residual life based on this parameter are very helpful in setting safe re-inspection periods. The parameter "A," originally developed by the Central Electricity Generation Board, England (CEGB), establishes a more precise quantification of residual lifetime through semiempirical relationships with the fraction of damaged grain boundaries. The parameter "A" is measured on replicas and photographs of the material with a 400X magnification. This magnification is a good balance between the need for resolution to identify microcavities (high magnification) and maintain a high number of grain boundaries in the visual field of the photograph taken (low magnification).

Lines transverse to the direction of maximum stress are drawn. As these lines intersect a grain boundary, the data point is classified according to the following rules:

1. An intersecting grain boundary is analyzed only between the intersection and the first triple point on both sides of the intersection.
2. A grain boundary is categorized as damaged if it contains one or more microcavities (or micro cracks) along its observable length.
3. Multiple intersections with the same grain boundary are counted as often as intersections occur.
4. A grain boundary intersected at a triple point is classified as damaged or not according to majority (that is, it is damaged if two or three of the grain boundaries forming the triple point are categorized as damaged).

The parameter A is the ratio of damaged grain boundaries to total grain boundaries. This parameter is correlated with the residual life of the component according to the following formula:

$$t_{\mathrm{rem}} = t_{\mathrm{exp}}\{[0.522/(A + 0.036)] - 1\} \tag{5.15}$$

where

t_{rem} = remaining life
t_{exp} = time of operation of the component
A = parameter "A"

The results obtained using these equations are considered conservative, considering normal scatter of experimental results. Systematically and periodically performing analysis to assess changes in the accumulation of damage prevents sudden deterioration and unnecessary repairs or replacements, and allows scheduling maintenance actions to extend the life of components subjected to creep damage.

References

1. ASTM A370-12.: Standard Test Methods and Definitions for Mechanical Testing of Steel Products. American Standards for Testing of Materials
2. Anderson, T.L.: Fracture mechanics: fundamentals and applications, 3rd edn. Taylor & Francis, Abingdon (2005)
3. Higdon, A.: Mechanics of Materials. Willey, New York (1967)
4. Pavlina, E.J., Van Tyne, C.J.: Correlation of yield strength and tensile strength with hardness for steels. ASM international, Materials Park (2008)
5. ASTM E1820-11.: Standard Test Method for Measurement of Fracture Toughness. American Standards for Testing of Materials
6. D.K. Felbeck, A.G. Atkins, *Strength and Fracture of Engineering Materials* (Prentice Hall, New Jersey, 1984)
7. Report GIE 7101-06-05.: Analysis of elongations in stand pipe. Petrobras PASA, 2005
8. Otegui, J L., Rubertis, E.: Cañerías y Recipientes a Presión. ISBN 978-987-1371-18-1. Eudem, Argentina (2008)
9. ASTM E647-11e1.: Standard Test Method for Measurement of Fatigue Crack Growth Rates. American Standards for Testing of Materials
10. NG-18 Report 146.: Test Method for Defining Susceptibility of Pipe Line Steels to Stress-Corrosion-Cracking. American Gas Association, Washington DC (1985)
11. Report GIE 2502-03/00.: Experimental Analysis of susceptibility to SCC in base and ERW materials of oil pipeline. Oldelval, Mar Del Plata (2001)
12. ASTM E139-11.: Standard Test Methods for Conducting Creep, Creep-Rupture, and Stress-Rupture Tests of Metallic Materials. American Standards for Testing of Materials

Chapter 6
Modeling Tools Applied to the Analysis of Mechanical Failures

6.1 Introduction

Most investigations of a breakdown or accident begin at the end of the story: after the explosion, when the fire is out, after the collapse, etc. At this point is when people ask: how did this happen, why? From this chronological end, the investigator collects, verifies, and evaluates the evidence to determine how and why the event took place.

The basis for failure investigations is the scientific method. Historically, there have been two versions of the basic scientific method. The first version involves collecting verifiable facts and observations about an effect, event, product, or topic. After these facts are evaluated, a general statement is postulated, which is consistent with the data. Empiricism accumulates observations and new developments. The general proposition can be modified or changed to be consistent with the known body of verifiable information. Rationalism allows preconceived notions to provide a framework for data and new facts. Some principles are considered inviolable; empiricists accept what is observed and try to reasons in order to explain it. Rationalists, on the other hand, propose what should be the order from the beginning and then organize observations and facts to fit within that framework.

The second and more modern version of the basic scientific method also involves the collection of data about a particular effect and the search for a principle common to the observations. However, after developing a working hypothesis, consistent with available data, then additional consequences or effects that have not yet been observed are anticipated. In short, each hypothesis is tested to determine if it has predictive value. These additional effects can then be verified by experimentation, by additional observations or perhaps re-examining tests, and data already collected. Often, a good working hypothesis allows predicting the presence of causes or effects that were present from the beginning but were overlooked, misunderstood, or considered unimportant during the initial probationary period. If additional consequences predicted by the hypothesis are

J. L. Otegui, *Failure Analysis*, DOI: 10.1007/978-3-319-03910-7_6,

confirmed, further confirmation of the hypothesis is obtained. The "model" defines what is possible and what is not possible.

As engineering systems grow, it is essential to understand them well enough to avoid unexpected failures. How do engineers design complex systems to remain reliable? Some of the most important tools for the design and optimization of such systems have been computer-aided engineering and computer-aided design. Engineers have widely used CAE or CAD tools for the past 40 years to draw engineering designs, and more recently, to simulate and test them.

Finite element methods (FEM), developed in the 1950s and 1960s, could do much more. By transforming a large computer model into many small elements, defining "nodes" between "elements," and then using mathematical expressions for physical properties, FEM can provide answers such as stress components at a specific location. This technique found immediate application in the aerospace industry and was instrumental in understanding the structural responses to stress. In the late 1970s, integration of graphical drawing and modeling software resulted in a better way to design and simultaneously test to improve the performance and hence reliability.

But to extend this concept to complex systems, the software had to be developed to simulate a wide range of responses, not only in the case of a static stress at a point, but also for motion, signal and energy transmission, and other input dynamics. Engineers needed to visualize how a system responds to variations in the design criteria, how it would work in whole or in part. Programs such as ABAQUS and ANSYS (both developed in the 1970s) were designed to allow engineers to expand FEM and CAD to solve the so-called multiphysics problems, including mechanical strength, thermal, electromagnetic, and vibration in a design. Then simulation software for electronics, fluid dynamics and electrical control came up, such as the simulation program with integrated circuit emphasis (SPICE). Using fluid dynamics computer programs, designers could not only understand the flow of fluids through a complex system, but also simulate how pressure increases, rapid temperature changes, and other dynamic changes could affect the overall system performance.

Once material properties, applied loads, and component geometry in the failure zone (including the geometry of previous defects) are sufficiently defined, mechanical modeling can be used to verify the relevance of causes which factual evidence may not identify. Mechanical modeling techniques are even more important when the experimental evidence does not ensure one single possible failure mode.

We have seen that mechanisms of ductile tearing or brittle fracture usually require the appearance, at some point during service, of stresses higher than historical. In other cases, damage mechanisms are activated when certain activation energy is exceeded, in most cases associated with the stress state in the component where these defects develop. Perhaps the only exception is in some cases of corrosion mechanisms, whose activation energy depends only on the electrochemical conditions and temperature of the medium. The stress state, in turn, defines the growth rate of in service damage.

Metallographic and fractographic tools of failure surfaces and material features have been discussed in Chaps. 3 and 5. These studies can verify, for example, if the mode of crack propagation is intergranular or transgranular, the shape and density of corrosion pits, etc. With these analyses, it is possible to assess the characteristics of damage or degradation, failure mode, existence of specific chemical species or previous in service damage, fatigue, etc. When the failure is related to initiation and growth of cracks from pre-existing defects, the evaluation is based on fracture mechanics analyses. Fracture mechanics uses an appropriate field parameter to relate three key elements of the analysis. These are the stresses applied to the region of the defect, material properties, and morphology and size of the defect. Fracture toughness of the material (as opposed to brittleness) is defined as the material's resistance to crack propagation, and is related to its ability to absorb energy.

Sometimes as part of a failure analysis and more often after a failure analysis has shown design problems, a mechanical redesign or evaluation of remaining life is performed. Finite element and fracture mechanics modeling are frequently used in these cases. The effect of operating loads on potentially critical points of a component is related to the generation of larger stresses that expected in design, e.g., dynamic load conditions and geometric stress concentrations. In a typical redesign, the complete geometry is modeled based on available drawings, service data, and abnormal conditions during operation.

Fracture mechanics analysis added to FEM models allow for assessing damage and failure mechanisms, and defining constructive and operational variables that could have contributed to those mechanisms. This is required when dealing with a time-dependent damage mechanism and remnant life, as in the case of fatigue, creep, and stress corrosion (which are discussed later in this chapter). Numerical models alone do not allow obtaining accurate life predictions, but allow reliable predictions of whether the degradation mechanisms may or may not lead to failure. In these cases, damage propagation rates strongly depend on the particular conditions of the materials involved (microstructure, temperature, medium, susceptibility, etc.).

The tools discussed below are also used for the design and integrity assessment of mechanical components; that is, models are predictive. For this purpose and in order to give reliable designs, the results are affected by safety coefficients. In failure analyses, on the other hand, these tools are used to re-enact the events that lead to mechanical failure. Thus, they are not intended to give "conservative" results, but results that are as accurate as possible. Some tools must therefore be used with care and specific knowledge of the analyst is required.

Mechanical stresses or strains in a solid are similar concepts to that of pressure in a liquid or gas. However, as the solid can transmit shear and tensile stresses as well as compression, in each of the particles in a solid there may be shear, tension, and compression components of stress. These stress components are defined with two subscripts, for example σ_{ij}. The first subscript identifies the direction normal to the plane in which the stress acts, and the second subscript indicates the direction in which such stress component acts.

Thus, two identical subscripts, for example σ_{yy}, denote a stress normal to the plane xz in the y direction, while two different subscripts, such as σ_{xz}, denote shear stress acting on the yz plane in the direction of z. The components (σ_{xx}, σ_{yy}, σ_{zz}, σ_{xy},

σ_{xz}, σ_{yx}, σ_{yz}, σ_{zy}, σ_{zx}) define the stress state at a given point *P*. Axes (*x*, *y*, *z*) are often designated with the letter *X* and each direction with a number: $x \rightarrow X1$, $y \rightarrow X2$; $z \rightarrow X3$. The nine stress components σ_{ij} are reduced to six, because $\sigma_{ij} = \sigma_{ji}$. Stress components can be further reduced to three normal (tensile or compressive) "principal" stresses, which are those defined in the "principal" directions (where shear stresses are zero). One of these principal stresses is the largest tensile stress at a given point, which is most important in the study of cracks and fractures. For those interested in mathematics, stress and strain components in a solid are characterized by second degree tensors in the three-dimensional orthogonal space (vectors such as loads and displacements are first degree tensors).

When performing a stress analysis, modeling involves determining the stress state at a certain point that is considered characteristic (typically, the most stressed) and then an equivalent stress is calculated, which is compared with the strength of the material. The "stress intensity" defined by ASME BPVC is similar to the equivalent stress, and is determined using a yield criterion. The two yield criteria are the maximum shear stress and the maximum elastic distortion energy, which are well known as criteria of Tresca and von Mises, respectively. For those interested readers, the literature detailing the basics of solid mechanics and the theory of elasticity is included in the reference list at the end of this chapter.

6.2 Numerical Modeling Tools

Numerical results are nowadays commonly used to define allowable operating stresses, for design of a component, integrity assessment, or redesign of nonconforming parts. Stress concentrations and residual stresses create very high stress gradients, that is, large variations of stress in a very short distance within the material. Analytical methods for stress analysis and design are based on simplifications regarding stress distribution that do not conform the reality in these cases. Precise determination of stress states requires direct measurements or numerical modeling. There are several modeling techniques, proven and reliable, which can be used. If a good analysis of the geometric boundary conditions and applied loads is made, so that the model accurately represents reality, the results are reliable enough to consider this method as a computational test, with a similar validity to an experimental test.

This section presents analytical techniques based on the through-thickness stress distribution in components, which use the results of finite element techniques incorporated into user-friendly commercial programs. These techniques allow interpretation of stress states in critical areas, such as those with high geometric stress concentrations, pressure and dead weight loads, dynamic loading, etc. Briefly mentioned in this section are some numerical methods that have been useful for the study of structures with mechanical and geometrical discontinuities, including cracks. The use of these numerical methods, or others that may develop in the future, is virtually mandatory outside the few cases where analytic solutions can be obtained.

The FEM is the by far the most widespread method, there being a number of commercial programs (FE-PIPE, CESAR, ALGOR, ABAQUS, COSMOS, LUSAS, etc.), various tools are integrated with computer-aided design (CAD), some even including design and computer-integrated manufacturing (CAD-CAM, or CIM). With the beginning of the new century, there has been a growing progress of the boundary element method (BEM), particularly in problems involving fracture. This method is a powerful calculation tool alternative to FEM. The most obvious attraction is that the BEM reduces problem dimensions to 2-D solutions, since solving 3-D problems only require modeling the surface of the volume. This reduces the size of the computer model, enables reduction of domain data and number of algebraic equations to be solved. The BEM method also allows calculation of stresses and displacements in the interior volume, but in this case FEM is more efficient.

Due to the increased availability of FEM-based commercial computing tools, we will only briefly describe the method. A disadvantage of most commercial programs is that they are limited to static conditions. That is, the model is maintained as a homogeneous solid; modeling conditions do not allow, for example, loosing parts, modeling paths of parts prior to an impact, etc. Specific tools are available for these cases, such as the commercial Working Model program and the like. These will be treated briefly at the end of the chapter.

FEM is called "finite element" because the volume of a given structure is divided into elements of finite size. These elements can be of different sizes and shapes, usually the "mesh" is refined near a region of interest. There are elements with one, two, and three dimensions; in any case study, the dimension of the structure exceeds that of the elements used. The elements are connected together at points called nodes, which ensure continuity of the displacement field. The method is usually applied in solid mechanics, to study the response of a structure with given constraints subjected to the action of specified loads. Elements build up 1-D, 2-D, or 3-D structures; for example, a beam fixed on a wall with lateral loads (1-D case), the shell of an amphitheater under the action of its own weight and wind (2-D case), or a solid volume with arbitrary loads on its surface (3-D case).

The formulation for a two-dimensional case provides an introduction to the method. To simplify the calculation when the problem domain is irregularly shaped (i.e., not rectangular or curved), modeling involves defining a global coordinate ($x - y$) system, and local and normalized ($\xi - \eta$) coordinates, associated with the element such that one of the local coordinates is 1 or (-1) at each element boundary. This facilitates numerical integration at fixed element points which are called Gauss points.

Consider the 8-node element depicted in Fig. 6.1; note both coordinate systems. Nodes are located at points (x_i, y_i), expressed in global coordinates and correspond with (ξ_i, η_i) local coordinates, with values -1, 0, and 1, as shown in the figure. Both coordinate systems are related by an interpolation function M, such that:

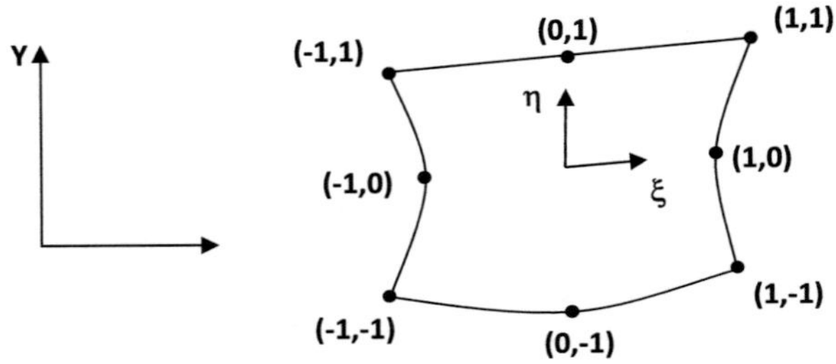

Fig. 6.1 Two dimensional FEM mesh

$$x = \sum_{i=1}^{m} M_i(\xi, \eta) \cdot x_i \tag{6.1a}$$

$$y = \sum_{i=1}^{m} M_i(\xi, \eta) \cdot y_i \tag{6.1b}$$

where the subscript "i" includes all "m" nodes of the element. Similarly, so-called "shape functions" allow interpolation within the element. Displacements in x and y directions at a given point in the element are:

$$u_e = \sum_{i=1}^{n} N_i(\xi, \eta) \cdot u_i \tag{6.2a}$$

$$v_e = \sum_{i=1}^{n} N_i(\xi, \eta) \cdot v_i \tag{6.2b}$$

where u_i and v_i are the displacements in the n points of the element. Both functions are bi-univocal and bi-continuous. When $m = n$ (points coincide with nodes) the element is called iso-parametric. If $N > M$ the element is sub-parametric and in the opposite case; super-parametric. For simplicity let us assume the element is iso-parametric, Fig. 6.1. The expressions u_e y v_e can be written in matrix form:

$$\begin{Bmatrix} u_e \\ v_e \end{Bmatrix} = [N] \cdot \{u\} \tag{6.3}$$

where [N] is the shape function matrix and {u} is the nodal displacement vector. Deformations are related to the displacements by the equation:

$$\{\varepsilon\} = \begin{Bmatrix} \varepsilon_{xx} \\ \varepsilon_{yy} \\ \varepsilon_{xy} \end{Bmatrix} = \begin{bmatrix} \partial/\partial x & 0 \\ 0 & \partial/\partial y \\ \partial/\partial y & \partial/\partial x \end{bmatrix} \cdot \begin{Bmatrix} u_e \\ v_e \end{Bmatrix} = [L] \cdot \begin{Bmatrix} u_e \\ v_e \end{Bmatrix} \tag{6.4}$$

where $\{\varepsilon\}$ is the strain matrix and [L] is a matrix operator. From the above two expressions,

$$\{\varepsilon\} = [\mathrm{L}]\,[\mathrm{N}]\,\{\mathrm{u}\} = [\mathrm{B}]\,\{\mathrm{u}\} \tag{6.5}$$

Matrix [B], customary in FEM notation, allows directly connecting the strain field in a given point of the element with nodal displacements. The matrix contains derivatives of functions N_i, with respect to (x, y), which are related to derivatives with respect to the local coordinates (ξ, η) through a Jacobian matrix [J].

The stress matrix is expressed according to Hooke's Law,

$$\{\sigma\} = [\mathrm{D}]\,\{\varepsilon\} \tag{6.6}$$

where [D] is the stress–strain constitutive matrix of the material. FEM is based on using variational principles or weighted residuals, where the nodal displacements are the unknowns to be solved. Forces and displacements in the nodes are related to each element by the equation:

$$\{\mathrm{f}\} = [\mathrm{k}]\,\{\mathrm{u}\} \tag{6.7}$$

where [k] is called the stiffness matrix, and is obtained by means of the integral

$$[\mathrm{k}] = \iint [\mathrm{B}]^{\mathrm{T}}\,[\mathrm{D}]\,[\mathrm{B}]\det[\mathrm{J}]\,\mathrm{d}\xi\mathrm{d}\eta \tag{6.8}$$

In many problems, it is possible to derive the above equation using the principle of minimum potential energy. In this method, the integrals are calculated numerically by using Gauss points in sufficient number to obtain the precision sought. For example, for 2-D elements it is common to use at least four Gauss points. When greater accuracy is required in a region of the problem, an optimal combination of the number of Gauss points and the size and type of element must be sought. Once the equations for each element are obtained, stiffness matrices are assembled to obtain the general equation of the problem

$$[\mathrm{F}] = [\mathrm{K}]\{\mathrm{U}\} \tag{6.9}$$

where [F] and {U} are respectively the global vectors of nodal forces and displacements, and [K] is the global stiffness matrix of the problem. Thus, displacements at nodes are calculated by computer, solving a system of linear equations. Some accuracy is lost when calculating stresses, since they are calculated from displacements. For nonlinear problems in which, for example, plasticity is to be considered, it is necessary to find the stresses and displacements incrementally, updating the matrix [D] in each load step. In these cases, and especially if the calculation is to be done in 3-D, the computational cost of the analysis steeply increases.

6.2.1 Example 6.A Modeling of Soil-Pipeline Interaction

This example shows a development in modeling a relatively recent problem: ruptures of pipelines buried in unstable soils. This case is a pipeline transporting liquid hydrocarbons, buried in a mountainous jungle area. The combination of

Fig. 6.A1 Buckles in a pipeline buried in unstable soil

steep slopes, up to 7,000 mm yearly rainfall (concentrated in a 3-month period), and topsoil with little consistency lying on bedrock, have contributed to the occurrence of some tensile and buckling failures, see for example Fig. 6.A1.

A comprehensive monitoring system of in-service pipe stresses was implemented. Also, a reliable modeling tool that estimates the most stressed pipe sections was developed in order to optimize the location of strain gauges and other devices. The FEM-based tool simulates the stress state imposed into the pipeline by ground displacements. Results of these studies support the system integrity management by the pipeline operator, which aims to operate reliably and maintain the pipeline transportation system, minimizing harm to people, environment, and facilities.

Two basic approaches are used for numerical modeling of these buried pipes: (1) standard, using beam elements to model the pipeline and a spring model to represent the surrounding soil and (2) advanced, using continuous solid type finite elements. Soil-pipeline interaction, in this second case, involves a nonlinear finite element model, using Abaqus software. The pipe is modeled with a special beam element, soil is simulated with continuous solid elements, and interaction between them is modeled by a shared node. This system works as a composite material and results show good agreement with actual physical behavior in the field. Also, alternative models that include shell elements for modeling the pipeline, continuous solid elements for the soil, and a contact algorithm to model soil—tube interaction were developed. But computing times are excessive to make this alternative a useful tool to help in decision making.

Figure 6.A2a shows a typical mesh for modeling the interaction between soil and a buried pipeline. In the detail of Fig. 6.A2b, some elements have been

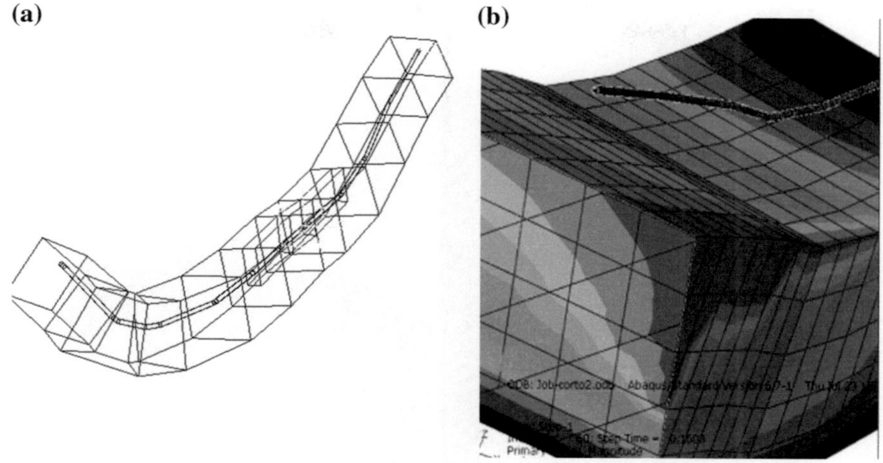

Fig. 6.A2 (**a**, **b**) Modeling of interaction between soil and buried pipeline

removed from the ground, in order to appreciate pipe elements. Shadow grades indicate predicted displacements both in the soil and in the pipe. These models are supplied with soil displacement data and information gathered by geotechnical specialists, usually obtained after a failure or through early detection by monitoring systems. These models are used to verify if movements induced into the pipe may reach a breakdown stress, thereby justifying failures. In one case, for example, it could be determined that a given soil displacement could cause the failure of a small diameter (14″) pipeline, while a larger diameter (30″) pipeline in the same right of way remained unaffected.

These models determine most stressed pipe sections, and thus allow defining, for example, where is the optimal location to place strain gauges for monitoring pipe stretch and preferential sites for rational location of displacement sensors into the ground. Another application is the definition of optimum positions for digging in order to relief stress in pipe sections with large axial strains due to soil movement, Fig. 6.A3. The location of these excavations is not easily defined, as it is common that sections with larger soil displacements do not correspond to the sections of larger strain applied to the pipe.

Maximum axial stress and strain in a buried section increase proportionally with the length of the tube that is subject to soil displacement. Figure 6.A4 shows model results to optimize the position of culverts in a buried pipe; these culverts are sometimes also used to detach the pipeline from soil displacement. The figure shows that maximum transverse displacements of the pipeline stabilize and become independent of the direction and magnitude of soil displacements. A rule of thumb for sizing the culverts in this particular case is that their base should be 2.5 times the pipeline diameter. Intelligently distributed intermittent sections of culverts have shown to be a good investment to ensure pipeline integrity.

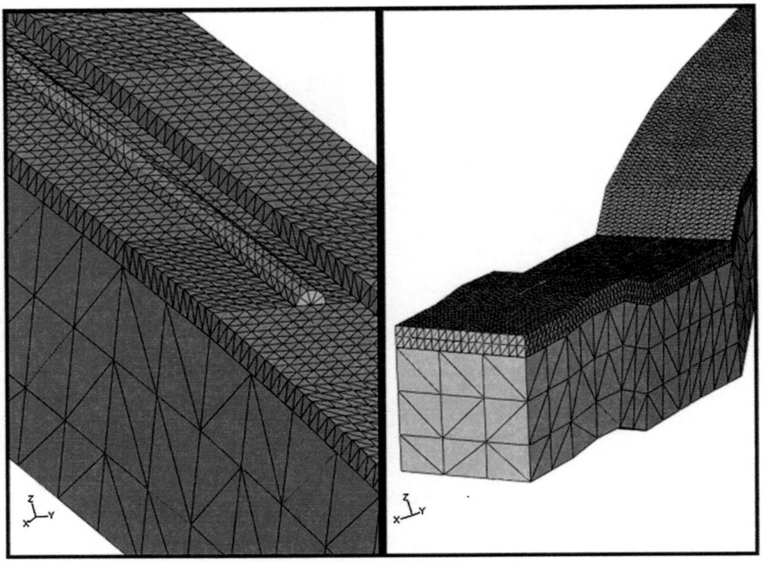

Fig. 6.A3 Model of intervention to relief stresses in a buried pipeline

6.3 Criteria for Modeling Pressure Components

A particular case for design, although very common, is when pressure is the most important external load in a component. As mentioned, design codes define allowable stresses as a fraction of the material yield strength or UTS. The ASME Boiler and Pressure Vessel Code (BPVC) refers to the Tresca criterion, while the code API RP 579/ASME FFS-1 "Fitness for Service" recommends the use of Von Mises criterion. To demonstrate structural integrity, both API and ASME Codes categorize stresses. Brief definitions of these categories are:

- General membrane primary stress intensity (PM) is the equivalent stress which is derived from the average value of the stress across the thickness of a section; these primary stresses are due to the design pressure and other specified mechanical loads but without considering peak or secondary stresses. Allowable equivalent stress is that of the material, Sm, multiplied by a service factor "k" that depends mainly on temperature of operation. API and ASME Codes have the same values of k for the same load combinations, see Table 6.1.
- Local primary membrane equivalent stress (PL) is derived from the average value across the thickness of a section, of the local primary stresses produced by the design pressure and specified mechanical loads but excluding all secondary and peak stresses. A region of stress in a component is considered as local if the distance over which the equivalent stress exceeds 1.1 Sm does not extend more

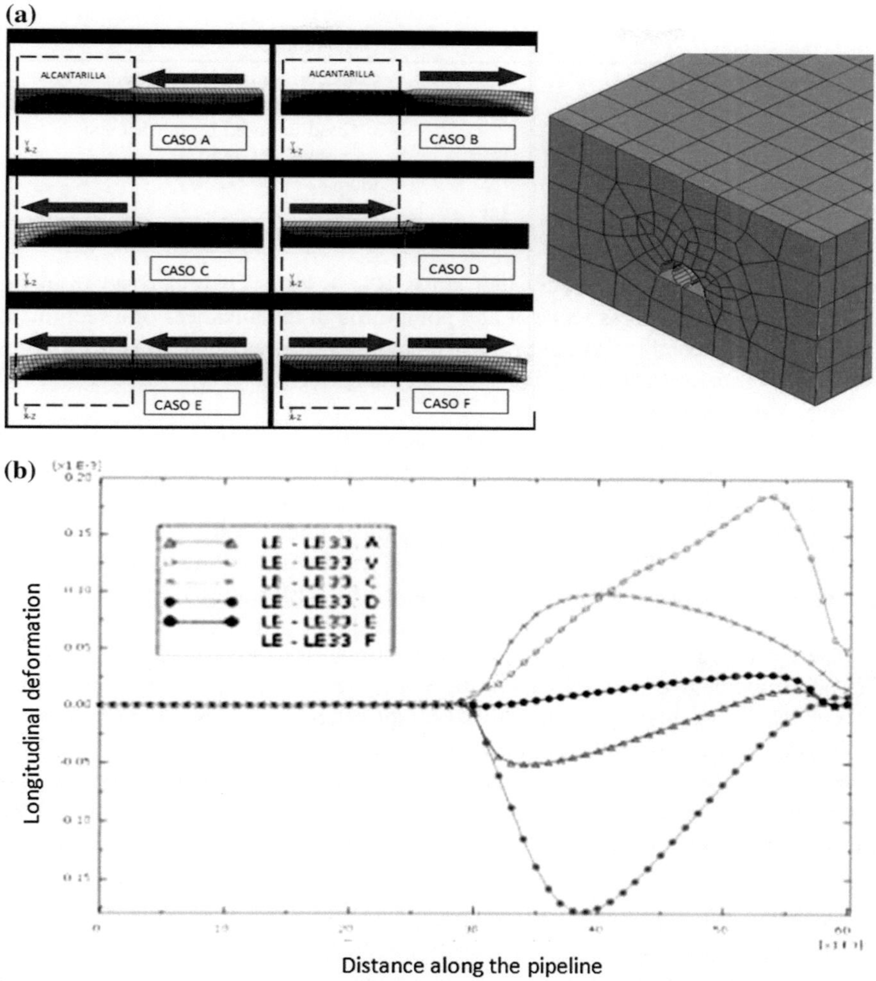

Fig. 6.A4 Model results to optimize the position of culverts in a buried pipe **a** Model detail. **b** Longitudinal deformation of pipe

than the pipe or vessel "size". This size is defined as the square root of the product of radius times thickness.

- Primary membrane (General or Local) plus Primary Bending Equivalent Stress (PL + Pb) is the equivalent stress, derived from the highest value across the thickness of a section, of the linearized general or local primary membrane stresses plus primary bending stresses produced by design pressure and other specified mechanical loads but excluding all secondary and peak stresses. The allowable equivalent stress value is 1.5 Sm.

Table 6.1 Stress factors

Load combination	k	Observations
Design loads	1	Consider corroded thickness and metal temperature, design pressure, weight, load imposed by mechanical equipment, etc.
Design and wind loads	1.2	Idem
Design and earthquake loads	1.2	Idem

- Primary plus secondary equivalent stress (PL + Pb + Q) is derived from the largest equivalent stress value at any point through the thickness of a section, as a result of the combination of primary membrane stresses (local or general), plus secondary stresses caused by local and thermal effects. The allowable equivalent stress value is 3 Sm.
- Primary plus secondary plus peak equivalent stress (Pl + Pb + Q + F) is derived from the largest equivalent stress value at any point through the thickness of a section for the combination of all primary, secondary, and peak stresses produced by operating pressures, mechanical loads, general and local thermal effects and the effects of global and local structural discontinuities. The allowable equivalent stress value is obtained by performing a fatigue analysis.

There is also a triaxial stress limit. The algebraic sum of the principal stresses at the point under investigation cannot be more than four times the Sm. Note that, except when considering cyclic stresses, equivalent stress should be compared with Sm, the allowable stress for the material.

Since the requirement for acceptability is that the equivalent stress does not exceed the specified allowable limits, and since these limits depend on the stress category for which we derive the equivalent stress, a procedure for the computation is established:

1. Determine the types of load to which the component is subjected.
2. The stress tensor is computed at each point along an assessment line. Then each set is assigned to one of the stress categories:
    - General primary membrane stresses Pm
    - Local primary membrane stresses Pl
    - Bending stresses Pb
    - Secondary stress Q
    - Peak stress F
3. Stress tensors assigned to each category of stress are added.
4. Principal stresses from the sum of the stress assigned to each category and determined, and equivalent stress calculated.
5. The equivalent stress is compared with allowance for each category.

Pm, Pl, Pb, and F do not represent numbers, but unique sets of six quantities representing the six components of the stress tensor. FEM results provide the combination for (Pl + Pb) or (Pl + Pb + Q) directly. For this reason, it is not necessary to determine Pl and Pb; the procedure was originally developed to perform manual calculations. ASME BPVC Code provides procedures for the design of components while API Recommended Practice summarizes useful and recognized procedures for determining fitness for service of components.

6.3.1 Example 6.B Stress Analysis of Damper Vessel

The purpose of this study is the analysis of the failure mode and causes in a dampening vessel at the discharge of the first stage of a large ethylene compressor at a petrochemical plant, Fig. 6.B1. The crack initiated in two places of the thick grained HAZ of the fillet weld, at the end of the gusset. One of the cracks beginning in the toe of the weld, the other one at the root, both propagated initially in the material of the weld. The welds of the gusset are single pass, geometric quality is very good. There is no evidence of manufacturing defects, or hard spots in HAZ.

The vessel was designed in 1979 according to ASME BPVC, section VIII div. 1. The procedure specifies discontinuous fillet welds; in this vessel the gusset was locally machined to reduce stress raisers. In the current design, the originally specified circumferential joint reinforcement was substituted by forged accesory called weldolet. The weldolet failed to increase the service life of the component, because the weld between the vertical gusset and the vessel was located directly on the wall of the vessel.

Fracture surfaces indicate fatigue propagation; cyclic stresses come to a large part from the transmission of vibrations from the compressor during operation. The evidence in this case and in the studies of previous failures shows that the vertical displacements of the vessel overcome thoroughly the limits tolerated by design. Analyses of vibration levels and frequency domain in the compressor did not allow detecting significant mechanical damage to the compressor or severe vibration induced by gas pulsation. Neither increases in recent vibration measurements nor loss in cohesion between frame, masonry, and foundation were found.

Numerical modeling verified that highest stresses are located in the crack area; see Fig. 6.B2, when using the reinforcement. The most relevant frequency of the vibration is 11 times the crankshaft rotation speed. Odd multiples of base frequencies are rare in this type of machine, so it is concluded that vibrations are not transmitted by the compressor. From this model results, it was determined that to reduce the frequency of reiteration of the flaws in future reconstructions of the vessel, it was recommended to modify its design: to replace the weldolet by a sleeve reinforcement, and to add other gusset plates, with an appropriate design.

It was concluded that changes were needed to the pads and possible cracks and rough spots in the weld reinforcement should be verified by NDT, to prevent the recurrence of cracking in this vessel. Vibrations could lead to cracking in other vessels, piping systems, and other components associated with this compressor. For the early detection of cracks in this vessel during operation, it was

Fig. 6.B1 Dampening vessel, cracked region shown

Fig. 6.B2 FEM model of dampening vessel

recommended to carry out frequent surface visual monitoring in the initiation area. The inspection of potentially dangerous welds in other connections and support systems was also recommended. The final solution to this reiterative problem is, however, to reduce the amplitude of the vibrations in this vessel and its associated pipes. To modify the frequency of pulsation of the fluid inside the damper requires a substantial modification of the length of the vessel.

6.4 Stress Analysis of Cracked Components

Design of components for use at room and cryogenic temperatures has adopted the concepts of toughness and transition temperature, as described in some detail in Chap. 5. Charpy impact test is used by several manufacturing codes. Around 1980 a new problem came up, as nuclear plants for power generation began to reach the end of their design lives. The nuclear industry was forced to demonstrate the integrity, reliability, and fitness for service of structures and pressure vessels containing defects, which were difficult to repair or replace due to the effect of radiation. The effort was later joined by other industries, which saw the opportunity to increase plant availability and efficiency, extend operation beyond expected useful life of components or increase operating parameters above design values. All this newly available technology helps minimize service costs and maximize the return on investment.

The CEGB procedure, developed by the Nuclear Electric UK, was originally used in the evaluation of components for the nuclear industry. It has been in continuous use and development in this harsh environment during the last three decades. It has been included in British, American, Swedish, and French codes. The "R6" methodology has contributed significantly to the safety and operational reliability in chemical, petrochemical, and energy industries. It is based on the failure assessment diagram (FAD), which is discussed in detail below.

The relevance of fracture mechanics is based on the fact that a structural component in the presence of a stress concentration may fail at a load substantially lower than allowed in design. Experimentally measured strength in a material such as ordinary glass is less than 1/100 of the theoretical strength of its atomic structure. This discrepancy led to postulate the existence of small cracks that cause the decrease of strength. The cracks act as stress concentrations such that the tip of those defects reaches the theoretical limit stress, but the average stress is kept much lower. Then, according to this concept, fracture occurs when the theoretical strength is reached at the tip of any crack or sharp defect. The reference list at the end of this chapter includes some literature that details the basic concepts of fracture mechanics.

For cracked bodies, subjected to external forces, it is possible to derive analytical expressions for the stresses in the vicinity of the crack. Figure 6.2 shows the three ways in which a crack within an elastic material can grow. In modes II and III, the cracked material is loaded by shear stresses, whereas in mode I, acting stresses are tensile. A cracked body can be loaded with any of these or a combination of two or three modes. But mode I represents over 90 % of failures of structures, so we will focus on mode I fracture propagation.

Consider the stress distribution in an elastic material adjacent to the tip of a through-thickness crack of length "2a," as shown in Fig. 6.3. The "crack tip stress intensity factor" when the cracked body is tensile loaded is called K_I, such that:

$$K_1 = \sigma(\pi a)^{1/2} \tag{6.10}$$

It can be shown (Westergaard's solution) that mechanical stresses in the adjacent material (Fig. 6.3) tend to infinity at the crack tip, but stresses far away from the crack tip are governed by remote conditions. If the cracked structure is subject to a uniform stress in the "y" direction, this stress σ_{yy} reaches a constant value σ^{∞}.

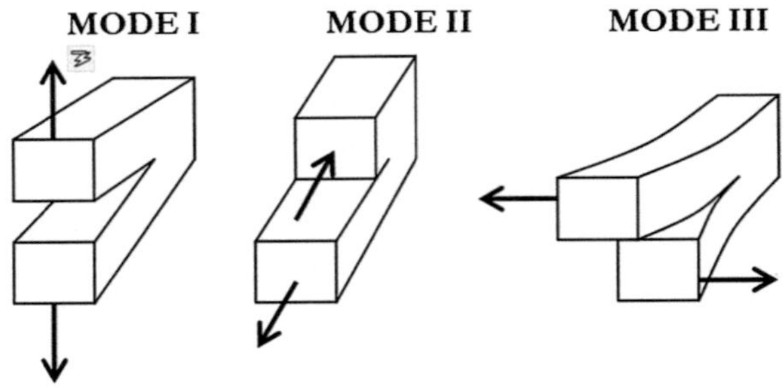

Fig. 6.2 Three load modes applicable to a crack

The crack tip stress intensity factor (or simply "K") defines the amplitude of the singularity at the crack tip, i.e., stresses near the crack tip increase in proportion to K. Furthermore, the stress intensity factor completely defines the conditions at the tip of the crack. This one-parameter description of the condition at the crack tip is one of the most important concepts of fracture mechanics.

The analysis carried out here to describe the stress and strain fields at the tip of a crack is based on the assumption that material′s behavior is perfectly linear elastic for any stress values at the crack tip. In reality, materials do not behave in this manner, and it is evident that the stress field near the tip of the crack must necessarily lead to the formation of a plastic zone located at the crack tip (Fig. 6.3). In what follows, we assume that the size "r_p" of this plastic zone is small enough to be neglected in the analysis.

6.5 Calculation of the Load Required for Brittle Fracture

Fort the stress intensity factor to be useful, one must be able to determine K from remote loads and geometry. Closed form solutions for K have been derived for several simple configurations. For more complex situations, the stress intensity factor can be estimated by experimental or numerical analyses.

Most configurations for which there is a closed form solution for K_I include a crack with a simple shape (a rectangle or ellipse, for example) in an infinite plate. Stated another way, crack dimensions are small compared with the size of the plate or component, and the conditions of the crack tip are not influenced by external limitations. As the size of the crack becomes comparable with thickness or other dimensions of the component, the edges begin to exert an influence on the crack tip. In such cases, a closed solution for the stress intensity is not normally possible.

Although solutions to the stress intensity factor are given in a variety of forms, K solutions can always be related with that of a through-thickness crack in an infinite plate through a suitable correction factor

$$K_{(I,II,III)} = Y\sigma\sqrt{\pi a} \tag{6.11}$$

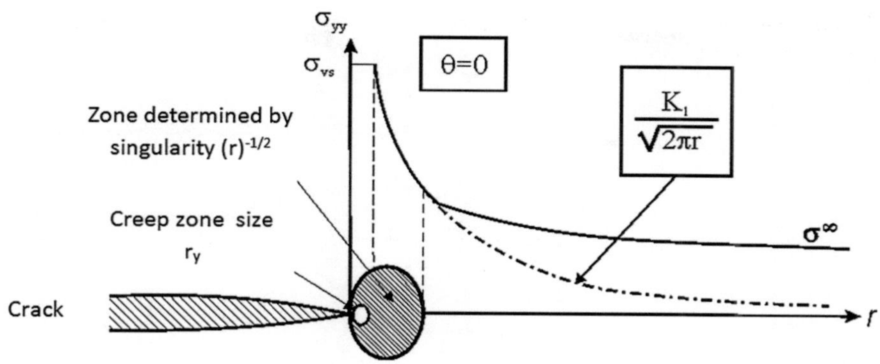

Fig. 6.3 Normal stresses to crack plane, opening mode I

where σ is the remote stress (the one in the crack site without considering the effect of the crack), "a" is a characteristic dimension of the crack, and Y is a dimensionless constant that depends on the geometry of the body and the crack, and loading mode. Figure 6.4 shows solutions for elliptical and semi-elliptical surface cracks, based on the Newman–Raju solutions.

$$K_I = \sigma\sqrt{\frac{\pi a}{Q}}f(\phi) \quad K_I = \lambda_s\sigma\sqrt{\frac{\pi a}{Q}}f(\phi)$$

$$f(\phi) = \left[sen^2\phi + \left(\frac{a}{c}\right)^2\cos^2\phi\right]^{\frac{1}{4}}$$

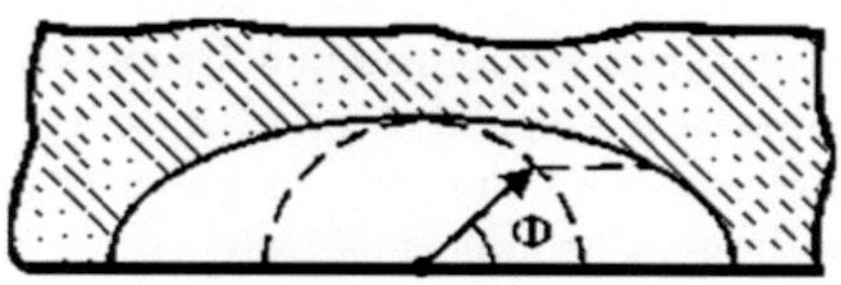

$$\lambda_s = \left[1.13 - 0.09\left(\frac{a}{c}\right)\right]\left[1 + 0.1(1 - sen\phi)^2\right]$$

$$Q = 1 + 1.464\left(\frac{a}{c}\right)^{1.65}$$

Fracture mechanics analysis of linear elastic structures becomes possible, once a solution of K for the geometry of interest it obtained. Solutions for stress intensity can come from several sources: published literature, experiments, or

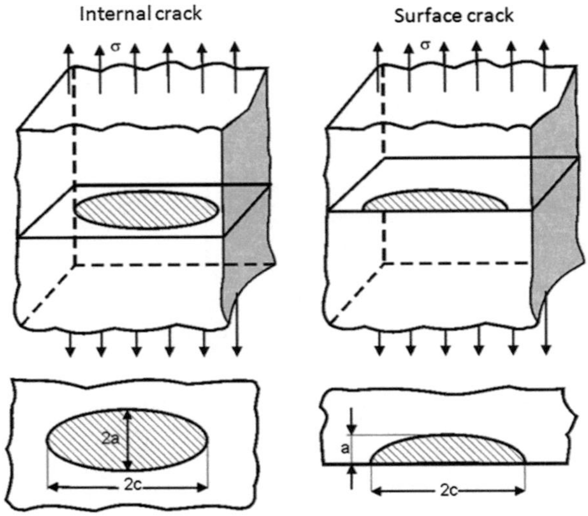

Fig. 6.4 Idealized geometries of internal and surface cracks

numeric analysis. A large number of stress intensity factor solutions have been published over the past 35 years. Several books contain collections of solutions for a variety of configurations. It is usually possible to find a solution of K for the geometry of interest (Fig. 6.5).

When a solution for K is not available, or the accuracy of the solution is in doubt, one can obtain the solution numerically or experimentally. To derive a closed form solution is not a viable alternative, as this is only possible with simple geometries, and almost all of these solutions have been published. An alternative is to use the principle of elastic superposition for building new solutions from known cases. Laboratory samples normally contain idealized cracks, but defects that occur naturally in structures have no obligation to maintain these ideals. Structural flaws are typically shallow and irregular. Moreover, severe stress gradients often arise in practical situations, as in the case of geometric stress concentrations (Fig. 6.6), whereas the samples undergo relatively simple loads.

Newman and Raju have published a series of solutions for K_I. Figure 6.5 illustrates some of the geometries: ellipse, half ellipse, and quarter ellipse, respectively. These solutions are applied to linear stress distributions, where the through-crack normal stress is resolved into components of bending and membrane, respectively. If the stress distribution in the thickness is not perfectly linear, bending and membrane stresses can be inferred as follows: the equivalent membrane stress is equal to the average stress integrated across the thickness, and the equivalent bending stress is inferred by computing the resultant moment (per unit length) and dividing by $6t^2$.

Newman and Raju solutions for non-through-thickness cracks subjected to membrane and bending stresses are expressed as follows:

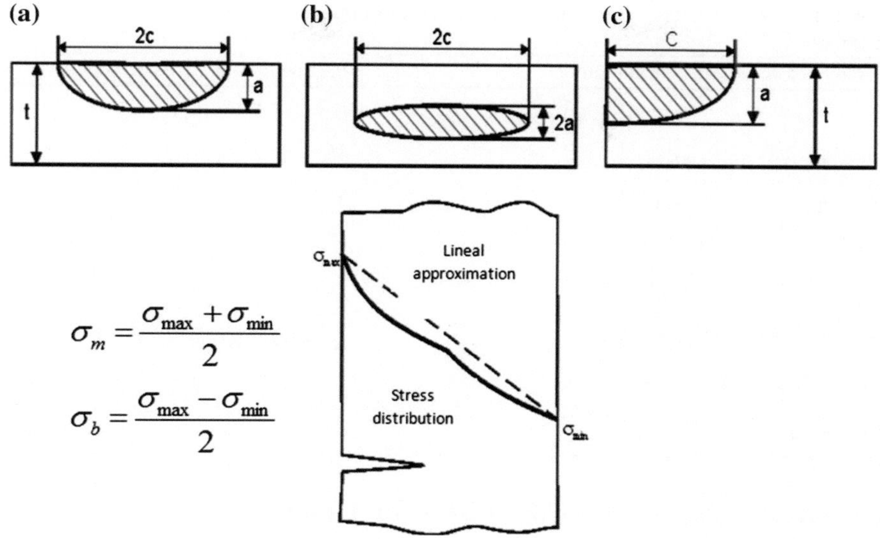

Fig. 6.5 Linearization of through the thickness stress distribution. **a** Semieliptic superficial crack. **b** Semieliptic superficial crack. **c** Corner crack

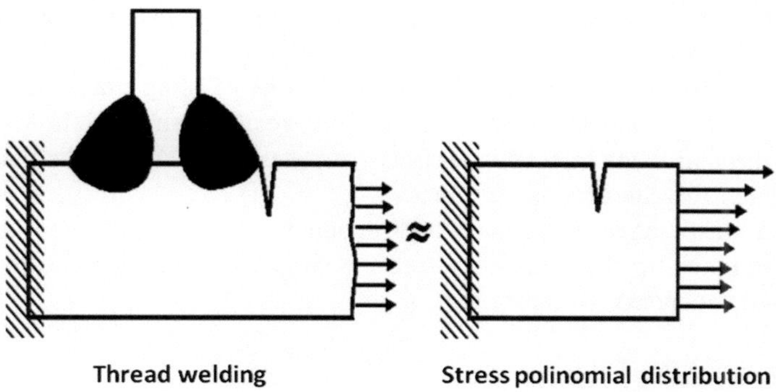

Fig. 6.6 Through the thickness stress distribution in a stress concentration

$$K_I = (\sigma_m + H\sigma_b)F\sqrt{\frac{\pi a}{Q}} \tag{6.12}$$

where Q is a shape parameter (Fig. 6.4) and F and H are geometric constants obtained from finite element analyses. Parameters F and H depend on a/c (crack depth to length ratio), a/t (crack depth to thickness ratio), and plate width. Table 6.2 lists solutions for F and H.

Table 6.2 Newman–Raju parameters for crack geometry and through the thickness stress distribution

Parameters for stress distribution	Parameters for CRACK geometry
$K_1 = (\sigma_m + H\sigma_b)\sqrt{\frac{\pi a}{Q}}F\left(\frac{a}{t},\frac{a}{c},\frac{c}{W},\Phi\right)$	$Q = 1 + 1.464\left(\frac{a}{C}\right)^{1.65}$
$F = \left[M_1 + M_2\left(\frac{a}{t}\right)^2 + M_3\left(\frac{a}{t}\right)^4\right] f_\Phi f_W g$	$M_1 = 1.13 - 0.09\left(\frac{a}{c}\right)$
$H = H_1 + (H_2 - H_1)(\sin\Phi)^P$	$M_2 = -0.54 + \frac{0.89}{0.2+\frac{a}{c}}$
$p = 0.2 + \frac{a}{c} + 0.6\left(\frac{a}{t}\right)$	$M_3 = 0.5 - \frac{1.0}{0.65+\frac{a}{c}} + 14\left(1.0 - \frac{a}{c}\right)^{24}$
$H_1 = 1 - 0.34\frac{a}{t} - 0.11\frac{a}{c}\left(\frac{a}{t}\right)$	$f_\Phi = \left[\left(\frac{a}{c}\right)^2\cos^2\Phi + \sin^2\Phi\right]^{1/4}$
$H_2 = 1 + G_1\left(\frac{a}{t}\right) + G_2\left(\frac{a}{t}\right)^2$	$f_w = \left[\sec\left(\frac{\pi c}{2W}\sqrt{\frac{a}{t}}\right)\right]^{1/2}$
$G_1 = -1.22 - 0.12\left(\frac{a}{c}\right)$	$g = 1 + \left[0.1 + 0.35\left(\frac{a}{t}\right)^2\right](1 - \sin\Phi)^2$
$G_2 = 0.55 - 1.05\left(\frac{a}{c}\right)^{0.75} + 0.47\left(\frac{a}{c}\right)^{1.5}$	

6.6 Calculation of the Conditions for Ductile Fracture

As was previously seen, mechanical loads in a structure can generate primary and secondary stresses. Primary stresses typically arise from externally applied loads and moments, while secondary stresses are localized and self-balanced across the section. Primary stresses, if they are sufficiently large, can lead to plastic collapse, but secondary stresses cannot cause plastic collapse. These, however, can contribute to fast fracture, that is, propagation of a pre-existing crack. Examples of secondary stresses include thermal and welding residual stresses. In some cases, however, thermal loads may cause primary stresses; a stress must be classified as primary when the category is not clear.

In linear elastic analysis, primary and secondary stresses are treated identically. The applied crack tip stress intensity factor is simply the aggregate of those due to primary and secondary components:

$$K_I^{\text{total}} = K_I^p + K_I^S \tag{6.13}$$

Superscripts p and s denote primary and secondary quantities, respectively. Primary and secondary stresses are treated differently in the elastic–plastic analysis, as discussed later.

Fracture mechanics equations can incorporate corrections in the estimation of stress intensity factors to take into account small amounts of plasticity at the crack tip. The effective length of the crack a_{eff} is defined as the sum of the sizes of the actual crack, a, and the plastic zone, r_p (see Fig. 6.3). The yield correction criterion for plastic zone has the effect of increasing the value of K to an effective value "K_{eff}" larger than the linear elastic K_I value as of Eq. 6.11. This correction becomes significant for applied stresses greater than approximately 0.5 σ_{ys} (half the material′s yield stress), but becomes inaccurate for applied stresses above 0.7 σ_{ys}.

In these cases, the correction for crack plasticity in plane stress is based on the analysis by Stone and Burdekin called strip yield, which models the behavior of the material at the tip of a growing crack as a series of small bars touching each other. These bars are treated as tensile test samples that yield and eventually collapse as the proximity of the crack tip makes local stress rise above yielding. Each configuration of component and crack requires a separate analysis. There is, however, an approximate method of generalizing the strip yield model with a single equation that describes all geometries:

$$K_{eff} = \sigma_{YS}\sqrt{\pi a}\left[\frac{8}{\pi^2}\ln\sec\left(\frac{\pi\sigma}{2\sigma_{YS}}\right)\right]^{1/2} \tag{6.14}$$

Failure to apply a correction for plasticity can result in an underestimation of the driving force over the crack, which would lead to nonconservative results.

The structures made of tough materials are usually not susceptible to brittle fracture, but may fail by plastic collapse if overloaded. The concept of the two-parameter Failure Assessment Diagram (FAD) allows describing the interaction between fracture and plastic collapse.

Equation (6.14) defines the effective stress intensity factor in a cracked infinite body. This relationship is asymptotic to yield stress, i.e., $K_{eff} \to \infty$ when $\sigma \to \sigma_{ys}$. Equation (6.14) can be modified for real structures by replacing σ_{ys} by the collapse stress σ_c of the structure. This ensures that the strip yield model predicts failure when the applied stress approaches the collapse stress. Work hardening can be taken into account by assuming a collapse stress as an average of yield and UTS, this material property is usually called flow stress. For tensile loaded structures, collapse occurs when the stress in the net section reaches the material collapse stress. Thus, σ_c depends on material properties and the crack size relative to the total section of the structure.

The equation for the FAD is defined as follows. First, the applied K_{eff} is divided by the linear elastic K_I value, giving rise to an equation that is independent of material, geometry, and loading

$$\frac{K_{eff}}{K_I} = \frac{\sigma_C}{\sigma}\left[\frac{8}{\pi^2}\ln\sec\left(\frac{\pi}{2}\frac{\sigma}{\sigma_C}\right)\right]^{1/2} \tag{6.15}$$

Then the stress ratio Sr and the K ratio Kr are defined as follows:

$$K_r = \frac{K_{eff}}{K_{IC}} \quad S_r = \frac{\sigma}{\sigma_C} \tag{6.16}$$

where K_{IC} is the material fracture toughness of the material (as defined in Chap. 5), and σ is the equivalent stress in the uncracked ligament. The FAD is then obtained by inserting the above definitions in Eq. (6.15) and taking the reciprocal

$$K_r = S_r\left[\frac{8}{\pi^2}\ln\sec\left(\frac{\pi}{2}S_r\right)\right]^{-1/2} \tag{6.17}$$

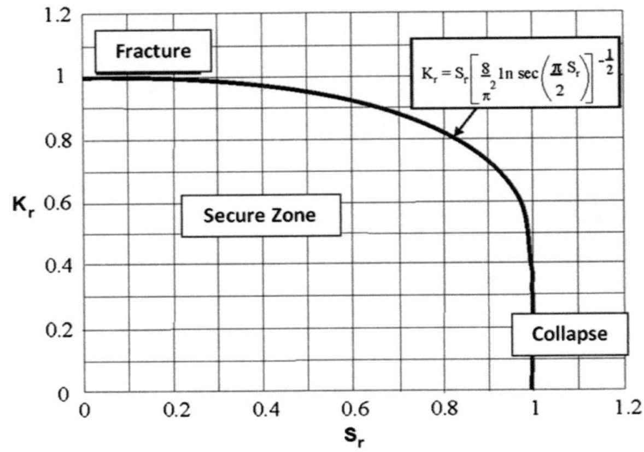

Fig. 6.7 Failure assessment diagram

Equation (6.17) is plotted in Fig. 6.7. This curve represents the predicted conditions for any failure. All points within the FAD are considered safe; the points outside the diagram are unsafe. Crack propagation (fracture) is expected if $Kr = 1$ ($K_I = K_{IC}$), while collapse of the ligament would occur when $Sr = 1.0$. In intermediate cases, collapse and fracture interact, and both *Kr* and *Sr* are lower than 1.0. To assess the importance of a particular crack in a structure, one must determine the applied values of *Kr* and *Sr*, and draw the flaw point in Fig. 6.7. The stress intensity ratio for the structure is given by

$$Kr_{\text{structure}} = \frac{K_I}{K_{IC}} \tag{6.18}$$

while the ratio of applied stress Sr is defined as the ratio of the applied to collapse stresses. Alternatively, Sr can be defined in terms of axial forces or moments. Only primary stresses are used to compute *Sr*, Secondary stresses, by definition, are self-equilibrating and do not contribute to collapse. Note that K_I is the linear elastic stress intensity; it does not include a plastic zone correction. Plasticity effects are already taken into account in the formulation of the FAD.

6.6.1 Example 6.C Calculation of Applied Stress at a Pre-existing Defect

This example is part of a failure analysis to determine the causes for a circumferential fracture in a 24 "diameter, ¼" thick pipeline, with more than 40 years of service. The metallurgical evaluation of the affected areas and analysis of evidence allowed defining that the failure occurred due to pre-existing discontinuities in the girth weld: lack of penetration, incomplete fusion, angular and radial misalignment, and lack of clearance at the root, see Fig. 6.C1.

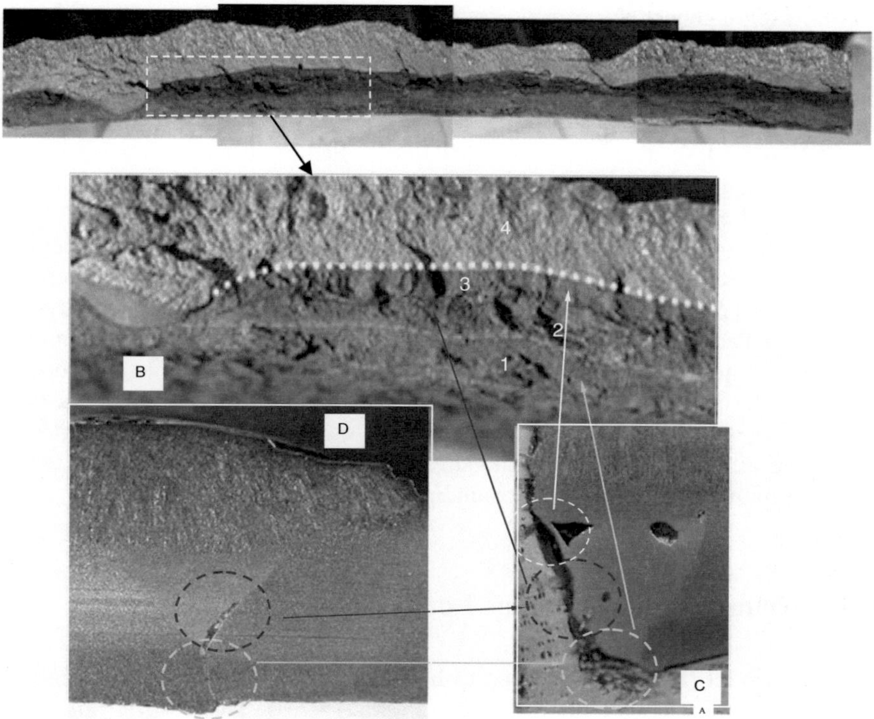

Fig. 6.C1 **a** Fractured pipe in circumferential weld, pre-existing defect, **b** Fracture detail, **c** Cross section in other zone of same weld, **d** Cross section at fracture initiation, correspondence of defects

The crack-like pre-existing flaw formed by the combination of lack of penetration and lack of fusion, resulting in a 4 mm deep, 84 mm long circumferential "crack" on the inner surface of the tube, (see darker area in Fig. 6.C1a. Similar construction defects were detected in other welds in the same pipeline. Crack shape and size, and material properties (tensile strength and Charpy impact toughness) were experimentally determined (as described in Chap. 5).

The evidence of some in-service propagation of these defects suggests that this section of the pipeline suffered axial deformations during service, due to soil movements. Much of the pipeline runs in areas with high density of subsoil use: under streets being paved, new crossings for sewer and water pipelines, etc. So soil settlements and slow motions of significant magnitude are likely. The mitigation plan requires defining the magnitude of the axial load that caused the failure. This was determined by the fracture mechanics methodology, using the FAD. The solution is plotted in Fig. 6.C2.

Fracture toughness K_{IC} of weld material was estimated from the correlation of Newton Roberts (see Chap. 5) with the value of energy absorbed in the impact test

$$K_{IC} = 8.43(CVN)^{0.63}$$

CVN = 50 J so that $K_{IC} = 100$ MPa m$^{1/2}$. Failure occurred by membrane tensile loading of the cracked section. This primary load includes internal pressure and imposed axial displacements.

The reduction of section is negligible; the axial load is distributed along the entire perimeter of the circumferential (girth) weld, of nominal thickness. Iterating in the FAD: $Sr = 0.99$, $Kr = 0.48$. Therefore, the total axial applied stress is $\sigma = 385$ MPa; failure is purely by plastic collapse. The stress due to internal pressure is significantly lower, so the condition for the observed fracture from the defect at the inner surface requires additional axial stresses. Welding residual stresses do not contribute to plastic collapse, as they are secondary and self-equilibrating. The origin of the extra stress is to be found in the tube axial strains imposed by movements of the surrounding terrain.

6.7 Mechanical Modeling of Longitudinal Cracks in Pipes

Circumferential (hoop) stresses due to internal pressure in pipes and cylindrical vessels are at least twice the longitudinal (axial) stress. That is why failures of pipes and other vessels caused by longitudinal crack propagation initiated in the inner or outer surfaces are much more common than fractures caused by defects in the circumferential direction (as the example shown before).

Brittle propagation of a defect can cause longitudinal fractures extending a great distance along the length of a pipe. A failure of a pipeline in Siberia in the 70 s resulted in a nearly 3 km long fracture. Pipes containing liquids tend to have much shorter longitudinal fractures, as the rapid reduction in pressure due to the leakage of fluid through the crack opening quickly reduces internal pressure. This in turn reduces the hoop stress and therefore the applied stress intensity at the tip of the propagating crack. In the case of compressible gases, this reduction is much slower and therefore longer cracks usually propagate.

There are specific analysis methods to determine failure conditions in the case of longitudinal cracks in the inner and outer surfaces of pipes and cylinders. To calculate applied K and equivalent stress for cracks that initiate at the inner surface, a stress equal to the internal pressure must be added to the hoop stress. This is because the fluid fills the crack mouth and pressure is generated inside of the crack. This pressure pushes both fracture surfaces in opposite directions and tends to open the crack, thus adding to the total applied stress.

To analyze cracks that initiate at the outer surface, outwards bulging of the surface crack tips must be considered, especially in thin-walled cylinders. This effect is due to the bending stress generated in the material in the mouth of the crack, which leads to a typical fish-mouth shape crack after failure, see for example Fig. 6.8.

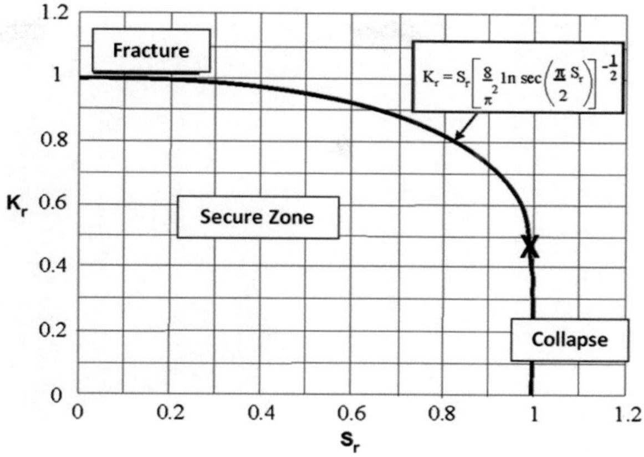

Fig. 6.C2 Position of failure condition in FAD

The effect of this secondary stress upon the driving force for crack propagation is quantified by a magnification factor called Folias factor, M, that multiplies the correction factor Y (Eq. 6.11). The exact form of this factor is an infinite series of terms; usually M is calculated as a partial development of the infinite series. A commonly used abbreviation:

$$M = \sqrt{1 + 1.255\frac{(L/2)^2}{(RT)} - 0.0135\frac{(L/2)^4}{(RT)^2}} \tag{6.19}$$

where:

T = wall thickness

L = length of the crack or defect projected on the longitudinal axis

R = tube radius

The Folias correction factor, as most all geometric magnitudes of discontinuities and defects in cylindrical vessels, are normalized with respect to the "size" of the vessel or pipe, given by the geometric mean value between diameter and thickness:

$$\text{Size} = (\text{Dt})^{1/2}$$

Because of the technological importance of longitudinal cracks in pipes, there are several specific methods to assess their criticality. In the previous section, we discussed in some detail the FAD method. When not used in graphic form, this method can be used directly applying Eq. (6.15), which defines the assessment criterion. This equation gives rise to the method "Logarithm Secant," used in North America, and the CDF (Crack Driving Force) method, as defined in the European code SINTAP (Structural Integrity Assessment Procedure). Calculations are made on the basis that effective K, as seen, fully characterizes the overall state of stress and strain at the tip of the crack.

Fig. 6.8 Fish-mouth shape crack

The FAD method is also used in the British procedure R6, in British Standard BS 7910 and in API STD 579–ASME FFS1 (which will be discussed in some detail in Chap. 9) and in procedures by EXXON, INSTA, and MPC. The method CDF is used in ETM and GE-EPRI (General Electric, Electric Power Research Institute, USA). These documents propose three or more assessment levels of increasing complexity and accuracy, different levels generate different expressions for the FAD and the CDF used in the analysis (the FAD described before corresponds to level 2). The assessment level is determined primarily by the degree of detail of the stress–strain data of the material used, and the reliability of estimates of the applied stresses and sizes of defects present in the component.

Pipeline integrity specialists have developed other methods for estimating the criticality of longitudinal cracks in pipes. CORLAS method is an assessment procedure similar to FAD, but presented differently. The Pipe Axial Flaw Failure Criterion (PAFFC) uses a graphical formulation. The graphics predict a failure hoop stress as a fraction of the yield stress of the material, depending on the surface length of the crack, with curves parameterized in function of a/c (depth/length of defect). Figure 6.9 shows an example of this type of computing, for a 30" diameter pipeline, which operating conditions are listed in the lower table. For small defects, failure stresses are much higher than the yield stress of the material. This indicates that the material will fail by plastic collapse rather than by fracture propagation.

6.8 Mechanical Modeling of Fatigue Crack Propagation

So far in this chapter we have seen the available tools to model final failure, based on material properties, applied loads, and geometry of previous defects. In some cases, mechanical modeling techniques are also useful to define the in-service

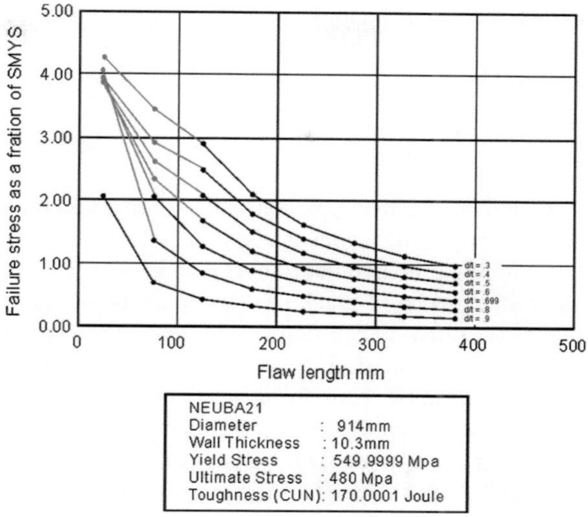

Fig. 6.9 Corlas predictions of failure pressures (LBB in *lighter grey*)

evolution of those defects that may ultimately lead to failure. That is, we can model subcritical propagation of these defects. In the case of fatigue crack growth the accuracy of numerical techniques is enough to be used as a prediction tool to predict when a detected defect would become critical, or when a defect that led to a failure was formed and began to grow.

Mechanical fatigue was defined in Chap. 4 as the process by which a defect (usually initially too small to be detected), grows during operation of the structure as a result of applied cyclic loading. We have seen in Chap. 5 that the fatigue process is divided into two phases: initiation of a crack from preexisting defects and crack propagation until it reaches a critical size for instantaneous fracture. The initiation process is very random and modeling tools are not sufficiently precise. However, crack propagation can, in some cases, be modeled with precision. Linear elastic fracture mechanics for constant amplitude cycle fatigue are well understood and are widely used for the analysis of cracked metal bodies.

More generally, fatigue crack propagation rate is a function of the applied cyclic K, $\Delta K = K_{max} - K_{min}$, and the mean load, which is usually defined by $R = K_{min}/K_{max}$, as an expression of the form:

$$\frac{da}{dN} = f(\Delta K, R) \tag{6.20}$$

where da/dN is crack propagation per cycle. The number of cycles needed to propagate a crack from an initial length a_0 to a final length a_f is:

$$N = \int_{a_0}^{a_f} \frac{da}{f(\Delta K, R)} \tag{6.21}$$

The functional relationship between *da*/*dN*, and ΔK usually exhibited by metals is that seen in Fig. 5.15. As discussed in Chap. 5, growth in region II is adequately expressed by the "Paris law":

$$\frac{da}{dN} = C \cdot \Delta K^m \tag{6.22}$$

C and m are material constants to be experimentally determined; Table 6.3 presents values for some steels.

Estimating the number of cycles from a certain crack size up to failure is based on an integration of crack propagation. The integration is between the maximum size of an initial defect (defined, for example, as the largest defect that could have escaped detection during pre-operational inspection) and the critical size for component failure (as defined in the previous sections).

As already seen, K corresponds to a general shape of the type: $\Delta K = Y\,\Delta\sigma\,\sqrt{\pi\,a}$ where σ is the applied stress, a is crack depth and Y is a geometry factor. Because propagation rate depends on crack size and geometry, a stepwise integration is required, and the differential equation is replaced by an incremental approximation:

$$\Delta a/\Delta n = C\,\Delta K^m \tag{6.23}$$

Rearranging terms and integrating both sides of Eq. (6.22), we have:

$$\int_{a_0}^{af} da = \int_0^{Nf} C(\Delta K)^m dN \tag{6.24}$$

Here, N_f is the number of cycles required to failure. As the crack grows, relationships *a*/*c* and *a*/*t* change and so do the correction coefficient *Y*. In these cases, one can resort to numerical integration:

$$\begin{aligned}\delta a &= C(\Delta K)_j^m(\delta N)\\ a_{j+1} &= a_j + \delta a_j\\ N_{j+1} &= N_j + \delta N_j\\ N_j &= \sum \delta N_j\end{aligned}$$

where are δN_j and δa_j are increments in crack length and N is the number of cycles. The procedure is to select a value of δN, then calculate ΔK using the corresponding value a_j and then get the next value of a. This is repeated until $a = a_f$.

Table 6.3 Paris equations and constants for some steels

Material	Equation and units
Ferritic-pearlitic steel	$\frac{da}{dN}(m/cycle) = 6.9(10^{-12})(\Delta K\ MPa\sqrt{m})^{3.0}$
Martensitic steels	$\frac{da}{dN}(m/cycle) = 1.35(10^{-10})(\Delta K\ MPa\sqrt{m})^{2.25}$
Stainless steels	$\frac{da}{dN}(m/cycle) = 5.6(10^{-12})(\Delta K\ MPa\sqrt{m})^{3.25}$
ERW welding	$\frac{da}{dN}(m/cycle) = 6(10^{-11})(\Delta K\ MPa\sqrt{m})^{5}$
Structural steels (conservative)	$\frac{da}{dN}(m/cycle) = (10^{-11})(\Delta K\ MPa\sqrt{m})^{3}$

6.8.1 Example 6.D Modeling of Fatigue Crack Growth in Compressor

The aim of the modeling presented here was to determine the origin and the reasons for the failure of a compressor in a petrochemical plant. The characteristics of various cracks and other defects present in the affected components were analyzed, material properties were defined and mechanical models and numerical were performed.

This is a two-cylinder reciprocating compressor. The failure was due to a perimeter crack developed between the two bolt holes in the crosshead of one of the cylinders. This crack is called main crack. Another crack discovered in the top of the crosshead is called remnant crack (Fig. 6.D1). During service, this part was dismounted several times to correct some abnormalities. Purpose of the modeling was to determine if it was possible to detect which of these maintenance activities led to the initiation of a fatigue crack, which eventually grew to cause the failure of the compressor.

A fabrication subsurface defect of 5 mm by 10 mm near the center of the hole of the crosshead was produced by a shrinkage cavity during solidification of the molten steel. This defect initially propagated to become an intergranular 9 mm by 12 mm corner crack at the surface of the central hole. Finite element modeling of the crosshead allowed defining the maximum stress (250 MPa), and the range of cyclic stress (127 MPa) applied to this initiation site in the central hole, under normal operation conditions. This initial fatigue crack grew in both sides of the hole, eventually creating the remnant crack.

Modeling of fatigue propagation was performed using Paris law and a method based on weight functions. This method allowed to predict how the remaining crack was being formed (Fig. 6.D2) and this was compared with beach marks left on the crack surfaces to verify consistency of the model.

Finite element and fracture mechanics modeling results defined a cyclic stress intensity factor above the fatigue threshold for the growth of fatigue cracks. Integrating Paris equation for fatigue propagation between initial and final crack sizes, a life of 50–80 million cycles was estimated, see Fig. 6.D3. This is less than a quarter of the total compressor operation time, estimated at 350 million cycles.

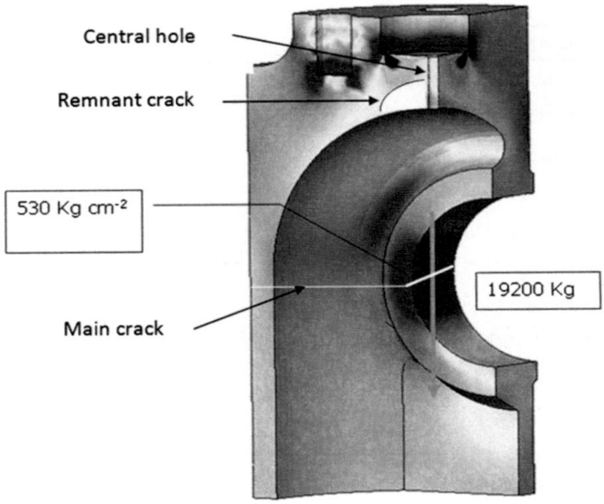

Fig. 6.D1 Model of failed crosshead

It was therefore possible to define with sufficient precision that the remnant crack had started to grow from an assembly problem after a maintenance operation, about 6 months before final failure. At the time of the failure, the remnant crack had grown up to 250 mm long, centered in the hole at the top of the crosshead. At this stage, the crack had entered a region of low cyclic stresses, so that its growth rate was very slow. The results of the failure analysis are beyond the purpose of this example, but the reader can find more details of the model developed in two of the references listed at the end of the chapter.

6.9 Determination of Fracture Toughness in Post-failure Samples

We have seen that knowing material toughness is essential for the analysis of cracked components. This material property significantly affects fracture behavior. Toughness is measured as the energy consumed by plastic deformation at the tip of a fracture at the time of its fast growth (Charpy impact, K_{IC}, and JIC tests were discussed in Chap. 5). We also saw that toughness is also related to ductility, whereby fracture resistance is largely related to the capacity of the material for plastic deformation, which can be measured in terms of a deformation (Crack Tip Opening Displacement or CTOD, $\delta 5$, etc.). All these methods require laboratory destructive testing of samples taken from the real component, machined with certain geometry.

A common property of many modern methods for the prediction of failures involves the existence of a material-depending characteristic length, commonly called critical distance. A variation of these techniques is the estimation of

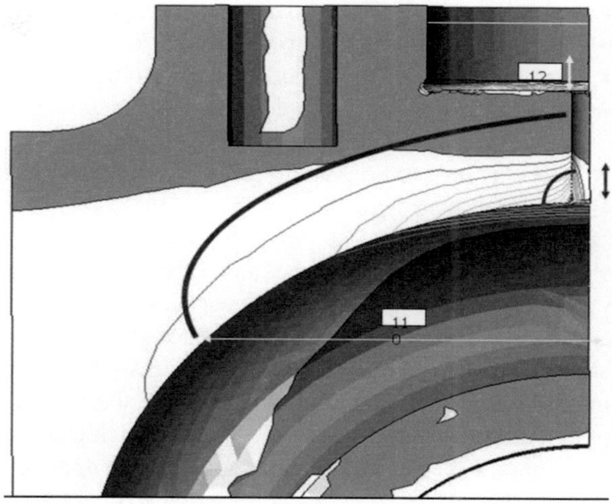

Fig. 6.D2 Numerical modeling of fatigue propagation in remnant crack

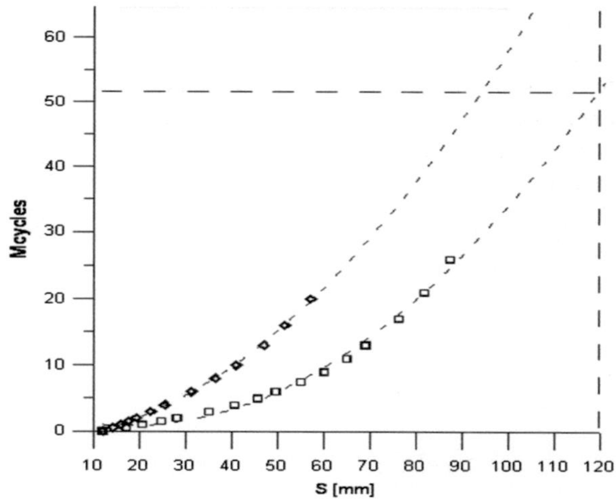

Fig. 6.D3 Numerical estimation of crack depth (s) as a function of fatigue life (million cycles)

material toughness from measurements of shear lips (BSL). We have seen in Chaps. 2 and 3 that these shear lips, located where the fracture surface reaches the surfaces of the cracked component, identify the transition between brittle fracture (near the center of the thickness) and ductile fracture adjacent to free surfaces, see for example Figs. 2.9 and 2.B1.

A failure event that involves propagation of a fracture is a real-life "test" for the material, in absolutely real conditions, and an excellent opportunity to take advantage of modeling techniques. Careful inspection and knowledge of fracture mechanics equations provide estimates of actual toughness of the material at the time of the event. This can be accomplished through some parameters:

- Lateral shrinkage during fracture propagation.
- Plastic zone size at the tip of the crack.
- Opening angle of the mouth of the fracture.
- Deformation at some characteristic distance from the tip of surface notches and other stress concentrators.
- Depth of indentations and correlation between hardness and toughness.
- Width of shear lips.

These methods are indirect and approximate formulas, validated within certain ranges using the equations of linear elastic fracture mechanics.

The correlation between fracture toughness and width of shear lips is of considerable importance in failures analyses. It has been theoretically and experimentally demonstrated that the size of the shear lips as measured at the free surface of a component is independent of its thickness, and is related to the fracture toughness, K_{IC}. This is because shear lips are approximately equal to the size of the plane stress plastic zone ahead of a crack tip, which has already been identified as r_p. The theoretical value of r_p is:

$$r_p = \pi/8(K_{IC}/\sigma_{YS})^2.$$

where σ_{ys} is the yield stress of the material. Using this theoretical expression and material data, we obtain:

$$\text{KIC} = (2.44\ \text{BSL})^{1/2}\sigma_{YS}$$

Recent applications of these correlations in failure analysis have resulted in reliably conservative estimates of toughness.

References

1. D.K. Felbeck, A.G. Atkins, *Strength and Fracture of Engineering Materials* (Prentice Hall, New Jersey, 1984)
2. T.L. Anderson, *Fracture Mechanics: Fundamentals and Applications*, 2nd edn. (CRC Press, New York, 1994)
3. D. Broek, *Elementary Engineering Fracture Mechanics*, 4th edn. (Kluwer, New Jersy, 1982)
4. ASTM E 1049-85 (reapproved 1990), *Standard Practices for Cycle Counting in Fatigue Analysis* (American Society for Testing and Materials, USA, 1998)
5. A. Higdon, *Mechanics of Materials*, 2nd edn. (Willey, NewYork, 1967)
6. API Standard 579, *Fitness for Service* (American Petroleum Institute, Washington, 2007)
7. ASME Boiler and Pressure Vessel Code Sec. VIII Div. 2, *Alternative Rules* (American Society of Mechanical Engineers, New York, 1998)
8. ASME Boiler and Pressure Vessel Code Sec. VIII Div. 1, *Rules for Construction of Pressure Vessels* (American Society of Mechanical Engineers, New York, 1995)
9. ALGOR Inc, FE-PIPE Pipe Stress Analysis and Design System, 1997
10. J.R. Farr, M.H. Jawad, *Guidebook for the Design of ASME Section VIII Pressure Vessels*, 2nd edn. (ASME Press, New York, 2001)
11. COADE Inc, CAESAR II pipe stress analysis and design code, 1997

12. Hibbitt, Karlson and Sorensen, Inc. *ABAQUS/Sandard User's Manual from CD ABAQUS 6.4*, Student edn. (Springer, Pawtucket, RI 2004)
13. Ibarra Pino, Modelado Mecánico de Recipientes a Presión en Plantas Industriales. Undergrad Thesis, University Nacional de Mar del Plata, Argentina, 2004
14. H. Shames, C.L. Dym, *Energy and Finite Element Methods in Structural Mechanics* (Hemisphere Pub, Washington, 1996)
15. O.C. Zienkiewicz, R.L. Taylor, *The Finite Element Method*, vol 2, 5th edn. (Butterworth-Heinemann, New Jersy, 2000)
16. API 618, *Reciprocating Compressors for Petroleum, Chemical and Gas Industry Services*, 4th edn. (American Petroleum Institute, Washington, 1995)
17. R.W. Hertzberg, *Deformation and Fracture Mechanics of Engineering Materials*, 4th edn. (Wiley, USA, 1989)
18. J. Malpartida, H Kunert, J.L Otegui, *Gestión de Integridad: Uso de la Modelación Computacional para Mitigar Amenazas por Fuerzas Externas en Ductos de Transporte de Hidrocarburos. VII INGEPET (GAS-3-JM-19-N)* (Lima, Perú, 2011)
19. H. Kunert, J.L. Otegui, A. Marquez, P. Fazzini, Nonlinear FEM strategies for modeling pipe–soil interaction. Eng. Fail. Anal. **24**, 46–56 (2012). ISSN 1350-630
20. H.G. Kunert, A. Marquez, J.L. Otegui, in *Numerical Tool Used to Model Pipeline Failures in Unstable Soil in Ecuador*, ed. by H.G. Kunert, P. Jaya, J.L. Otegui, 5th International Conference on Engineering Failure Analysis (ICEFA 5), Netherlands, 2012
21. Technical Report GIE P2302-03-11, Root cause analysis of leak in NGL Camisea pipeline, PK 91. (Coga, Peru, 2011)
22. J.L Otegui, H. López Montenegro, A. Márquez, Numerical life estimation after fatigue failure of a complex component. Fatigue Fract. Eng. Mater. Struct. **28,** 547–556 (2005). ISSN 8756-758X
23. P.G. Fazzini, A.A. Márquez, J.L Otegui, P Barcia, Cause and effect assessment after a complex failure of a large Ethylene compressor. Eng. Fail. Anal. **13**, 1358–1369 (2006). ISSN1350-6307
24. Pipeline Research Council International, Inc. *Extended Model for Pipe Soil Interaction*, PR-271-0184, (2003)
25. Report Gie 2303-09-09, *Análisis de Causas Raíz de Falla en Soldadura Circunferencial en Gasoducto Metrogas Tramo VII. Metrogas* (Argentina, 2009)
26. ASTM E-1820: ASTM E 1820-01: Standard Test Method for Measurement of Fracture Toughness, (ASTM International, West Conshohocken, 2001)
27. L. Susmel, The theory of critical distances: a review of its applications in fatigue. Eng. Fract. Mech. **75**, 1706–1724 (2008)
28. D. Taylor, S. Kasiri, A comparison of critical distance methods for fracture prediction, Department of Mechanical Engineering, Trinity College, Dublin 2. Irel. Int. J.Mech Sci **50**, 1075–1081 (2008)
29. Lai, M.O. (Man On), The relationship between toughness and shear lip size Ph.D. Thesis, University of Auckland, 1977

Chapter 7
Root Cause Analyses

7.1 Introduction

The scientific investigation of an accident, catastrophic event, or failure should be structured like a pyramid. There must be a large base of verifiable facts and evidence in the bottom. These facts form the basis of analyses based on scientific principles and proven methods, that links facts and analytical results of the various statements of witnesses and participants; it links what is possible and what is not possible. If the statement of a witness would not fit within these limits, it is not reliable. The logic provides order and consistency to all statements, facts, principles, and methodologies brought to bear on a particular case. For a first-rate research, the logic that connects the pieces must be rational.

The facts and analyses, taken together, support a small number of conclusions which form the apex of the pyramid. The conclusions should be directly based on the facts and analyses and not on other conclusions, assumptions, or hypotheses. If the facts are logically and systematically organized, the findings should be almost obvious. Root cause analysis (RCA) is often used in relation to the human (or management) factor, and how procedures and management techniques can be improved to prevent the problem from occurring again. The RCA is often used in association with large systems, such as plants and projects, where there is a great emphasis on safety and quality assurance through formal procedures. It is presumed that if true causes are identified and then corrected, the event will not be repeated, at least not in the same way.

After the root causes for an event have been determined, the intent of any corrective action is to reduce the likelihood of recurrence for the same reason, or to mitigate consequences if it returns.

Forensic engineering is a term usually related to the investigation of an event that is somehow referred to law, courts, legal debate, or disclosure. We have seen (Chap. 1) that when litigious consequences are foreseen, the analysis departs from some of the RCA principles discussed in this chapter. We will return to this again in Chap. 10. In what follows, we make a short historical review of recent failures that led to what was called the NASA method for RCA.

J. L. Otegui, *Failure Analysis*, DOI: 10.1007/978-3-319-03910-7_7,

7.2 The Explosion of the Challenger Space Shuttle

When the space age dawned in the second half of the twentieth century, the basic physical laws needed to design and fly a spaceship were thought to be clear. Otherwise, manned flight into orbit and beyond would have been just a crazy dream. The trick was to exploit the laws correctly. But knowing the laws of nature is not enough for a team to compete with them successfully. It takes a creative engineering genius to design a ship that not only successfully launches, but also orbits the Earth, re-enters our atmosphere, and glides to a safe landing. Successful integration requires a lot of specialized knowledge and accomplished engineering teams involved in the intricacies of rockets, combustion, structures, aerodynamics, life support, heat transfer, computer control, and a host of other specialties.

The space shuttle program (USA, 1980–2010) required a lot of engineers and managers to fulfill its mission. Each shuttle involved more than a million pieces, suggesting the degree of complexity of the entire hardware, software, and operation system. In the 1980s, the directors of the National Aeronautics and Space Administration (NASA) estimated that flights would be 99.999 % reliable. This represents a failure rate of 1 in 100,000 and would imply that one could launch a shuttle into space every day for 300 years and expect to lose just one. Engineers, who are more familiar with the shuttle itself and with machines in general, predicted a success rate of "only" 99 %, i.e., one error out of 100 launches. A safety officer in charge, who personally observed tests during the development of rocket engines, predicted a failure rate of 1 in 25. The Challenger accident in 1986 showed that this estimate was the actual failure rate and, as a result, a success rate of 96 % after just 25 launches.

In a retrospective analysis published in early 2011, NASA summarized previous shuttle records as "lucky." The possibility that a catastrophic failure occurred in the first nine shuttle missions was, in fact, as high as 1 in 9. In the following 16 missions, including the 1986 Challenger mission, the chances of failure were 1 in 10.

The American Space Shuttle used two solid rocket boosters (SRB). Figure 7.1a shows a picture of the vehicle in its launch condition: the ship with two booster rockets on each side, and the huge orange liquid fuel tank, the largest volume at launch. SRB rockets were long and thin tanks, to be constructed in sections. They were manufactured by a company called Thiokol in Utah. The connection between tank sections included a series of fasteners and seals, including two rubber "O-rings," Fig. 7.1b.

That space shuttles struggle against the force of gravity is evident in the slow, agonizing seconds of takeoff during each launch from Cape Canaveral. The failure of the Space Shuttle Challenger occurred on January 28, 1986 (mission STS-51-L), 6 s after takeoff the main engine was ignited. When the engine reached its maximum power, the two mounted rocket boosters ignited. 36 s later, the Challenger crossed the sound barrier. Suddenly the shuttle was beaten by violent winds. By detecting severe fluctuations in the flight path, the control system directed the

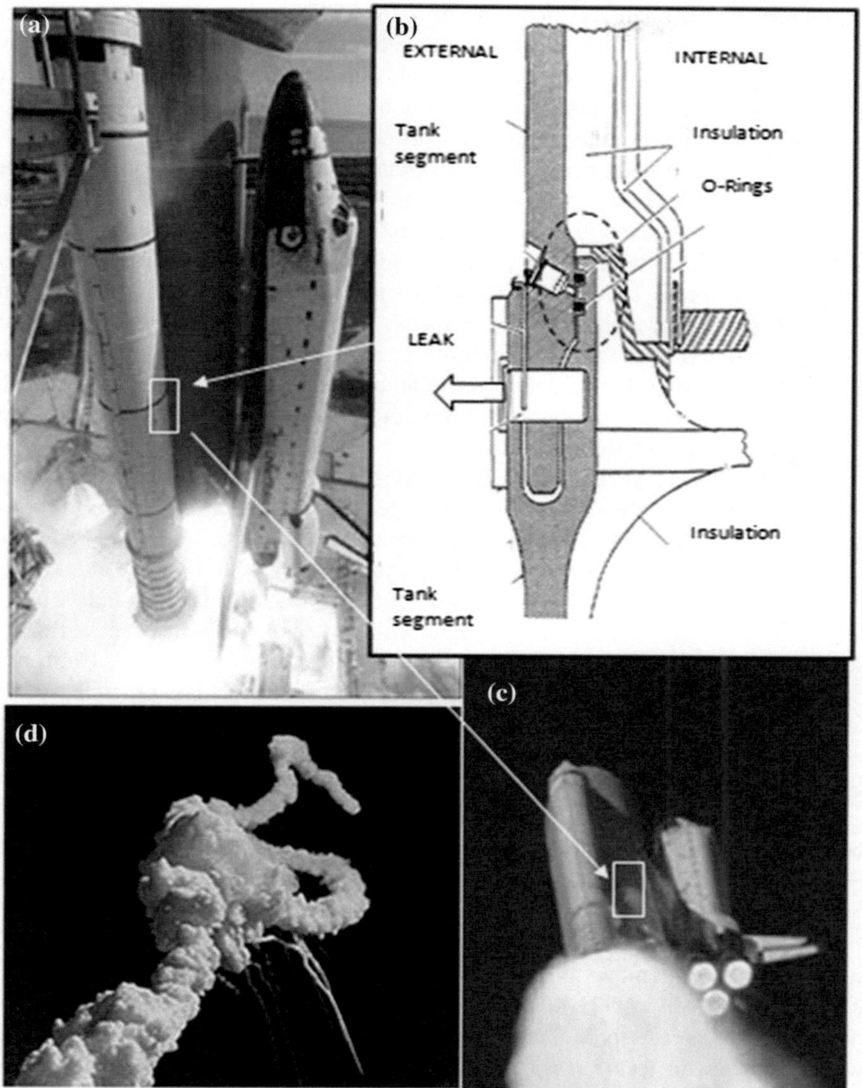

Fig. 7.1 **a**, **b**, **c**, **d** Immediate causes for the challenger failure

power system of the rocket boosters to compensate. Thus, the shuttle's engines power increased by 104 %.

At that moment, a tongue of fire that had not been detected started licking the starboard booster rocket joints, Fig. 7.1c. After a 72 s flight, the hydrogen that leaked from the fuel tank ignited. The booster rocket turned on its moorings and punctured the fuel tank, which was torn from top to bottom. In the resulting explosion, hundreds of tons of liquid wrapped the shuttle in a ball of fire. The two booster rockets were thrown; then the explosion destroyed the shuttle, Fig. 7.1d. Everything happened so fast that there was no time to implement any emergency maneuver. In any case, the shuttle was not equipped with ejection seats.

The commander had opened the radio channel, but had no time to say anything; he was heard exclaiming, "Oh, oh." As the shuttle disintegrated in the air, some members tried to activate their emergency oxygen supplies. Though the cabin was preserved virtually intact until rushing into the Atlantic Ocean, 14 km down, aerodynamic forces killed all those who survived the initial blast. Family and friends of the crew watched the sky in horror. Although it was clear that something terrible had happened, the PA system was providing data on the altitude and the speed of the shuttle. Finally, the spell was broken. The phlegmatic voice in the public address system announced: "Flight controllers are analyzing the situation. Obviously a major malfunction."

President Reagan set up a commission to investigate the crash. Shuttle launches were canceled while the committee deliberated and NASA raised the possibility of launching satellites using "old" unmanned rockets. Unmanned launches take place today with rockets from Russia and other countries. In fact, since the shuttle fleet was discharged in July 2011 until the time of the writing of this book, only Russian Soyuz rockets are capable of exchange with the international space station for sending personnel and supplies. A privately held company based in California, SpaceX, is using the capabilities of NASA to test launch Falcon rockets with unmanned Dragon capsules, bound for the space base.

The commission reviewed videos of the flight and the flames; clouds of smoke coming out of the joints of the starboard SRB booster rocket were detected. The commission concluded that the rubber rings that should have sealed the joint between the segments of the booster rocket had failed on takeoff, as highlighted in Fig. 7.1a. The failure was due to a faulty design; of which Thiokol had previously warned NASA. The rubber seals could fail at low temperatures. Some company engineers had even advised to suspend the launch until a solution to the problem was found. Two members of the presidential commission commented that NASA was playing "a kind of Russian roulette every time they launched a shuttle." Old rubber rings were discarded and a new type of mechanism for sealing the joints was designed and tested. In 1988, when NASA scientists were satisfied, space flights resumed.

The failure of the Challenger forced to rethink the details of the design and operation of the shuttle, and changes were made based on lessons learned. After a 20-month pause NASA resumed missions and the shuttle fleet flew successfully to the 113th mission, which ended with the disintegration of the Columbia as it entered the Earth's atmosphere in 2003. The improvements made in those 20 months reduced the probability of failure to 1 in 38. The odds were changing during the program because constant modifications were made to the system. For example, when the Environmental Protection Agency (EPA) banned the use of Freon, NASA had to stop using it to blow foam insulation on the external fuel tank.

The compound used to replace Freon did not allow the foam to adhere to the tank, and as a result more foam spilled during takeoff and flight. This increased the risk of accidents such as the one that finally destroyed the Columbia shuttle. For the nine missions that followed the ban on Freon, the chances of disaster increased from 1 in 38 to 1 in 21.

One of the most important consequences of the Challenger accident was the creation by NASA of technologies for root cause analysis (RCA), which is discussed in this chapter. Apart from the deficiencies in design, RCA showed that serious organizational shortcomings led to failure. Understanding these shortcomings has led, among others, to the development of failure tree techniques (events, conditions and exceeded barriers) for the determination of root causes.

The nuclear electricity generation industry is undoubtedly the one with highest consequences in case of an accident; a radioactive release has consequences for all biological species, including, of course, humans. In the U.S., nuclear power generation plants are controlled by the Department of Energy (DOE). In 1992, the agency issued the standard DOE-NE-STD-1004-92, a guide for RCA specified by DOE Order 5000.3A.

7.3 Methodology for Root Cause Analysis

Root cause analysis (RCA) describes the results of detailed evaluations of various aspects of the background and conditions that led to a "failure," or, more generally, an "undesired result." Examples of undesirable results: failure, anomaly, broken equipment, defective product, problem, incident, accident, etc.

When performing an RCA, detecting only the immediately visible cause (the "immediate cause") is not enough. Root causes are harder to see; however, they can contribute significantly to an unwanted result and, if not corrected, will continue to create similar types of problems. Each branch developed during the analysis would reach approximately the same level of causality.

Immediate causes: the occurred event and any condition that existed immediately before the undesired outcome, which directly resulted in its occurrence and, if been deleted or modified, the undesired outcome would have been prevented. They are also known as direct causes.

Root causes: each of the multiple factors (events, conditions or organic factors) that created or helped create an immediate cause and subsequent undesired outcome. If been eliminated or modified, the undesired outcome would have been prevented. Typically, multiple root causes contribute to an undesired result.

Root cause analysis (RCA): structured method of assessment that identifies the root causes of an unwanted outcome and appropriate actions to prevent recurrence. RCA continues until all organic factors have been identified or until data are exhausted. The RCA is a method that helps professionals determine the three questions defined in the foreword: what happened, how it happened, and why it happened. It also allows learning from problems, past failures, and accidents.

The purpose of a RCA is to identify root causes, so that latent failures can be eliminated or modified, and future occurrences of similar problems or accidents can be prevented. If the RCA is not performed and the analyst identifies and manages only the immediate causes, then the underlying causes may continue to produce similar problems in the same or related areas.

The RCA is to identify systemic problems, such as lack of budget for maintenance, and correct them so problems or accidents do not to happen again. RCA bases are:

- Clearly define the undesired event.
- Collect data, including a list of all potential causes.
- Create a tree of events and causal factors.
- Continue asking "why" to identify root causes.
- Check the logic and remove items that are not causes.
- Generate solutions that address immediate and root causes.

Each cause can be of three types:

1. A condition of the system, prior to the failure (e.g., inappropriate material, manufacturing defect, errors in control systems, etc.).
2. An event that occurred at the time of failure and contributed substantially to it (operating parameters above allowable or nominal, overpressure, operation errors in start up, damage by third parties, natural events such as earthquakes or tsunami, etc.).
3. An exceeded barrier, that is, a safety device that upon the occurrence of an event should have avoided a major failure, but did not (examples: a relief valve after an operating overload, nondestructive evaluation that does not detect a manufacturing defect, etc.).

DOE regulations define that each RCA report should include five distinct phases, while recognizing there may be some overlap among them:

Phase I Data collection. The information to be collected consists of conditions before, during, and after the incident (event, accident): staff participation, environmental factors, and any other relevant information.

Phase II Evaluation. It includes the following steps:

a. Identify the problem.
b. Determine the importance of the problem.
c. Identify the causes (conditions or actions) immediately preceding.
d. Identify the reasons for these causes, until reaching root causes.

Phase III Corrective actions. The effective implementation of corrective actions for each cause reduces the likelihood of a repeated problem and improves reliability and safety.

Phase IV Reports. Reports and processing system (ORPS) are part of the process required by DOE. Also included are discussing and explaining the results of the analysis with management and involved staff, including corrective actions.

Phase V Monitoring includes determining whether corrective action has been effective in solving problems. A review of the effectiveness is essential to ensure that corrective actions have been implemented and can prevent recurrence.

In this chapter we will discuss in some detail phases I and II, in Chap. 10 we will return to phases III, IV, and V.

7.3.1 Example 7.A Is a Blown Fuse a Failure?

This example is often used in textbooks and courses of RCA. A fuse in the electrical circuit burns and causes the lights to go out in a building. You can identify the immediate cause, "the fuse burned," and replace the fuse. You can also identify the intermediate cause: a "short circuit" and repair the failed wiring. However, if you do not identify and correct the organic factor that led the fuse to jump (e.g., lack of maintenance of the wiring because the maintenance budget was insufficient), other systems may have similar failures due to lack of maintenance.

But this example starts with a faulty premise: the fuse is placed on the electrical system to function as a barrier. That is, an overload causes breakage of this element, and this avoids the occurrence of a failure of major consequences, as in the case of a fire in the wiring. It is common to consider the operation of a barrier as a failure, because the system is designed for this not to occur.

Other cases where barriers are NOT exceeded are, for instance, a breakage during test loading a new structural element, or when damage is detected during inspections or tests, etc. A recent case, discussed later, is the emergence of leaks in a buried pipeline when a well was dag at a section where a buckle or loss of thickness was suspected, after nondestructive evaluation and geological studies. When removing the certain weight of top soil some elastic recovery occurs in the buried pipe, so (pre-existing) cracks can open up and cause leakage. In all these cases, the "failure" occurs in controlled conditions, and its consequences are much lower than would be expected if it had occurred at some later point during service.

7.4 The Cause–Effect Tree

Some formal tools may be used for assembling cause and effect trees, some of which will be seen below. The basic instructions in the assembly of this tree are:

- Put the unwanted outcome on the top of the tree.
- Add all the events, conditions, and exceeded or failed barriers that occurred immediately before the undesired outcome and what could have caused them (Fig. 7.2).

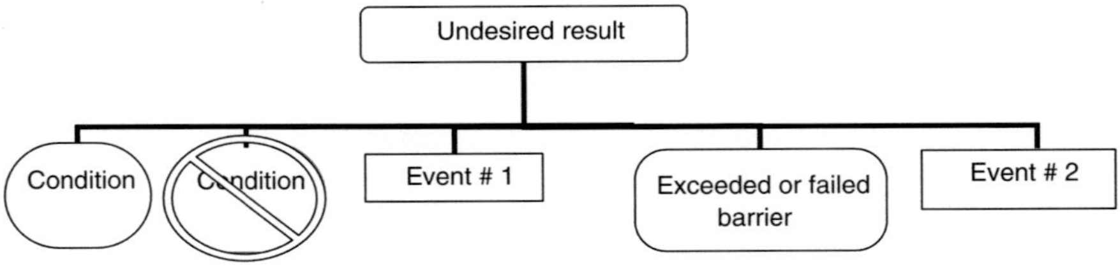

Fig. 7.2 Example of cause–effect tree: immediate causes

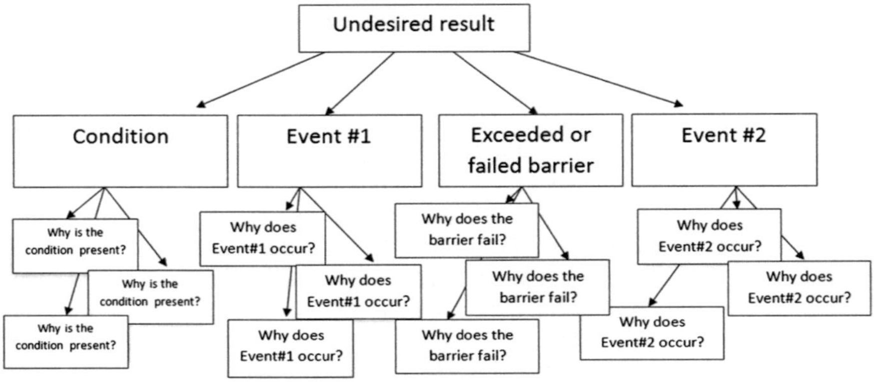

Fig. 7.3 Cause–effect tree: intermediate causes and root causes

- Make brainstorming to ensure that all possible causes are included, not just those that you are sure are involved.
- Be sure to consider people, hardware, software, policies, procedures, and environment.
- If you have solid data to indicate that any of the possible causes is not applicable, it can be removed from the tree. Caution: do not rush to eliminate causes. If there is a possibility that it is a causal factor, leave it and delete it later when more information becomes available.
- After you have identified all possible causes, ask yourself why each may have occurred.
- Be sure to keep your questions focused on the original problem, Fig. 7.3. For example, "Why was this condition present?" "Why was the event?" "Why was the barrier exceeded?" or "Why did the barrier fail?".
- Individual failure trees can be used to determine the causes of failure of any subsystem involved in the basic event (e.g., components) and can then be incorporated into the event tree.
- The resulting tree of questions and answers should lead to a complete picture of potential causes for an undesired event.

- Check your logic with a detailed review of each potential cause. Compare each cause with verified facts.
- If the action, deficiency, or decision in question had been corrected, eliminated, or avoided, could have prevented or avoided the unwanted result? If the answer is yes, then it is a cause. Keep it in the tree. If this is not the case, then remove it from the tree.

To ensure that data is not lost, it is important to start the data collection phase for RCA immediately after the identification of occurrence. Taking care not to affect safety during recovery, it is recommended, if possible, to collect data even during an incident. The information to be collected consists of conditions before, during, and after the occurrence of the failure, the participation of personnel (including actions taken), environmental factors, and other relevant information to the condition or problem. Photographing the affected area from multiple views is very useful during and after the investigation. Every effort should be made to preserve physical evidence, such as failed components (see Chap. 2), fires, spills, partial implementation of work orders and procedures, etc. This should be done despite the operational pressures to restore the equipment or service.

Once you have gathered all the data associated with the occurrence, the data should be checked for accuracy. The investigation can be improved by retaining all the physical evidence, establishing a quarantine area, for example. Some areas to consider when determining what information must be included:

- Activities related to the appearance.
- Initial or recurrent problems.
- Hardware (equipment) and software (program type issues) associated with the occurrence.
- Recent changes to program or administrative team.
- Physical environment and circumstances.

7.5 Data Collection: Interviews and Statements

Interviews should be conducted to find information and failures (more on this in Chap. 10). Questions should be prepared before any interview. This is essential to ensure that all necessary information is obtained. The worksheets for the cause and effect tree can be used as a tool to help gather information. Interviews should be done preferably in person, to those more familiar with the problem, although this is sometimes impractical in terms of the number of people involved. Interviews can be documented by any format desired by the interviewer.

Although interview preparation is important, it should not delay contact with participants and witnesses. The first interview may consist solely in hearing the story; you can arrange a second more detailed interview, if necessary. The interviewer should always consider the objectivity and the frame of reference of the respondent. You should consider interviewing other officers conducting work in the past.

Eyewitnesses are important sources of information, but must be carefully examined and evaluated. Often the witnesses do not report directly their observations; they tend to report personal findings as internally trying to make sense of what they have observed. Over time, rethinking things and hearing comments may make witnesses to report seeing things that they did not really witness. Also, witnesses can over time forget things that they really saw; they would remember what they or others think is interesting. The perception of the event is often conditioned by their education and training, experience, physical condition with regard to vision, hearing, and social or cultural prejudices. Skillful questioning of witnesses can separate personal factual observations from assumptions.

That an RCA failure analysis is not an attempt to lay blame is important for two reasons. First, the research team must understand that the real benefit of this analytical method is the improvement of plants. Second, those involved in the incident will generally adopt an attitude of self-defense and assume that the investigation aims to find and punish the person or persons responsible for the incident. Therefore, it is important for researchers to dispel fear and replace it with positive team effort to solve the problem.

The interview process is a key part of any investigation. It is crucial for the investigator to be a good listener with good diplomatic skills. To get a complete picture of significant incidents, he must meet all the staff who may have direct or indirect knowledge of the event. It is necessary to verify the information obtained in the interview. A prepared list of questions or topics to discuss helps keep the interview organized and prevents the interviewer from forgetting questions about key issues:

- What happened? Clarifying what really happened is an essential requirement. The natural tendency is to define perceptions rather than the actual event.
- Where did it happen? In addition to the location, determine whether the event also occurred in similar locations or systems.
- When did it happen? When you know the real-time "snapshots" of an event, it is much easier to quantify the process, operations, and other variables that may have contributed to the event. What changed? Failures and deviations always occur due to specific variables. Therefore, it is essential to define the changes that occurred in conjunction with the event.
- Who was involved? This is a controversial question. Errors and events are often the result of human error or insufficient knowledge. However, the purpose of the investigation is to solve the problem, not assigning blame. Therefore, any comments or statements must be totally objective, impersonal. Each staff involved in the incident should be assigned a code number or other identifier, such as "operator A, B, C," etc. This helps reduce the fear of punishment and reduces prejudice or preconceived opinions about individuals within the organization.

The simplest way is that the investigator interviews each person. This should be done in private, without distractions. Group interviews may be held in the workspace. When investigating controversial or complex incidents, it is advisable to have two interviewers: one may ask while the other records the information. Interviewers must coordinate their interrogations and avoid overwhelming or intimidating the interviewee. Interviewers can also compare their impressions and reach a consensus on their views.

The group interview is advantageous in some cases. The exercise of group problem solving promotes the exchange of ideas, particularly with people from different disciplines (maintenance, production, engineering, etc.). Such an exchange can help solve a problem or event. This approach is particularly useful when the investigator has completed his evaluation and wants to review the findings with those involved in the incident. The investigator in this case acts as a facilitator in the process of problem solving and uses a tree diagram or cause–effect sequence of events as a working tool for the meeting. Groups meetings do not work in a hostile environment. If the event is controversial or political, personal agendas of the participants generally prevent positive results.

7.5.1 Example 7.B Causal Tree in a "Hot Tap" Repair

In this example, the failure or undesired event was the loss of the coupon and wick guide for "hot tap" drilling a 30″ pipeline for a derivation. The cost of this failure seems small but involves the loss of the entire device, the need to redo the process (in an area possibly not as appropriate), and a high risk for recurrence. In this example, the testimonial evidence was predominant for the resolution of the RCA.

This type of derivation is performed with the pipeline in operation, and is based on the use of a device that allows drilling in a local environment free from air, to avoid explosive mixture. A sketch of the operation is shown in Fig. 7.B1 (1, 2, 3, 4). In this case, the hot tap was applied to the replacement of a stretch of defective pipeline.

Cutting is performed with a system consisting of a front cutter, Fig. 7.B2a, which carries a drill or roving guide (wick), Fig. 7.B2b. Pegs lie near the tip of the wick guide which open when the wick has passed through the tube wall, so that the cut piece (coupon) is lifted when the cutter is removed after the cut is completed.

Figure 7.B3 shows the top of the cause and effect tree, possible immediate causes are indicated. A first inspection verified that the wick guide did not remain attached to the drill when withdrawing the drill. The wick detached from its seat in the cutter, and thus remained attached to the coupon inside the pipeline. Coupon loss is important because it stays inside the pipe and prevents proper placement of the line stop seal to isolate the section. These first conclusions are reported in a so-called Appendix A, as shown in Fig. 7.B3. Therefore, possible causes are branded, verified, and eventually deleted from the tree.

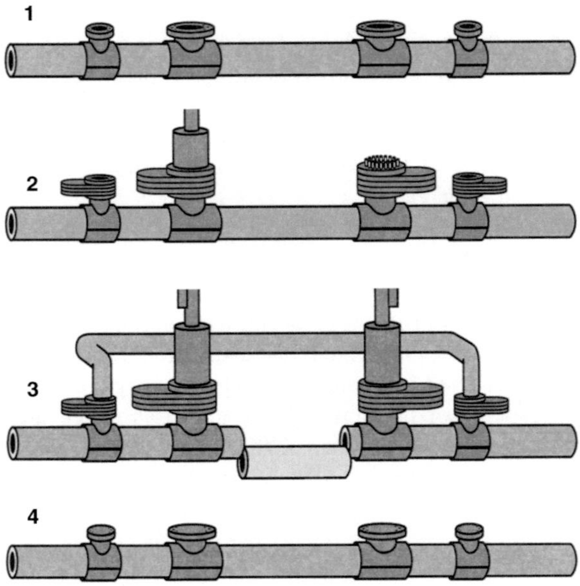

Fig. 7.B1 (*1*, *2*, *3*, *4*) Hot tap drilling in pipeline

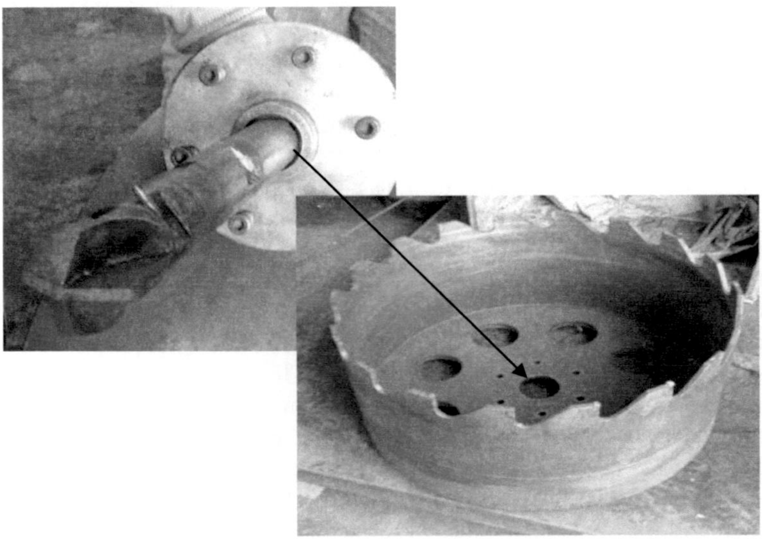

Fig. 7.B2 **a**, **b** Drill and guide (wick) in hot tapping machine

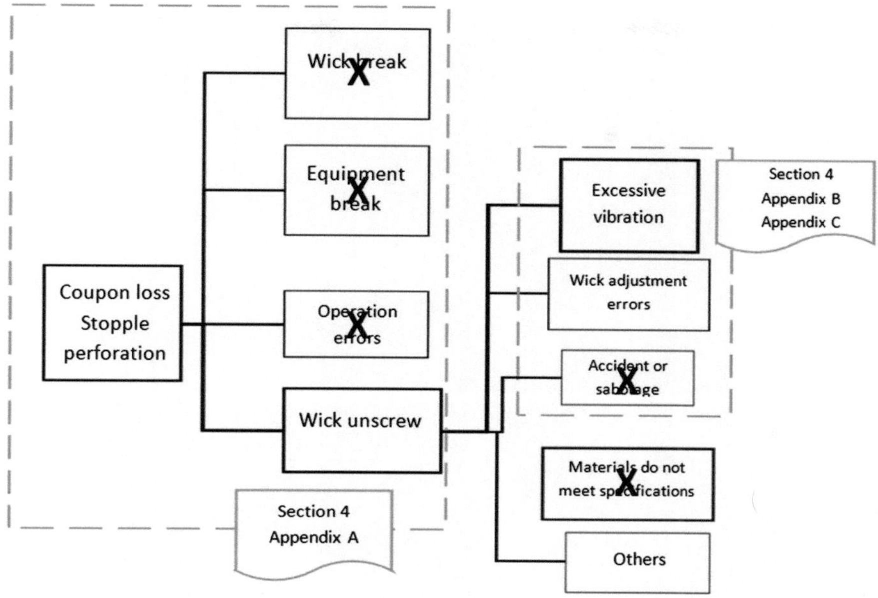

Fig. 7.B3 Upper part of the cause–effect tree

In Appendices B and C statements by operators are analyzed. They indicated that unusual vibrations in the machine had been detected during this cut. These vibrations may have been due to:

- Lack of maintenance.
- Lack of alignment of the machine.
- Deficiency in the edges of wick and cutter.
- Errors in fixing the wick to the cutter frame ("anchor").

It was not until cutting the section of the pipeline and the recovery of the failed hot tap that more information was available. Figure 7.B4 shows (a) intermittent rubbing marks on the side surfaces of the inner surface of the flange and (b) wear at the sides of the cutter teeth. Furthermore, Fig. 7.B5 shows that the wick guide anchor to the cutter body is a threaded joint with a tightening mirror. It is reasonable that vibrations during this failed operation had loosened the tightness, and led to unscrew the guide wick.

With this, the immediate causes of the failure got clear and let us see root causes. What caused wear of the inner surface of the flange? A dimensional analysis revealed a certain lack of roundness of the flange, due to shrinkage of the longitudinal welds during its placement in the pipe. This lack of roundness exceeds tolerances defined by the drill supplier. Additionally, the anchoring system of the wick was found to be inadequate. A recess and bolt system would be much more resistant to vibration. These two causes are consequences of other circumstances, as outlined in the complete causal tree shown in Fig. 7.B6.

(a)

(b)

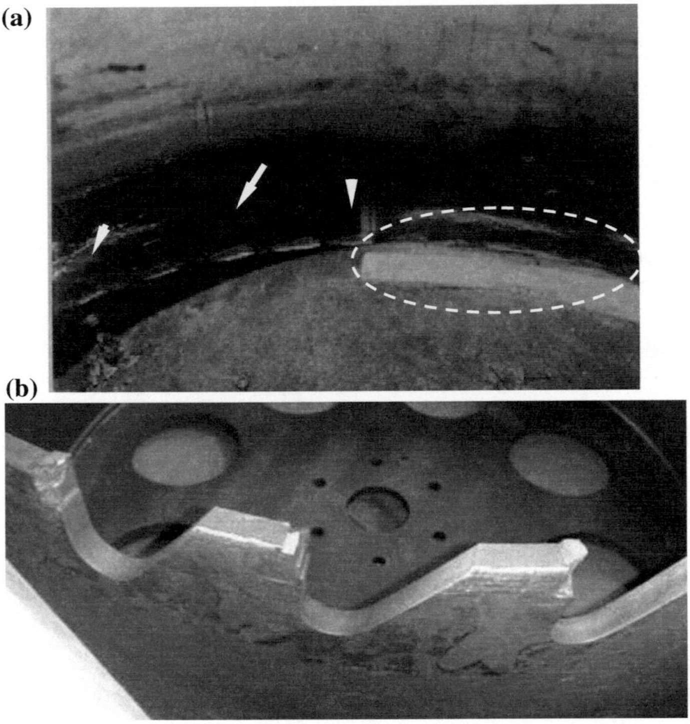

Fig. 7.B4 **a** Intermittent rubbing marks on the side surfaces of the flange and **b** Wear at the sides of cutting teeth

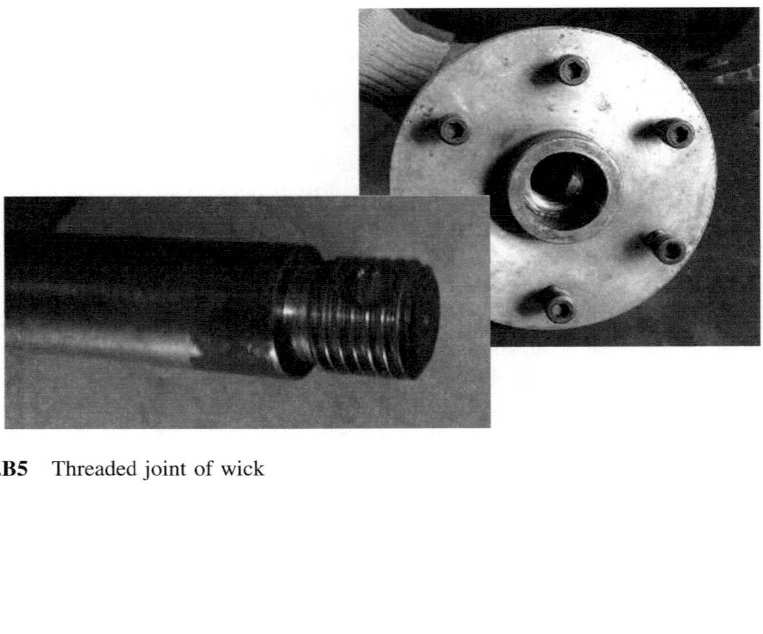

Fig. 7.B5 Threaded joint of wick

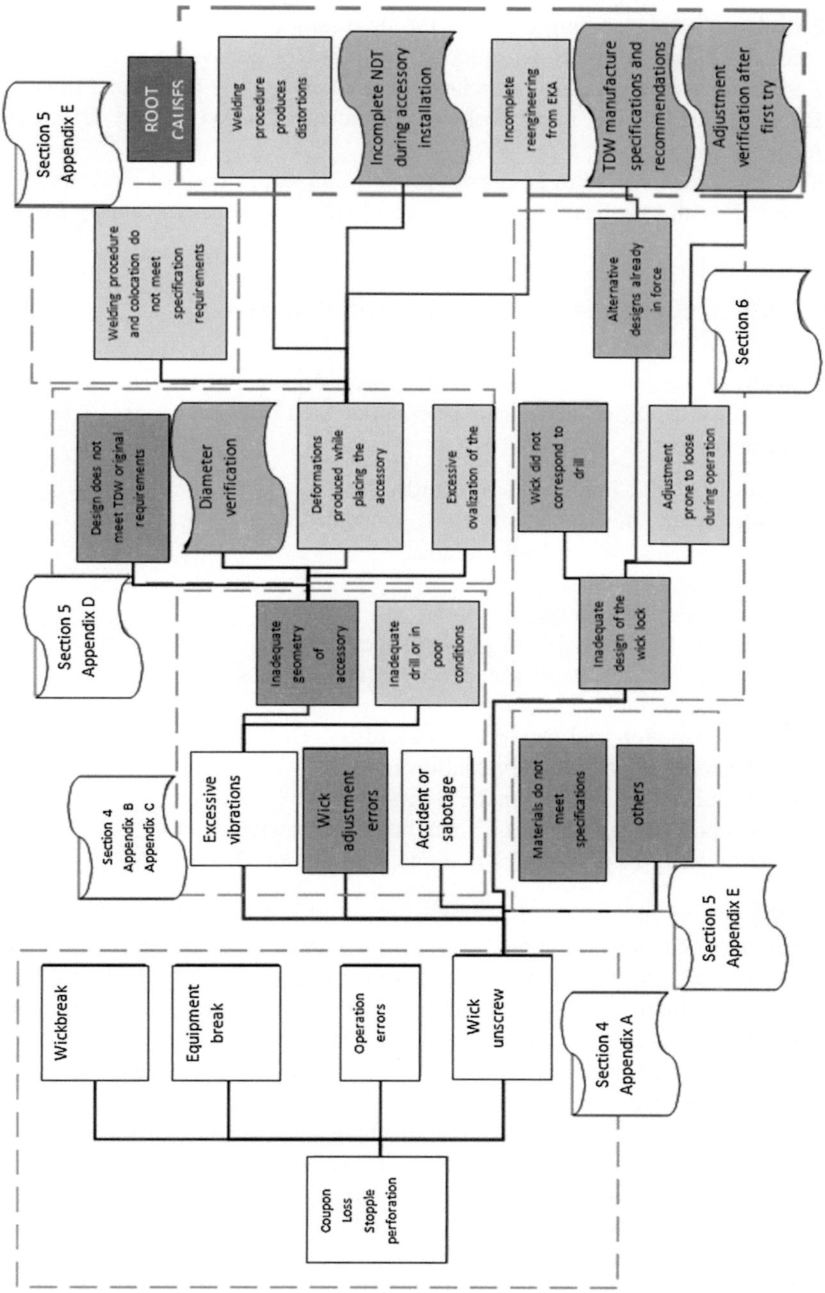

Fig. 7.B6 Cause–effect tree (root causes are indicated)

We conclude that the failure is the added a result of:

1. Shrinkage during cooling down of the field longitudinal weld distorted (ovalized) the body of the fitting, reducing its diameter transverse to the pipe.
2. Post-weld inspection did not include an accurate survey of internal diameters.
3. Rubbing against the inner surface generated vibration in the drill, which loosened the wick.
4. The wick anchor system is unsafe, prone to loosen in case of vibrations. This system had already been replaced in other versions of the same boring machine.

7.6 Data Collection: Documents and Records

The review of records and documents, or parts of documents, relevant to the failure analysis, allow defining origin and dates associated with revisions. Examples of relevant documents:

- Operating records, maintenance records.
- Correspondence, minutes of meetings, work orders.
- Records of inspection and surveillance.
- Equipment process data.
- Procedures and instructions.
- Manufacturer manuals, drawings, and specifications.
- Specification and results of preoperative qualifications.
- Records of equipment history.
- Design information and technical standards.
- Related quality control and evaluation reports.
- Safety requirements, safety analysis reports.
- System reports, trend diagrams, installation parameters.
- Results of analyses of samples (chemical, radiological, etc.).

In addition to the analytical methods discussed in Chaps. 2, 3, and 5, an RCA evaluator must consider additional information related to the physical design of the system. Another important aspect related to the expected results of the analysis is to determine if there is other information or operational expertise in similar events in other facilities. It is convenient to review records from the manufacturer or supplier of the involved equipment to determine if there is further information regarding the problem.

Design parameters: it is essential to understand the design parameters and specifications of the systems associated to a failure. The purpose of design review is to establish the specific operating characteristics of the system. The level of detail required for a design review varies with the type of event, but cannot be omitted in any investigation. Information can be obtained from four sources: equipment identification plates, purchase specifications, vendor specifications, and operation and maintenance manuals. The troubleshooting instructions by the supplier provide information on the most common causes of abnormal behavior. The purpose of design review is to determine design limitations, acceptable operating conditions, possible errors, and indices to quantify the operating condition.

Installation: each machine and system has specific installation criteria that must be met. Acceptable deviations and best practices, as defined in the reference documents, should be noted.

Operating environment: the best approach is to determine all variables and limits used in normal production. For example, define speed range, flow rates, product variation, etc.

Operating procedures: these data should be reviewed to determine actual production practices. The monitoring and control systems provide the best automated database for this part of the evaluation. Two reference sources are needed to complete this task: the design review report and provider manuals. Transients, such as startup, changeup, and shutoff procedures should be carefully evaluated. These procedures often cause deviations in quality and have a direct impact on the reliability of the equipment.

Operational practices: the normal trend is that operators resorting to shortcuts in procedures frequently lead to many problems. The investigation should fully assess actual practices used to operate the machine or system.

Maintenance records: a complete history of actual and scheduled maintenance should be developed, including inspections, lubrication, etc. The main details required include frequencies and types of repair, frequency and types of preventive maintenance, failure history, and any other facts that help in the investigation.

Maintenance practices and procedures: Procedures should be compared with maintenance requirements defined by the design and vendor manuals, and then determine if these procedures are performed in a timely manner and if appropriate techniques are used.

7.7 Rebuttal, Iteration, Coincidence, and Causality

One of the most colorful individuals who advanced investigation methodologies in the nineteenth century was a French detective named Francois Vidocq (1775–1857). Vidocq revolutionized criminal investigation, emphasizing the systematic collection of evidence, preservation of evidence, and analysis techniques. Apparently, he killed a man in a fencing duel when he was 14; later, as a soldier in the Bourbon army, he boasted that he had 15 duels and was repeatedly jailed for gambling, counterfeiting, and similar offenses. Later in his life, Vidocq used his knowledge to invent the method of plaster casts of footprints and impressions at the scene of crime, criminal records, ballistics, disguises, and subterfuge in tracking suspects.

Vidocq created then the first private detective agency, and inspired many writers. Sir Conan Doyle based "Sherlock Holmes" on the exploits of the private detective Vidocq, particularly in relation to his technique for linking the physical evidence. In Les Miserables, by Victor Hugo, Jean Val Jean and Inspector Javert are also based on Vidocq. The same occurs with Edgar Allen Poe's "Murders in the Rue Morgue," the first modern detective story.

In the 1930s, Karl Popper in "The Logic of Scientific Discovery" described how the laws of the physical sciences evolved through a sequence of rebutted hypotheses. He noted that an initial hypothesis with some explanatory value is rebutted (that is, is proven to be false) by verifiable experiences, and replaced by a new one that is better able to explain the known facts. Interestingly, he noted that each new situation that is most explanatory has greatest potential for rebuttal. Popper proposed that a hypothesis is scientific if it is refutable. That is, there must be an experiment or observation to show that the hypothesis is incorrect. If a proposition or hypothesis cannot be refuted by any empirical measure, it is not scientific.

Moral feelings as "murder is a mortal sin" are not scientific because they are not subject to empirical rebuttal. Such statements are simply philosophical, moral, religious, or personal judgments. There should not be any point in the investigation of an accident, failure, event, or phenomenon where a working hypothesis cannot be verified by the facts and evidence cannot withstand scrutiny by rebuttal.

The attributes of a good working hypothesis are:

- All underlying data should be objectively verifiable.
- Hypotheses must be compatible with all relevant data, not only any selected data.
- Scientific principles underlying the hypothesis must be verifiable and repeatable.
- Hypotheses should provide some predictive value.
- Hypotheses must support genuine efforts at rebuttal.

Investigators often get caught up in wanting to confirm their own opinions. They can establish laboratory tests, recreations of the event, or additional tests which results point to support a certain hypothesis. Likewise, they may tend to avoid analyzing data or results that may contradict their hypothesis. This is especially true when time and energy have been already invested in the investigation of certain assumptions, and deadlines and budgets predispose the investigator against the start of a new line of investigation. The expression "no other explanation" and the like are signs that a disguised argument is being used.

A confirmation bias can develop over time and eventually be institutionalized through repeated reinforcements. Investigation funded by the tobacco industry that seemed to support the hypothesis that smoking did not cause health problems was given to advertising, but concealed investigations gave opposite results.

If a person accepts only the facts that support his particular scenario, he can convince himself that the world is flat, that the Holocaust never happened, that astronauts never landed on the moon, that fossil remains of mollusks at the top of the mountains actually grew there, that Bigfoot roams the Rocky Mountains and just poses for photographers who cannot focus a camera, or that the planets and the Sun travel in a perfectly circular orbit around the Earth. A perfectly logical case can be made to support all previous arguments and many more, excluding a few contradictory facts. The defender will try to exclude information contrary to a particular argument by:

- Discrediting the information.
- Discrediting the carrier.
- Drawing up the circumstances that explain the information.

The "sin of a priori," as defined by R. K. Noon, can occur when an investigator is committed to a particular theory of how the failure occurred and then set to test his hypothesis. In other words, he has already decided the answer and simply seeks to obtain sufficient evidence to comply with the forms or regulations.

An important feature associated with the modern version of the scientific method is iteration. The initial hypothesis iteratively evolves as new information is discovered, verified, considered, and compared to the existing body of information. This is an example of why many investigators do not reveal anything out of their team while the investigation is ongoing. Only when investigators are confident that the evaluation is completed, most significant conclusions may be disclosed publicly. A hypothesis is considered a complete reconstruction when it holds that:

- The hypothesis accounts for all verified observations.
- When possible, the hypothesis accurately predicts the existence of additional, yet not known, evidence.
- The hypothesis is consistent with accepted scientific principles, knowledge, and methodologies.

The veracity of a scenario can be considerably enhanced if several independent methods to reach the same conclusion are used simultaneously. Each test is totally independent of each other, but produces the same conclusion.

Occam was a fourteenth century priest, who is mentioned by Noon as the ancestor of the expression: "Keep it simple, stupid!". This means that when two or more scenarios explaining how an event could have occurred, the simplest scenario that fits the facts is usually the correct one. Regarding failure analyses, a person applying Occam's law reduces redundant or unnecessary steps in the scenario, which cannot be verified or refuted. The theory of coherence of truth states that a scenario can only be true if it is logically consistent with other proven scientific theories and verifiable facts.

A cause and effect relationship has a time-ordered sequence, but the converse is not necessarily true. Sometimes two events occur so close in time that it is assumed that the first event, in some way, been the cause of the second event. An apparent orderly time sequence, i.e., coincidence, is not sufficient in itself to demonstrate a cause and effect relationship. The match is essentially a random effect that includes events that occur independently.

It is sometimes argued that there should be a causal link, observing how unlikely a coincidence is. But the law of large numbers says that something that is possible but has a very low probability of occurrence in a given time, will eventually occur if given a sufficiently large number of appearances. To show that something is not just a coincidence, there must be a verifiable causal link between the two events. The burden of proof is to show that the two events are not a coincidence, not the other way around.

The next step from coincidence is correlation. The correlation occurs when there is a demonstrable relationship between two events. When event A occurs, event B also seems to occur with some regularity. Indeed, the correlation between A and B can provide tested and repeatable predictions. The correlations may indicate that there is a direct cause and effect relationship between two events, but may also indicate that the two events simply share a common factor. For example, R. K. Noon mentions correlations between failures in air conditioning and pregnancy. While one obviously does not cause the other, what they have in common are the blackouts. Voltage fluctuations and failures occur in air conditioning compressors, and increase the rate of major failures. Also, blackouts that leave people in the dark promote caresses and other displays of affection, which in turn result in a greater number of births 9 months after the event.

7.7.1 Example 7.C RCA of Failure in Directional River Crossing

An analysis in which the search for factual evidence was predominant for the resolution of an RCA is presented in this example. This is a horizontal directional drilling (HDD) crossing of a 36″ pipeline under a large river. Nominal thickness for this pipeline is 8 mm; following specifications, thickness of the pipe string to be introduced in tunnel was 12 mm. The crossing includes the creation of a tunnel more than 500 m long. The cost of the failure was the total loss of the crossing, and the need to repeat the process in another location, resulting in the loss of a few million dollars. In this case, the RCA is a consequence of a project whose initial objective was to determine whether the crossing could be recovered after the failure, and thus to reduce the economic consequences of the failure. Therefore, this project involved three activities:

- Inspection and audit to ascertain the condition of the crossing.
- Definition of causes that led to the event.
- Definition of potential remedial alternatives.

We will not discuss the process of rebuttal, due to the complexity of the causal tree; let us just present the most important conclusions. Figure 7.C1a shows a schematic elevation of the proposed tunnel, the horizontal scale is greatly reduced. It indicates the riverbed, the trace of the tunnel and the forward direction of the

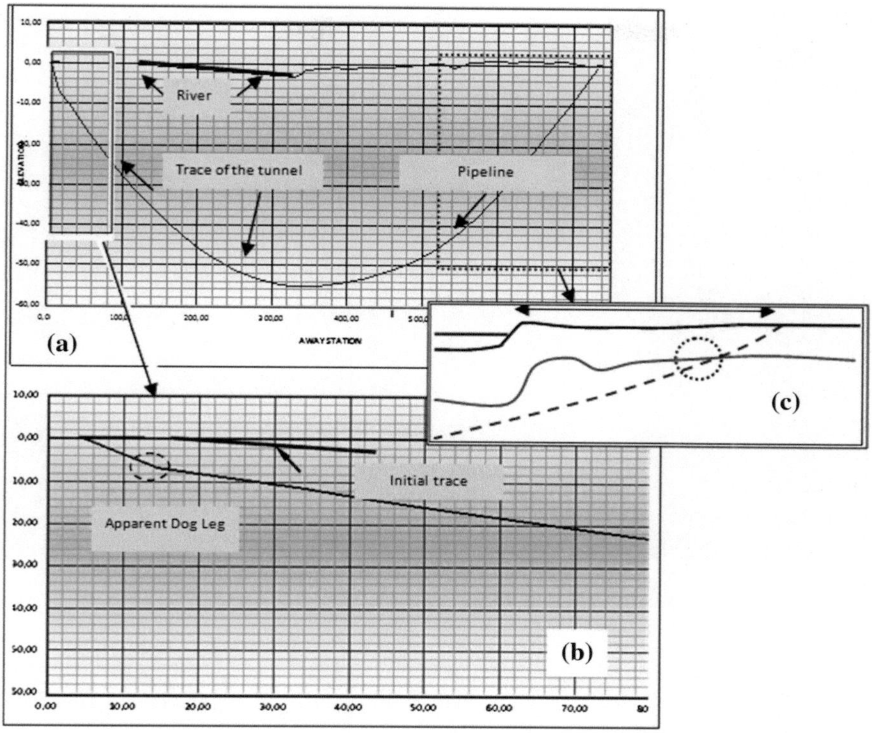

Fig. 7.C1 **a** Sketch in elevation of the HDD tunnel (scales in m), **b** detail at south bank, **c** detail at north bank

pipe being pulled once the tunnel was completed. Figure 7.C1b shows a detail of the boxed area, where digging of the tunnel began. Figure 7.C1c shows a detail of the delineated area, where the pipe was introduced into the finished tunnel.

Figure 7.C2 shows the drilling machine and the drill shaft. The machine generates an axial thrust and rotation of the drill ("reamer"). When operations started, the machine drew off due to bad conditions of the topsoil, which is why the initial trace was abandoned and the operation was restarted from a place several meters further from the south river bank (see detail in Fig. 7.C1b).

The first tunnel (smallest diameter) is the most difficult. It is directed by the directionality of the mud jet nozzles located injected at the perimeter of the drill, which in turn rotates and advances. Figure 7.C3a shows in some detail the ending of the second tunnel boring, now with the shaft in tension. While a centering device tries to keep each new diameter coaxial with the former (Fig. 7.C3a), gravity and tunnel curvature force the section of the tunnel accept a certain degree of lack of vertical roundness, as shown in Fig. 7.C3b. Once the tunnel was finished, the pipe was pulled from the opposite bank of the pipe, where the pipe string had been previously girth welded and placed in rollers and winches, see Fig. 7.C4.

Fig. 7.C2 Drilling machine and drill shaft

Fig. 7.C3 Finishing second step (diameter) of tunnel. Insert: Scheme of gradual enlargement of the tunnel

Fig. 7.C4 Pipe string in the opposite side of the river

Figure 7.C5 shows (finally!) the undesired event: once the pipe came up at the end of the tunnel on the south bank, it was observed that it had collapsed by buckling. First thing was the attempt to save the operation, so, a method to determine the extent of the collapse in the length of the pipe within the tunnel was defined. Geo-referenced data indicated an excessive curve ("dogleg") near the exit point of the pipe, see Fig. 7.C1b. If the problem had begun there, it would have been possible to recover the work with an excavation in the south bank. An internal gauge was introduced from the opposite end, but it got stuck a few meters from the north bank end. The circle in Fig. 7.C1c shows this position, which coincides with the interface between the two layers of soil: clay layer (lower) and loose gravel and sand (top), product of fluvial debris over centuries. This interface was the problem that led to the dogleg on the other side, too.

So that there was a high probability that almost the entire length of the pipe were collapsed. Numerical modeling was developed, which allowed determining that the conditions for bending, torsion and axial tension in that area of the pipe were sufficient to cause local collapse, see Fig. 7.C6. The onset of lateral collapsing strongly depends on the thickness of the tube. Once started, the collapse probably was spreading as the tube moved. The evidence observed at the site of the crossing and mechanical models developed allowed to define the collapse of the pipe by combined effects of external pressure and bending and tensile loads in a large localized tunnel curvature as the immediate cause of the incident. While there is no physical evidence (pipe and tunnel had to be abandoned) it is considered that the entire length of the tube from the point shown in Fig. 7.C1c is collapsed.

Fig. 7.C5 Undesired event: collapse by lateral buckling at the end of the tunnel

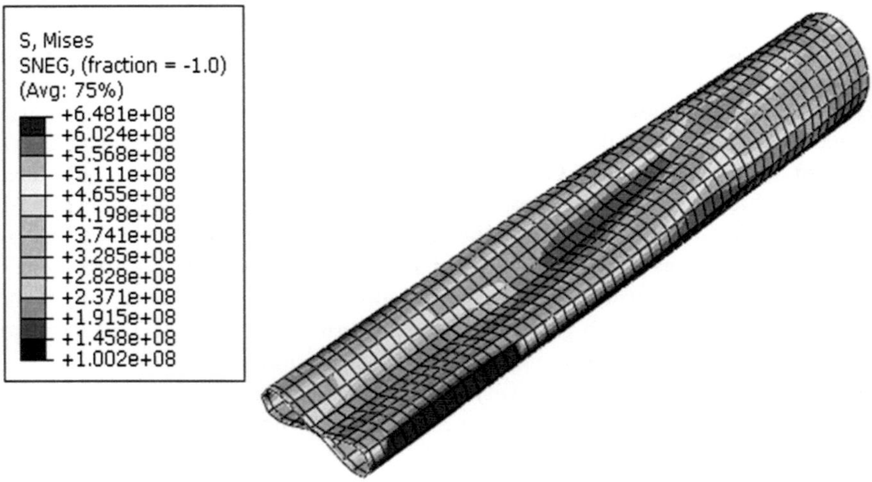

Fig. 7.C6 Modeling of local buckling and post-collapse propagation

Witnesses and data were vetted to define the likely origin of the discontinuities in the tunnel that led to buckling collapse of the pipe. When insertion of the pipe string into the tunnel at the north end was first attempted, a rupture occurred in one of the brackets of the column (Fig. 7.C4). The fall of the pipe caused a small soil slide in the mouth of the tunnel, which had to be cleaned by passing an internal cleaning tool (a reamer).

It was also discovered that increasing the length of the tunnel (Fig. 7.C1b) forced to place two additional tubes on the string. No more 12 mm thick tubes were stocked on site, so it was rush decided to use two 8 mm thick tubes. These were placed one on each extreme of the string, meaning that the first tube entering the tunnel was thinner than recommended by engineering. Numerical modeling found that buckling stress in this thinner first tube was two and half times more critical than in the rest of the column. Modeling also proved that the influence of the (12 m long) thinner tube in the resistance to collapse was equivalent to the entire column being of reduced thickness.

The position on the tunnel of the collapse site, as found by probing after failure, roughly coincides with the site of collapse predicted by the mechanical model. This is also the approximate stop position for the cleaning tool introduced to open up again the tunnel, as witnesses reported. All this evidence indicates that at this position stones may have accumulated at the bottom of the tunnel, after having been pushed during "cleaning." This position is consistent with the highest peak of mud pressure during pull back, and with the highest value of torque in the reamer. This section also recorded peak tensile load.

Vertical tunnel ovalization allowed the flattened tube to accommodate in a vertical plane within the tunnel. Because of this and because lubrication conditions were very good, the collapse was not detected by an increase of torque, mud pressure, and tension above allowable limits.

Figure 7.C7 shows the complete cause and effect tree, and the definition of root causes. After evaluating management and construction practices, the geometry of the tunnel, drilling records of torque, mud, and tension, there are no data to show that the construction was not in line with industry standards for HDD.

Two previous incidents had a definite influence in the chain of events that led to the failure, both related to the difficult surface layer of gravel. First, the need to restart the pilot bore which led to the addition of one pipe at each end of the string. Second, the failure of a tube hanger made the pipe front to impact the earth so close to the entrance of the tunnel that a small collapse of the gravel occurred. This then led to loose gravel or other discontinuities being left in the tunnel, which led to the increase in stresses in the pipe.

These are fortuitous accidents, but there are two stages at which quick decisions were made, in the heat of the process, that did not follow prescribed engineering procedures. The first and more obvious is the use of two sacrifice pipes, which was related to the need for more pipes than originally calculated. The decisions to use the available thinner pipes, and then to place one in each end of the string, were made ignoring the criticality of this condition.

The second failed decision, not so obvious but also important, was to pass the reamer along a short length of the tunnel, after the small collapse at the entrance. It is clear now and should have been clear then, that any debris pushed by the reamer

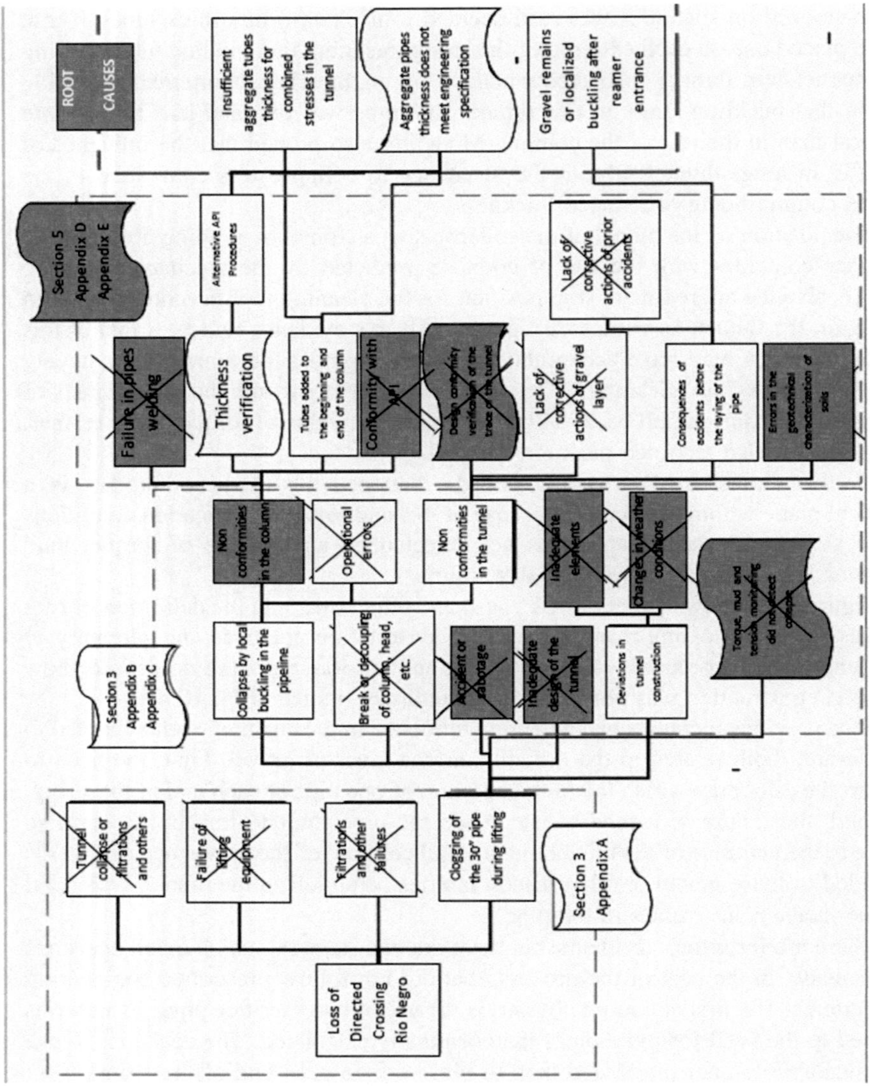

Fig. 7.C7 Cause–effect tree, and definition of root causes

would accumulate at the site where the reamer would be stopped. Once launched, the reaming should always be pushed all the way through the other side. Both decisions were made in order to keep a timetable. As is many times the case, engineers are pressed within their companies to reduce costs and times, but it is their duty to avoid conditions that could violate standards or proven procedures.

7.8 Levels in Failure Analyses

A thorough RCA of a failure should be limited to cases that really justify the cost. Even a simple analysis requires an investigator assigned to the project until it is resolved. Furthermore, the analysis requires the participation of all plant personnel directly or indirectly involved in the incident. Generally, the investigator must make numerous interviews, and must review many documents to extract relevant information.

RCA analysis is usually performed on random or non-recurring events. The problems that often justify the use of this method include equipment, machinery, or system failures, and deviations from operational and economic performance, safety and regulatory issues. Its use should be limited to those incidents or events that have a measurable negative impact in terms of performance, personal safety or compliance.

Some questions to ask here are: What is the impact? And will it happen again? Although the logical question of how we prevent recurrence usually cannot be answered until the analysis is complete, we may just have to learn how to minimize the impact if correcting the problem is not economically viable.

A recommended practice is being developed by API (RP 585, "Pressure Equipment Integrity Investigation") that will be applicable in Petrochemical, oil and gas plants. This recommended practice is to be published in 2014. Its objectives are to clarify the general elements of a Failure Investigation Program and complement existing programs. Not limited to a single RCA methodology, this RP provides guidelines applicable to all methods. The RP defines a list of target equipment in a process plant, related with pressure vessels, piping, rotating equipment, and other components under pressure. Three types of events are defined:

- Failures.
- Quasi-accidents.
- Deficiencies in equipment.

A central aspect of incidents is that they differ in frequency and severity of their consequences; they can be chronic or sporadic. Failures are defined as incidents of greater severity. Incident definitions include:

- Leak, fire, structural failure, etc.
- Quasi-accident (near-miss).
- Increased damage than expected, discovered in a routine inspection.

- Loss of thickness greater than allowable (no leakage).
- Pressure Relief Device out of service, detected in routine inspection.
- Etc.

The concept is that every incident should be investigated, but the depth of the analysis depends on its importance. Thus, a risk is defined depending on the consequences (Real and Potential); and the risk defines the level of analysis:

Level 3: A single event with large consequences (actual or potential).
Level 2: Isolated event with consequences of lesser magnitude, or group of incidents with high potential consequences.
Level 1: Quasi-accidents (less detail).

API RP 585 defines that every organization must have a classification matrix, as shown in Table 7.1. This will allow each member to know, in advance, how to proceed with an accident. API RP 585 defines the formation of the task team and the type of analyses to be performed, for each level of analysis: team leader, team members, sponsor, methods, investigation extension, type and extent of report.

For Levels 2 and 3, which require evaluation of physical evidence, API RP 585 develops recommendations for the preservation of evidence, who should collect it, when and how. The order of collection depends on how perishable the information is. Collecting information should NOT include trying to figure out why the accident happened. Following types of evidence are defined in order of priority for collecting testimonies, physical evidence, and documentation.

Testimony: this evidence is perishable. The evidence should be collected before people begin to talk to each other. As mentioned in Chap. 2, first eye witnesses should if possible remain isolated until questioning; they could write what happened in the meantime.

Physical evidence: provides general guidelines, as discussed in Chap. 2:

- Preserve the damaged equipment.
- Do not distribute or manipulate fracture surfaces.
- Do not clean damaged parts unless strictly necessary, to allow chemical analysis.

Depending on the findings, RP585 makes recommendations aimed at preventing future events. It is emphasized that the commission does not analyze the cost of each recommendation, this is undertaken by management. The commission assigns a responsible for the implementation of each recommendation and a deadline. The final report should then contain answers to the three questions of the prologue of this book: What happened? How? Why? As we have seen, the report should clearly define Immediate Causes, Root Causes, and Recommendations.

Table 7.1 Draft classification matrix

API RP 585	Level 1 investigations	Level 2 investigations	Level 3 investigations
Incident characteristics:	1. Condition or damage found that if it had been allowed to progress would have led to loss of containment before the next scheduled outage or inspection interval 2. Discovered pressure equipment integrity damage significantly beyond expectation, no loss of containment 3. Very small leaks from pressure equipment or joints that were easily contained	1. Leak from pressure equipment that resulted in small fire or equipment damage only to the immediate area 2. Unexpected pressure equipment failure from damage mechanisms 3. Unexpected pressure equipment damage discovered that required unit shutdown 4. Repetitive Level 1 type incidents in the same process or system	1. Leak or rupture from pressure equipment that resulted in a large fire or equipment damage. 2. Leak or loss of containment from pressure equipment that was beyond regulatory reporting requirements
Investigation characteristics:	Near miss report. Investigate using simple root cause tools	Investigate using company or department formal root cause analysis	Investigate using company or department formal root cause analysis
Team makeup recommended For Level 2, 3 Leader must be formally trained in a RCA method	Investigated by the pressure equipment integrity personnel from the affected area No team required, can be investigated by one person Involve subject matter expert as needed	Leader would be someone trained form the affected area Team members would include 1 or 2 others of different disciplines from the area; include a pressure equipment integrity person Involve subject matter expert as needed	Leader would be someone trained and from another area of the plant or business unit Team members should be at least 3 and from different disciplines such as inspection, operations, process safety Subject matter expert should be included on the team
Initiation	Immediately	Within 48 h	Within 48 h
Sponsorship	Supervisor of leader	Department head	Plant manager
Reporting	Near miss report format	Formal report using prescribed form	Formal report using prescribed form

Finally, this Recommended Practice defines methodologies for dissemination of the findings and a practice for auditing and continuous improvement of the program.

7.9 Alternative Formats for a Failure Investigation

Table 7.2 summarizes the most common methods for causal analysis. Each is advisable to certain circumstances, as described below.

Analysis of events and causal factors: identifies the sequence of a series of tasks or actions and the surrounding circumstances that led to an occurrence. The results are shown in a graph that provides a picture of the relationships of events and causal factors.

Change analysis: used when the problem is obscure. It is a systematic process that is generally used for a single occurrence and focuses on the elements that have changed.

Barrier analysis: is a systematic process that can be used to identify physical, administrative, and procedural obstacles or controls that should have prevented the occurrence.

Management oversight and risk tree analysis (MORT): MORT (and Mini-MORT) are used to identify deficiencies in barriers and controls, support functions and management. MORT identifies the specific factors associated with the development of an event and identifies management factors that allowed these factors to exist.

Human performance assessment: identifies the factors that influence performance in tasks. The focus of this analysis method is in operation, working environment and management factors. Methods to study human–machine interface to improve performance take precedence over disciplinary measures.

Kepner–Tregoe: for problems and decision-making, provides a systematic framework to collect, organize, and evaluate information and applies to all phases of the investigation process. Its focus on each phase helps keep them separated. The root cause phase is similar to the change analysis. The Kepner–Tregoe method, commonly called K–T method is a rational structure that maximizes the critical thinking skills of key people. K–T includes:

- Assessment of the situation.
- Analysis of problems.
- Decision Analysis.
- Analysis of potential problems.

The remarkable success in marketing the K–T method has generated many competitors. There seems not to be a best not to be a method. Each has its good points and weak points, depending on the particular application, the instructor and the ability of the person to apply the particular method. Some seem to work better for management problems, and others for equipment problems.

There are many packaged and sold investigation methodologies. Most commercial methods come with training manuals, and with their own jargon. There are professionals who teach the methods. These investigation methodologies offer formatted flowcharts and worksheets for interim and final reports and a series of publications that help the novice in a particular methodology. Some also offer

Table 7.2 Most common methods for causal analysis

Method	When to use	Advantages	Disadvantages	Remarks
Events and causal factor analysis	Use for multi-faceted problems with long or complex causal factor chain	Provides visual display of analysis process. Identifies probable contributors to the condition	Time-consuming and requires familiarity with process to be effective	Requires a broad perspective of the event to identify unrelated problems. Helps to identify where deviations occurred from acceptable methods
Change analysis	Use when cause is obscure. Especially useful in evaluating equipment failures	Simple 6-step process	Limited value because of the danger of accepting wrong, "obvious" answer	A singular problem technique that can be used in support of a larger investigation. All root causes may not be identified
Barrier analysis	Use to identify barrier and equipment failures and procedural or administrative problems	Provides systematic approach	Requires familiarity with process to be effective	This process is based on the MORT hazard/target concept
MORT/Mini-MORT	Use when there is a shortage of experts to ask the right questions and whenever the problem is a recurring one. Helpful in solving programmatic problems	Can be used with limited prior training. Provides a list of questions for specific control and management factors	May only identify area of cause, not specific causes	If this process fails to identify problem areas, seeks additional help or use cause-and-effect analysis
Human performance evaluations (HPE)	Use whenever people have been identified as being involved in the problem cause	Thorough analysis	None if process is closely followed	Requires HPE training
Kepner–Tregoe	Use for major concerns where all aspects need thorough analysis	Highly structured approach focuses on all aspects of the occurrence and problem resolution	More comprehensive than may be needed	Requires Kepner–Tregoe training

processing flowcharts and producing comprehensive reports, formatted according to the particular methodology. Some also provide some advanced training classes where a person can become a certified trainer in a particular method.

Most popular investigation methodologies commercialized today are Proact®, Kepner–Tregoe® (i.e., Project eThink® and Logic®), Apollo® (i.e., RealityCharting®), and Tap Root®. Other training materials for RCA are known by the name of the instructor who developed the materials, for example, FPI Chiu Chong International®. However, this is just a small list. There are too many methods to mention by name.

It is advisable to avoid proprietary investigation methods that do not use the scientific method, or when assumptions implicit in their analytical techniques do not pursue rebuttal tests and do not allow to easily define initial hypotheses that can evolve as evidence is collected. Without these fundamentals, the report resulting from the investigation can be a disjointed collection of facts and comments strictly logically packaged into a nice format.

7.9.1 Example 7.D RCA of Fire in a Heater at a Petrochemical Plant

This example describes a case in which most failure causes are not conditions or events related to the materials and operation, but with the control systems. This is a cabin-type furnace used for heating a hydrocarbon flowing through tubes around the walls and the "roof" of the furnace. This furnace suffered an explosion followed by fire after less than 2 years of operation, see Fig. 7.D1a, b. Figure 7.D2 shows the initial part of the causal tree.

Background and testimonial data verified that the blast was the most important event of a number of problems related to control of the flame and re-ignition and even earlier explosions, though smaller and with less impact. The furnace had problems with the gas supply from the beginning; it was designed for natural gas. Early on, the idea ("myth"?) was conceived that potential pockets of CO_2 in the well gas used as fuel could generate flame extinction for lack of fuel.

The control of general operation of the furnace, sequence of tasks and events was made via a program executed by a PLC. Modifications to the original programming of the PLC began early in the life of the equipment. Modifications were made to program procedures for re-igniting, startup, and control. PLC programming was modified to allow immediate reigniting the furnace without the prior venting and gas purge, as recommended by the manufacturer. The version of the control software as analyzed after the failure did not match the version of the program that was supposedly installed after last modification. Additionally, it was found that, at the time of failure, some sensors were out of service.

Fig. 7.D1 **a**, **b** Furnace that suffered explosion followed by fire

Fig. 7.D2 Initial part of the cause–effect tree

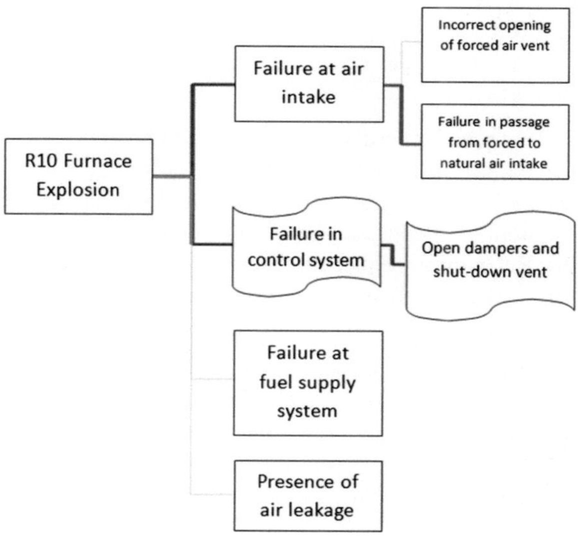

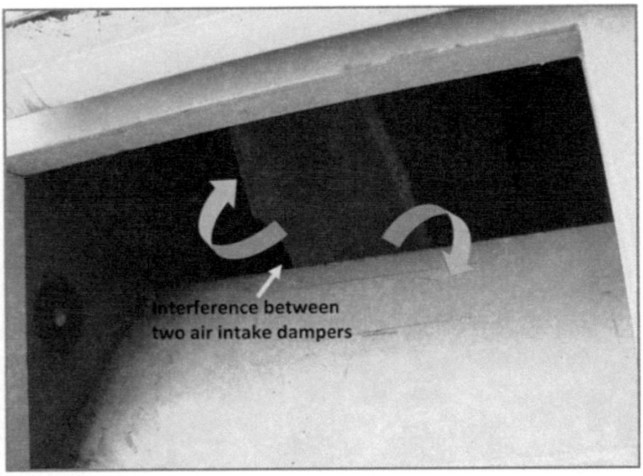

Fig. 7.D3 Gates of natural and forced air interfere between themselves

Let us go straight to Monday's newspaper (that is, conclusions of the RCA): A constructive detail was key in the sequence of events leading to the failure. This is a system of "dampers" or gates for combustion air intake. The furnace had a system to toggle forced air or natural air intakes. The two gates are illustrated in Fig. 7.D3. When both dampers were in an intermediate position between open and closed, there was interference. This was "corrected" during assembly by grinding the edge of one of the dampers (see arrow in Fig. 7.D3). But this recess was insufficient. Now the gates not only collided but could be locked in an intermediate position.

The immediate cause of the incident is supported by the occurrence of the following:

(a) Frequent off calls due to poor combustion.
(b) Failure of forced air intake fan.
(c) PLC programming error that inhibited automatic passage from forced to natural air intake in case of fan failure.
(d) Re-ignition sequence unsafe in case of lack of flame (10 attempts for 1 min).
(e) Lack of control, alarm, and stop in case of dangerous conditions.

The complete causal tree and the investigation of all rebutted causes will not be detailed now; interested reader can go to Ref. [19].

The combination of several faults led to the failure. The faulty passage from forced to natural air intake led to detonation because control and safety systems also failed. The successive changes in the programs in the PLC resulted in the following inconsistencies:

- Error switching from forced to natural intake.
- Burner gas pressure was at maximum during restart.
- Pilot ignition sequence was different from design.
- Sensing error in natural air inlet damper.

In previous versions, the PLC program was designed to arrest in the case of a flame failure during operation. Later versions included a restart sequence that energized the ignition transformer to immediately recover the flame.

Due to a programming error, the execution of the restart sequence caused the deactivation of forced to natural air intake. As a consequence, in case of flame off, the furnace was rekindled. Then, a failure in the forced air intake caused fuel intake to continue to operate without air; thus inhibiting the formation of the flame as air–fuel mixture was too rich, above the upper limit of ignition.

Under these conditions, with no flame and full throttle, gas accumulated in the furnace during re-ignition attempts. This situation was dangerous because the atmosphere inside the furnace could enter the unstable range with an unexpected air intake (for example if the force air failure was intermittent).

References

1. H. Petroski, *To Forgive Design: Understanding Failure* (Harvard University Press, 2012)
2. R.K. Noon, *Scientific Method, Applications in Failure Investigation and Forensic Science* (CRC Press/Taylor and Francis, Boca Raton, 2009), ISBN 978–1–4200–9280–6
3. R.K. Mobley, *Root Cause Failure Analysis* (Butterworth-Heinemann, Oxford, 1999), ISBN 0-7506-7 158-0
4. U.S. Department of Energy, Office of Nuclear Safety Policy and Standards, *Root Cause Analysis: Guidance Document*, (1992), DOE-NE-STD-1004-92
5. American Petroleum Institute, *Pressure Equipment Integrity Investigation*, Draft 8, API RP 585 (2011)
6. U.S. Department of Energy, *Occurrence Reporting and Processing of Operations Information*, DOE Order 5000.3A, May 30 (1990)
7. J.L. Burton, *Method Identifies Root Causes of Nuclear Reactor Scrams* (Power Engineering, October 1987)
8. D.L. Gano, *Root Cause and How to Find It* (Nuclear News, August 1987)
9. R.J. Nertney, J.D. Cornelison, W.A. Trost, *Root Cause Analysis of Performance Indicators, WP-21* (System Safety Development Center, EG and G Idaho Inc., Idaho Falls, 1989)
10. J.D. Cornelison, *MORT-Based Root Cause Analysis, WP-27* (System Safety Development Center, EG&G Idaho Inc., Idaho Falls, 1989)
11. J.R. Buys, J.L. Clark, *Events and Causal Factors Charting, SSDC-14* (System Safety Development Center, EG&G Idaho Inc., Idaho Falls, 1978)
12. W.A. Trost, R.J. Nertney, *Barrier Analysis, SSDC-29* (System Safety Development Center, EG&G Idaho Inc., Idaho Falls, 1985)
13. M.G. Bullock, *Change Control and Analysis, SSDC-21* (System Safety Development Center, EG&G Idaho Inc., Idaho Falls, 1981)
14. G.J. Briscoe, *MORT-Based Risk Management, WP-28 (SSDC)* (EG&G Idaho Inc., Idaho Falls, 1991)
15. J.L. Harbour, S.G. Hill, *HSYS, A Methodology for Analyzing Human Performance in Operational Settings*, Draft, EGG-HFRU-3306 (1990)
16. C.H. Kepner, B.B. Tregoe, *The New Rational Manager* (Princeton Investigation Press, Princeton, 1981)

17. R.J. Latino, K.C. Latino, *Root Cause Analysis: Improving Performance for Bottom-Line Results*, 3rd edn. (CRC/Taylor Francis, Boca Raton, 2010)
18. P. Wilson, *Root Cause Analysis* (Oxford Press, 1999), ISBN 970613526X
19. J. Booman, M.O. Gonzalez, J.L. Otegui, Failures in cabin type hydrocarbon heaters due to inadequate fuel control. Eng. Fail. Anal. **21**, 31–39 (2012), ISSN 1350-630

Chapter 8
Damage and Failure Mechanisms in Machinery

8.1 Definition of Failure in Machinery

A machine (from Latin *machina*), within the broad concept used in this text, is a set of parts or fixed and mobile elements which operation permits to use, direct, regulate or transform energy. Machines consist of sets of mechanisms of a greater or lesser degree of complexity, which together present a single degree of freedom, so that machine operation is uniquely determined by a resulting action. Thus, machines have mobile and fixed elements supported or connected to a fixed structure to secure adequate rigidity and stability. In a machine, as in other more or less complex systems, the definition of failure is given from:

- *Function*: when the machine fails to function successfully because it has a decreased performance in terms of production capacity, quality, or opportunity.
- *Reliability*: when the machine performs its duty but does so with recurrence of events that determine detentions or operation has risks above levels considered acceptable for operation. One of the parameters to measure the reliability of a machine is the mean time between failures (MTBF). From this point of view a failure is considered a detention of the machine due to causes that could be solved by replacement or repair of parts and require workers with skills to recognize and solve the problem. These are corrective actions, involving the intervention of the machine to a greater or lesser degree. Thus failures differ from unscheduled stops, which are recovered by the intervention or assistance of operating or maintenance personnel without replacement of parts beyond normal consumables.
- *Structural integrity of the machine*: when damage is detected in structural parts or the machine, and if progresses would affect the operation and possibly lead to a catastrophic failure that would affect most important and costly parts of the equipment.

Machines are more or less complex elements, consisting of assemblies and subassemblies, mechanical, hydraulic, pneumatic, electrical, and electronic systems, and are constructed in many different metallic and nonmetallic materials.

J. L. Otegui, *Failure Analysis*, DOI: 10.1007/978-3-319-03910-7_8,

This, and the fact that the different parts are in contact with the most diverse fluids and environmental factors, make it impossible to approach the study of failure modes in individual machines. However, there is a set of elements common to most machines that regardless their application have failure modes that are unique to them, and can therefore be studied in a generic way. These elements are, for example, shafts, transmission components (gears, chains, and belts), bushings and bearings, fasteners (threaded bolts, rivets, pins and keys), rotors in rotating machines such as centrifugal pumps, rotary compressors, etc.

There are techniques to detect premature failures and prevent further damage, called predictive maintenance techniques, which are analyzed in some detail in Chap. 9. These techniques include real-time monitoring of operating variables, vibration analyses, scheduled and properly documented visual inspections. In case of a total or partial failure in a machine, the comprehensive assessment of the failed system, that is, a complete failure analysis, is extremely important to determine the mechanism of failure and its causes.

8.2 Modes of Failure in Shafts

We consider here "shafts" as rotating elements which are subjected to torsion and bending moments. Shafts are linked to the main mechanism through transmission elements and to the support structure by bearings. For structural and economic reasons, gears, pulleys, and arms may be a continuous part of the shaft itself. Crankshafts and camshafts are exception, since they are not merely cylindrical shapes because journals and cams are an integral part of the overall machine kinematics. Shafts that are subject primarily to torsion loads are elements mainly designed to transmit power.

The most critical points on shafts are geometric discontinuities that are generated at diameter changes, keyways, slots for seals and gaskets, and oil holes. Figure 8.1 shows the subcritical propagation of fatigue cracks from geometric concentrators in a shaft keyway and in the fillet of a crankshaft. More characteristic failure modes in shafts are summarized in Table 8.1. The column "Ref" indicates typical locations of occurrence of each type of failure, according to the scheme shown in Fig. 8.2.

8.2.1 Example 8.A Failure of Shaft in a Hydraulic Pump

A fault was detected in the control shaft of a hydraulic pump that powers fishing net maneuvers in a fishing vessel. Inspection showed clear evidence of fatigue propagation of a crack originating from the bottom of the keyway (Fig. 8.A1). From this initiation site, beach marks are clearly indicative of successive stages of crack propagation. The shaft also had signs of material loss by fretting, both in the

Fig. 8.1 Fatigue crack propagation from geometrical concentrators

shaft and in the mating part in the cube. A rundown of several millimeters in diameter is a sign of having been working loose for long.

The shaft failure occurred by fatigue initiated at the bottom of the keyway. This way of fracture propagation is typical in such assemblies and well documented in the literature (Fig. 8.A2). As seen in Chap. 4, fatigue is due to the combination of variable loads and a high stress state due to the presence of the stress concentrator, in this case, the keyway. Both the shaft and the hub have evidence of having worked with excessive clearance, exaggerated in the later stages of damage by material loss due to fretting, Fig. 8.A3.

8.3 Failures of Bearings

Bearings are mechanical elements that have evolved greatly, achieving a very wide range of applications in all industries with high reliability. The service performance of bearings depends strongly upon the conditions of assembly, lubrication, sealing of the rolling element chamber, and environmental service conditions. Therefore, failure analyses of bearings should consider not only the conditions of the bearing, but also the context of all the above factors and machine parts related to their function. Failure modes of bearings are typified in different standards, such as ISO 243, which defines damage as a function of visual characterization, in each case making reference to a number of causes. Manuals from bearing manufacturers provide extensive documentation with images of different failure modes (Table 8.2).

Failure modes that are characteristic to bearings, which do not occur in other machine elements, are electrical erosion, false brinelling, and indentation. The growing use of variable speed drives and programmable control electric motors have produced an increase in sparking (or arcing) failures in bearings, due to high frequency currents that circulate between the rotor and the housing through the rolling elements.

Table 8.1 Characteristic failure modes in shafts

Failure mode	Cause	Ref.	Characteristic
Bending brittle fracture	Overloading in relation to material strength	1	Smooth and granular in appearance, transverse to the shaft, without plastic deformation
Torsional brittle fracture	Overloading in relation to material strength	2	Helicoidally shaped, no plastic deformation
Torsional ductile fracture	Overloading in relation to material strength	3	Fibrous, circumferential direction, transverse to the shaft
Plastic deformation by bending and torsion	Break of parts related to the shaft, abrupt stop, or catastrophic failure of the machine	4	Visible to the naked eye, there are broken parts still in place
Fatigue Torsion	Alternative loads, stress concentrators, metallurgic, or geometric initiators	5	Beach marks, to 45°, with a section of brittle or ductile fracture in the remaining ligament.
Deflection			Beach marks, transverse plane, with a section of brittle or ductile fracture in the remaining ligament.
Buckling by torsion or deflection	Very thin wall section axes, resonance, combined states overload	6	Plastic collapse of the tube wall of the shaft. Wall fractures, oblique or longitudinal. It occurs in materials with very thin walls.
Abrasive and/or adhesive wear	In sliding parts, lubrication failure or contamination, low speed, overload, improper bushing material	7	Grated surface or with material adhesion, signs of heat damage, corrosion materials. There are usually cracks associated with this damage
Fretting	Parts in contact with relative displacement, lack of interference during assembly, high alternating loads.	8	Erosion in surfaces with material transfer, powder waste or mixed with lubricant, reddish or brown
Corrosion, pitting	Part of the axes in contact with corrosive fluids, confined spaces, or degraded lubricant	9	
Surface fatigue or pitting	Shaft surfaces bearing rolling elements, the shaft acts as bearing inner race	10	
Plastic deformation in multiple keyways	Worn striation overload, poor heat treatment	11	Striations are distorted in the contact limit plane
Keyway fatigue	Lack of fit between the shaft and the transmission element, small radius at the bottom of the keyway	12	Fracture progresses by fatigue parallel to the surface, in the form of peeling

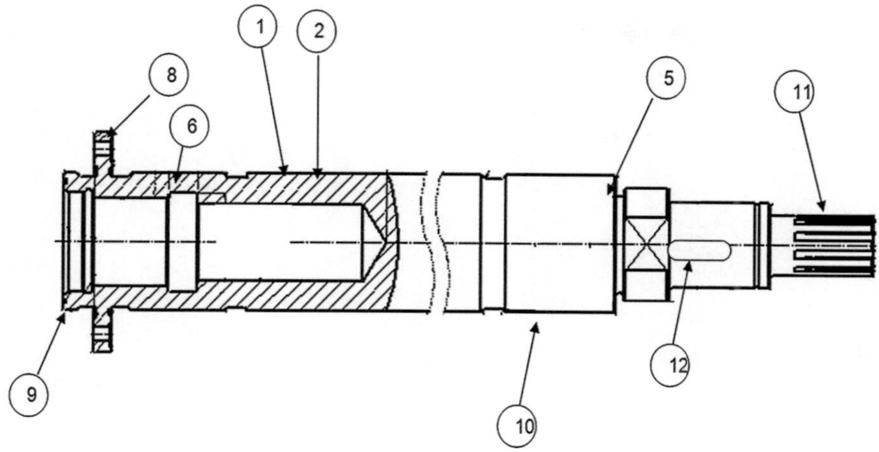

Fig. 8.2 Typical locations for occurrence of failures listed in Table 8.1

Fig. 8.A1 Failure in a control shaft of a hydraulic pump in a fishing vessel

Fig. 8.A2 Metals handbook reference for this type of crack: fatigue initiation at the *bottom* of the keyway

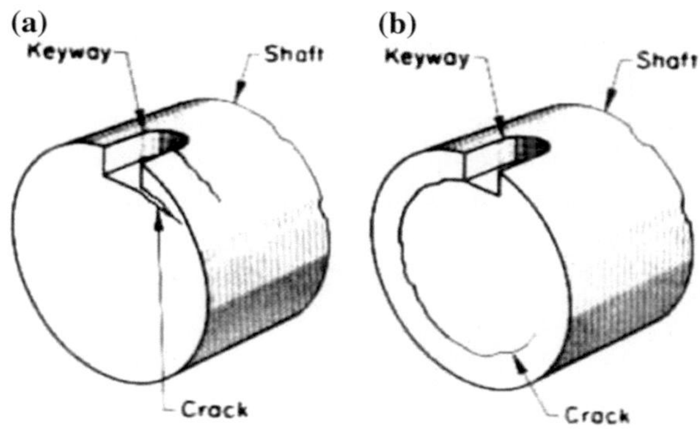

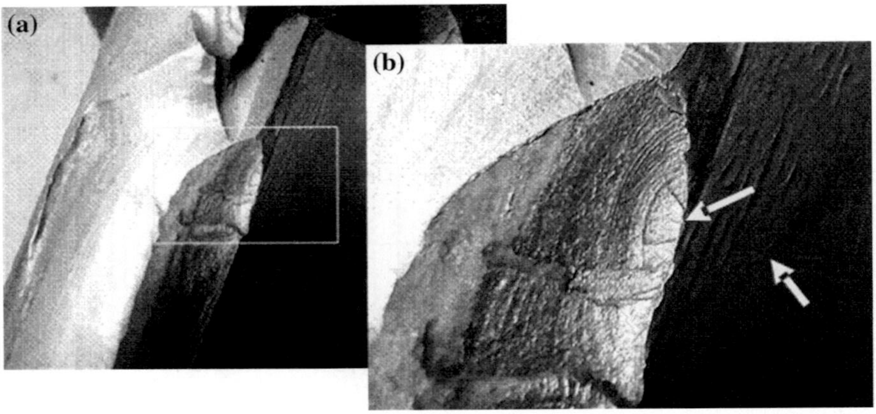

Fig. 8.A3 **a** Crack initiation site and further fatigue propagation **b** Marks left by crack in the keyway

Table 8.2 Failure modes in bearings

1. Fatigue	1.1 Subsurface fatigue	
	1.2 Surface-initiated fatigue	
2. Wear	2.1 Abrasive wear	
	2.2 Adhesive wear	
3. Corrosion	3.1 Moisture corrosion	
	3.2 Frictional corrosion	3.2.1 Fretting corrosion
		3.2.2 False brinelling
4. Electrical erosion	4.1 Excessive voltage	
	4.2 Current leakage	
5. Plastic deformation	5.1 Overload	
	5.2 Indentación, brinelling	5.2.1 Indentation from debris
		5.2.2 Indentation by handling
6. Fracture	6.1 Forced fracture	
	6.2 Fatigue fracture	
	6.3 Thermal cracking	

False brinelling is erosion of the races corresponding to the location of the balls or rollers. It is produced by the action of vibrations without the machine being turning. Vibrations displace the lubricant film, allowing contact between the metal parts. This can occur during transport or in service, due to the vibration of adjacent machines. True Brinelling or indentation occurs under static overload, mainly due to malpractice during assembly or manipulation; these are always associated with significant plastic deformation (Figs. 8.3 and 8.B1).

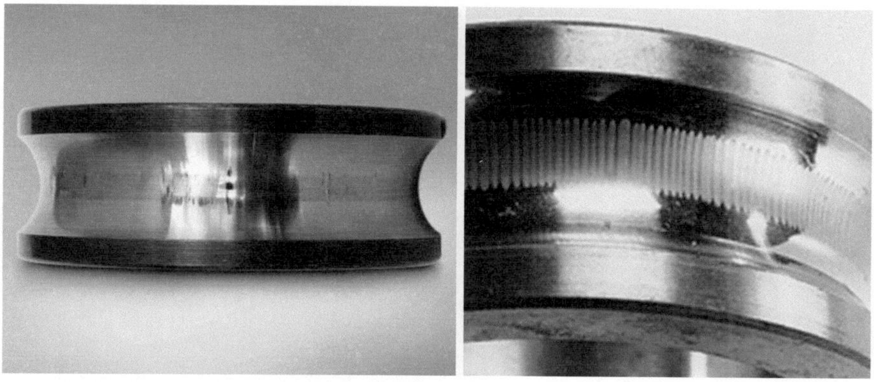

Fig. 8.3 Marks of false brinelling

Fig. 8.B1 Damage to the generator shaft. Adhesive erosion can be appreciated with material loss **a** by slide of the inner race of the bearing **b** by contact with seal caps

8.3.1 Example 8.B Failure of Bearing in A.C. Generator

A group of similar generators presented a recurring failure in their bearings. Catastrophic failure occurred in several of them and in others damage was detected in the early stages of the problem. Inspection of the bearings and the machinery made it possible to define the main cause of failure as arcing caused by failure of the insulation of the outer race (Fig. 8.B2). The same problem induced degradation in the lubricant, causing loss of fluidity (Fig. 8.B3)

The failures occurred shortly after performing a relubrication operation. During this operation, the entry of fresh grease displaced degraded grease into the race, which generated an above normal temperature rise and increased drag torque. Both factors caused loosening of the race and its rotation with respect to the shaft. This rotation produced adhesive erosion damage (Fig. 8.B1) and eventually the

Fig. 8.B2 Slide bearing cover. Insulating sheets are observed in the outer races and the discharge ports of excess grease. Grease is degraded by the effect of the electric arc

Fig. 8.B3 Grease retained between rolling elements, severe degradation is observed, with hardened fibrous appearance, typical of arcing temperature elevation

destruction of the bearing. An inadequate fixation between the inner race of the bearing and the shaft was identified as a secondary cause for failure.

To prevent recurrence of the failure, failed bearings were replaced by bearings which are electrically insulated in the outer race; the design of fixing the inner race was modified by adding a piece that provides the race an axial compression against the shaft.

8.4 Failure in Sliding or Plain Bearings

Sliding bearings are designed to transmit radial force between surfaces, generally cylindrical supports in shafts. These bearings are built with materials with low coefficient of friction and high wear resistance, with generally much lower hardness than the shaft in the contact zone. These bearings can be constructed to work with or without lubrication, obtaining in the first case a reduction of friction and service lives several orders of magnitude larger than those for the nonlubricated. Lubricated bearings operate by the principle of the oil wedge; this is formed by rotating one part over the other, separating the parts in contact. So bearing performance varies with the type of contact lubrication:

- Hydrodynamic contact, when the lubricant layer completely separates the contact surfaces.
- Elasto-hydrodynamic contact, when the surfaces can come into contact but roughness profiles do not support loads above the material's elastic limit.
- Mixed, when the load is supported partly by the surfaces.
- Dry contact, when there is no or very little lubricant.

The materials used in the construction of slide bearings are varied and different technologies exist for the construction of the bearings. Journal bearings are generally constructed so as to be easily replaceable in the form of bushings or shaped in halves and are composed by many metal layers. The layers that are in contact with the support or structure of the machine are heavy-duty steel and accommodate some elastic interference, giving the final shape to the bearing. The layers in contact with moving pieces are made of low friction alloys; there might be one or more contact layers, according to the technology used. Antifriction materials are alloys based on aluminum, copper, and tin.

Table 8.3 summarizes the damage mechanisms antagonistic to properties that define anti-friction materials. These are compatibility, formability, fatigue strength, indentability, embedibility, hardness, and corrosion resistance. These properties are referred to the service behavior and tribological system and are strongly related to the thickness and arrangement of the different layers.

In particular, the "formability" is referred to the ability of the material to deform plastically to redistribute the stresses in the presence of geometric abnormalities in the bearing or the assembly. The "embedibility" is the ability to incorporate and immobilize particles into the mass of the bearing without causing further damage or minimizing the consequences.

Table 8.3 Damage mechanisms antagonistic to properties that define anti-friction materials

Tribological property of the material	Associated damage mechanism	Failure causes
Compatibility	Adhesive erosion	Roughness, inadequate oil film
Conformability	Adhesive erosion, superficial fatigue erosion	Misalignment, assembly defect, geometric deficiencies
Fatigue resistance	Superficial fatigue	Excessive dynamic loading, localized loads (misalignment, ovalization, etc.)
Embeddability	Abrasive erosion	Particles in oil, contamination
Hardness	Extrusion, plastic deformation	Cavitaion and/or overheating
Corrosion resistance	Corrosive erosion	Degraded lubricant (overheating), contaminated lubricant (water, fuel, coolant)

8.4.1 Example 8.C Failure of Sliding Bearing in a Large Gas Engine

A catastrophic failure occurred in several bearings of a 5,000 HP gas piston engine. After operators detected abnormal operating noises, they found a lot of bearing metal in the bottom of the oil pan, so that the unit went out of service. The main fault was detected during disassembly: damage in several rod bearings, and erosion (adhesive and abrasive) in the corresponding crankshaft journal.

The engine had been in service for 50,000 h and was soon to be subjected to a planned overhaul; the manufacturer had offered a replacement without charge of all rods and bearings. Failed bearings were found completely destroyed so did not provide evidence of the origin of the failure. The rest of the bearings showed fatigue erosion damage (Fig. 8.C1) with a marked loss of material from the back side and crushing fretting. Wear copied roughness and shape of the crank housing (Fig. 8.C2).

Fractographic and metallographic analyses allow concluding that the loss of material on the sliding surface was due to subsurface fatigue in the aluminum-based friction coating. This caused the shedding of particles which were entrained in the oil flow to other parts of the bearing and out of the system (Fig. 8.C3).

One of the main causes of the onset of fatigue is attributable to the loss of support of the bearing in its housing, due to crushing of the initial roughness of the bearing housing in the rod, and subsequent fretting. The catastrophic failure was due to an escalation of the damage described, which degenerated into the total destruction of the bearings. Replacement rods and bearings provided by the manufacturer incorporated modifications in rod roughness and the method for positioning and fixing the bearing.

Fig. 8.C1 Fatigue erosion damage with a marked loss of material in the antifriction layerof bearing (aluminum based babbitt alloy)

Fig. 8.C2 Backside of bearing. Erosion and fretting can be seen. Wear-copied roughness and shape of the crank housing

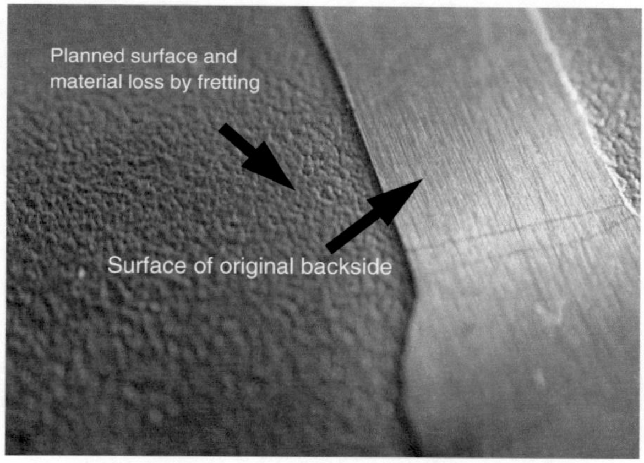

Fig. 8.C3 Micrograph of the bearing material; subsurface cracks typical of fatigue damage can be seen (X100)

8.5 Failure of Transmission Elements: Gears and Couplings

8.5.1 Gears

Causes for gear failure often come from foreign elements getting into the gear, so that the initial inspection is most important to determine the main cause of failure. Inspection should be performed without disassembling the joint, seals, and keyways, as modifications during disassembly may induce erroneous traces or signals into the analysis. Similarly, recording the tooth contact patterns should be done in relation to the other components. It is also important to check the condition of the lubricant, to take oil samples before the first cleaning and check for signs of corrosion, contamination, or overheating. Inspection of the bearings in the shafts is crucial because gear failure is very often the cause of bearing failure. Bearing wear can generate increased center distance or misalignments. The general condition of the bearings may give clues to lack of lubrication, contamination, or overload.

Verification of the gear design can confirm the existence of overload or fatigue if operating conditions had been changed. In all cases it is necessary to confirm microstructural and mechanical properties of the materials used in the gear assembly. The calculations and checks can be made according to AGMA 2001-B88, Standard 1010-1095, Nomenclature of Gear Tooth Failure Modes. The main causes of failure are excessive bending stresses at the base of the tooth and excessive contact stresses in the support zone of the tooth. A more complete description of gear failure modes can be seen in Table 8.4.

Many of the terms used to define the state or appearance of a gear should not be interpreted outside the context of the standard; transcribed below are specific definitions.

- **Burnish**: an alteration of the original manufactured surface ranging from a dull to a brightly polished condition.
- **Wear**: the removal of metal, without evidence of surface fatigue or scoring, resulting in partial or complete elimination of tool or grinding marks, and/or development of a discernible shoulder ridge at the bottom of the contact area near the root or at the toe or heel end of the pinion tooth contact area.
- **Abrasive Wear**: wear caused by lapping of mating surfaces by fine particles suspended in lubricant, fuel, air, or imbedded in a surface.
- **Adhesive Wear**: wear caused by shearing of junctions formed between operating surfaces in direct metal-to-metal contact; sheared-off particles either remain affixed to either the mating surfaces or act as wear particles between the surfaces.
- **Rippling**: an alteration of the tooth surface to give an appearance of a more or less regular pattern resembling ripples on water, or ash scales. Note: examination of the surface with oblique lighting usually is necessary to show this condition.

Table 8.4 Gear tooth failure modes (ISO 10825)

1. Indications of surface disturbances	1.1 Sliding wear	1.1.1 Normal wear (running in wear)
		1.1.1.1 Moderate wear
		1.1.1.2 Polishing
	1.2 Corrosion	1.1.2 Abrasive wear
		1.1.3 Excessive wear
		1.1.4 Moderate scratching (scoring)
		1.1.5 Severe scratching
		1.1.6 Interference wear
		1.2.1 Chemical corrosion
		1.2.2 Fretting corrosion
		1.2.3 Scaling
	1.3 Overheating	
	1.4 Erosion	1.4.1 Cavitation erosion
		1.4.2 Hydraulic erosion
	1.5 Electric erosion	
2. Scuffing		
3. Permanent deformations	3.1 Indentation	
	3.2 Plastic deformation	3.2.1 Plastic deformation by rolling
		3.2.2 Plastic deformation by tooth hammer
	3.3 Rippling	
	3.4 Ridging	
	3.5 Burns	
4. Surface fatigue phenomena	4.1 Pitting	4.1.1 Initial pitting
		4.1.2 Progressive pitting
		4.1.3 Micro pitting
	4.2 Flake pitting	
	4.3 Spalling	
	4.4 Case crushing	
5. Fissures and cracks	5.1 Quench cracks	
	5.2 Grinding cracks	
	5.3 Fatigue cracks	
6. Tooth breakage	6.1 Overload breakage	6.1.1 Brittle fracture
		6.1.2 Ductile fracture
		6.1.3 Semi-brittle fracture
	6.2 Tooth shear	
	6.3 Breakage after plastic deformation (smeared fracture)	
	6.4 Fatigue breakage	6.4.1 Bending fatigue
		6.4.2 Tooth breakage

- **Ridging**: alteration of the tooth surface. The result is a series of parallel raised and polished ridges running in the direction of the sliding motion, either partially or completely across the tooth surfaces.

Fig. 8.4 Gear failure modes: Progressive pitting (left) Spalling (right) Subsurface fatigue, initiated *below* the cement layer

- **Pitting**: small irregular cavities in the tooth surface, less than 1 square millimeter, resulting from the breaking out of surface metal (Fig. 8.4. left).
- **Spalling**: the breaking out of flakes or irregular area of the tooth surface, 1 square millimeter or larger; a condition more extensive than pitting (Fig. 8.4. right).
- **Scoring**: displacement of metal by local momentary weld of the gear tooth, resulting in the development of a matte or frosted, dull surface.
- **Discoloration**: any alteration in the normal color of rated surfaces.
- **Corrosion**: a general alteration of the finished surfaces by discoloration, accompanied by roughening not attributable to mechanical action. Rust is a special case of corrosion.
- **Chipping**: ring and pinion gears. A condition caused in the manufacturing processing in which a small irregular cavity is present only at the face/crown edge interface. The edge chipping phenomenon occurs when sufficient fatigue cycles accumulate after tooth surface wear relieves the compressive residual stress on the tooth proffle′s side of the profile-to-topland interface. Chipping within 1 mm of the face/crown edge interface is to be called chipping, not pitting or spalling.

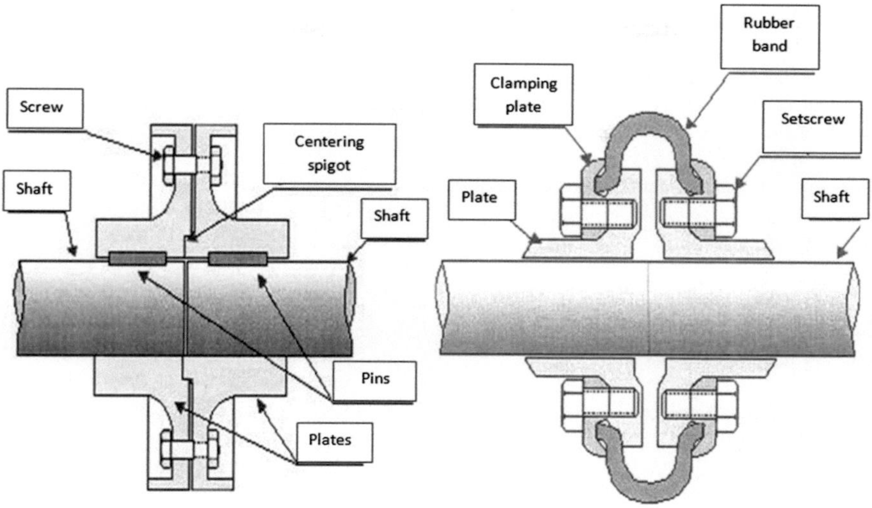

Fig. 8.5 Elements in rigid and flexible couplings

8.5.2 Couplings

The mission of couplings is to join rotating shafts of a mechanical system, to transfer the rotational motion and torque. Couplings are designed to allow some degree of misalignment and absorb torsional vibrations. There is another requirement for couplings: providing a break point between the drive shaft and the driven, acting as a fuse to the occurrence of severe torsion overload in the system (Fig. 8.4).

Couplings are designed to meet certain operational parameters, including limits to rotational speed and transmitted torque, probable misalignment of the shafts, vibration, admissible free movement in case of reversing (back lash), and environmental factors and conditions. If all these factors were evaluated in the selection of the coupling, and the coupling is placed, used and maintained properly, there should not be failure events throughout the expected life. If premature failure occurs it can result in major economic loss and a serious risk to the facilities and people. Many coupling systems include a design that minimizes damage in case of catastrophic failure (Fig. 8.5).

When the stiffness of the coupling is similar to that of the shafts attached, said coupling is called rigid. More common are flexible couplings, which are designed to have a low stiffness against displacement and misalignment between shafts, while maintaining high torsion rigidity. Flexible couplings have a metal or polymer portion that is designed to absorb the deformation.

Tabla 8.5 Failure modes in couplings

Failure mode	Cause of failure
Severe wear or erosion in the flexible elements or bearings of the axes	Excessive misalignment
Break of the polymeric flexible element or cut bolts	Instantaneous torsional overload
Fatigue of flexible elements, bolts, or metal discs	Torsional vibrations Excessive stops and starts
Bearing failure	Lack of lubricant Excessive misalignment Excess temperature in service
Cubes or plates loose on the shaft	Instant or variable torsional overload
Wear in jagged elements	Lack of lubrication Instant or variable torsional overload
Degradation of elastomers, severe corrosion in axes and cubes	High temperature Chemical attack, corrosive atmosphere

The methods that are used to secure the coupling to the two shafts depend upon the transmitted power. For low power and small diameter shafts sliding couplings are fit with a keyway, and cubes or plates are positioned with screws which press against the shaft. For medium to large powers, the cubes are fixed to the shafts by interference press fit. Determining the appropriate interference is decisive in the reliability of the coupling. If the interference is low, the cubes can loosen and generate fretting failures in the hub shaft. If, however, the interference is very high, high stresses can be generated in the shaft, which are added to the normal service stresses and can cause fatigue failure of the shaft. While there is a lot of coupling systems, typical failure modes can be classified according to Table 8.5.

8.5.3 Example 8.D Failure of a Flexible Coupling

The coupling was placed between shafts of a high power internal combustion engine and a reciprocating compressor for natural gas transport. This flexible coupling is made of two steel sheet packs and was specified for maximum 1600 RPM speed and torque of 45.000 Kgm. Nominal working torque is 30.000 Kgm. Flexible disk packs are mounted to the ends of a spacer and to both shaft hubs, by means of bolts (Fig. 8.D1).

The rupture of the disk packs produced a catastrophic failure. Of four bolts, which fix the disc package with the spacer, two were cut to the height of the first thread (Fig. 8.D2) and the other two had severe plastic deformation in the nuts. Only one sector of the flexible disc pack remained attached to one of the bolts (Fig. 8.D3).

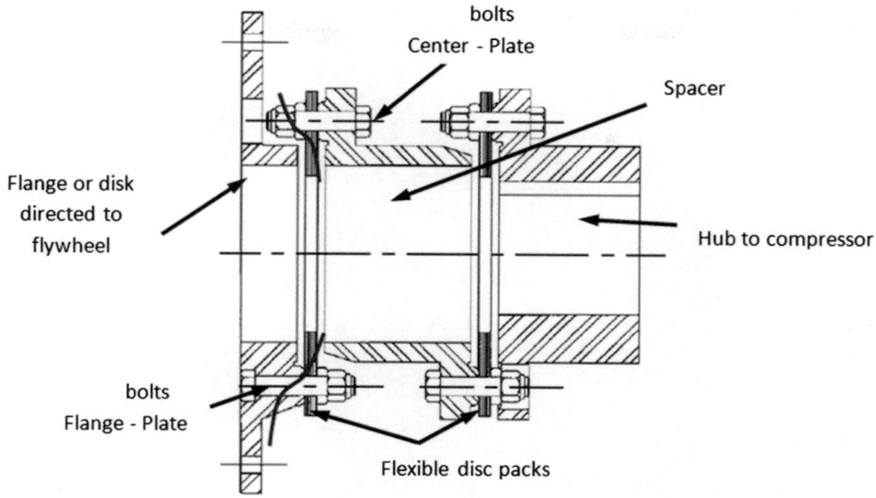

Fig. 8.D1 Cross-section of the coupling. The position of the breakage of the disc pack is indicated

Fig. 8.D2 Binding disc to the flywheel

Although the disk pack was completely destroyed, the relative position of all parts of the coupling disc and flexible disc package was rebuilt. Older fractures (Fig. 8.D4) and marks left by the fracture on adjacent sheets were identified (Fig. 8.D5).

Fig. 8.D3 Spacer and part of the disk pack

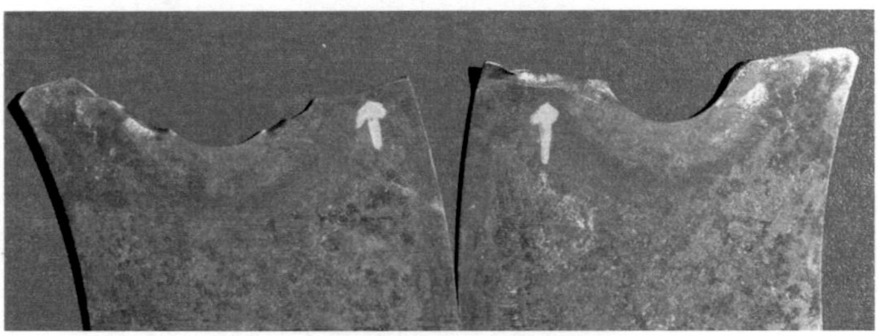

Fig. 8.D4 Consecutive discs were placed by corrosion signs. The mark left by the broken edge of one over another is indicated

Ultimate failure was caused by the failure of successive discs due to fatigue crack propagation. These cracks were initiated in the bolt holes. The failure of the discs was progressive, and may not have generated variations in the vibration level of the machines until the last moments of failure. The appearance of this type of damage is typical of lack of bolt preloading, which was confirmed by erosion and fretting marks left by the discs in the bolts and on the contacting surfaces of the spacer. The presence of corrosion on the plates could have favored fatigue initiation. Rupture of the connecting flange occurred as a result of the collapse of the disc pack, partly by unbalanced bending loads at the time of failure, and partly by the impact of the bolts attached to the separator piece (Fig. 8.D6).

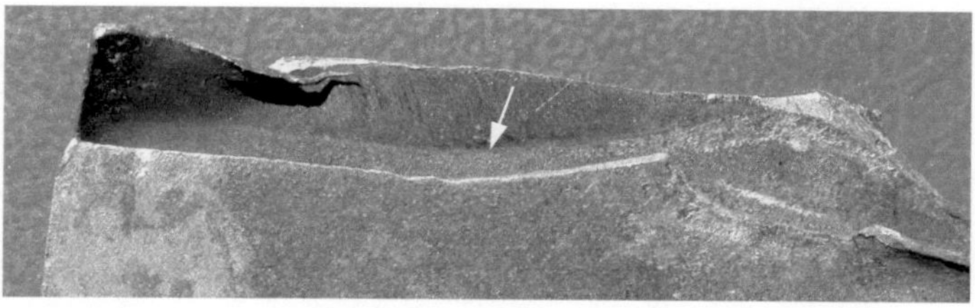

Fig. 8.D5 Disc detail. The imprint left by the upper disc is indicted

Fig. 8.D6 Contact marks left by the plate pack in the bolts next to the failure

8.6 Failure of Fasteners, Bolts, and Other Threaded Elements

The helical thread screws have been one of the most important inventions in mechanics. Screws serve for transforming rotary movements into linear displacements, and for developing mechanisms such as mechanical presses, and most important, threaded fasteners. There is a lot of non-threaded fasteners and new types of fasteners are constantly developing, such as rivets, pins, annular seals, spring seals, self-tapping screws; more recently adhesives are becoming important, mainly for bonding composites. However, threaded studs or bolts allow disassembling the joined elements practically without damaging, facilitating inspection, maintenance, and replacement of parts.

One of the main objectives of mechanical design is to reduce the number of screws and bolts. However, there will always be a need for them for the reasons listed. For example, Boeing 747 jumbo jets require up to 2.5 million fasteners, some of which cost several dollars each. To cut costs, new designs are constantly made to decrease the number of connecting elements, facilitating and simplifying installation and tooling techniques. The choice of methods for joining pieces is extremely important in quality engineering design. So it is essential to have a thorough understanding of the principles involved to ensure joint designs that satisfy service conditions.

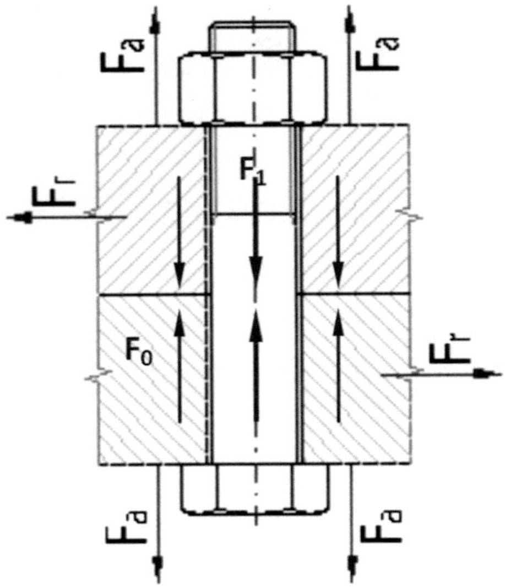

Fig. 8.6 Diagram of loads acting in a bolted connection

Bolts and threaded joints can be subjected to all kinds of loading conditions. Loads in bolts are mainly axial and transverse. The axial load come from preloading of the element to be assembled and is modified in service by external forces. Preloading is the tensile force in the screw shank which keeps the parts together by compression.

Transverse loads are transmitted by the structure when the friction force generated by the axial preload is not sufficient; in this case the screw shank portion must withstand sear stresses. Overall compressive forces generated in the parts to be joined must be enough to make the transverse or shear loads minimal or nil. The net compression force between the joined parts is the difference between the preload of bolts "F1" and the external load "Fa" applied to the joint (Fig. 8.6).

The classical approach for calculating bolted joints is based on the determination of the pure axial load on the bolt as a result of preload (or pre-stress) and external forces. In many applications, such as flanges, plates joining structures, anchors etc., the deformation of the parts to be joined makes the load supported by the bolts not perfectly axial, but with significant bending components. This causes that stresses in the stem are greater than those calculated by the classical approach. When there are eccentric loads there may appear fatigue or overload failures within the bolt head rather than in the apparently more stressed threads.

During tightening (pre-stress) of the joint, the bolt sustains an elongation while the parts that are fastened get compressed. The relationship between the deformation of the pin and the deformation of the joined parts is given by the ratio of their respective rigidity. Then, when the operational force is applied to the connection, loading in the joined parts is reduced and the load in the connection bolt increases.

The assembly load diagram is used for determining the load carried in service by the bolt (Fig. 8.7); it is constructed with the initial bolt preload F_0 and rigidities of parts being joined.

Notation in Fig. 8.7 is:

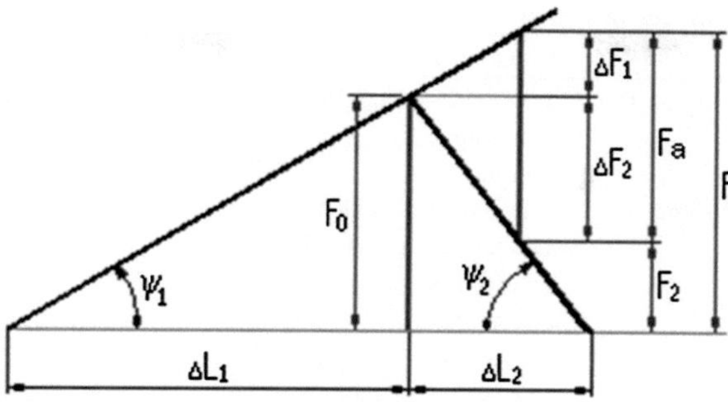

Fig. 8.7 Load diagram of the assembly of a bolted joint

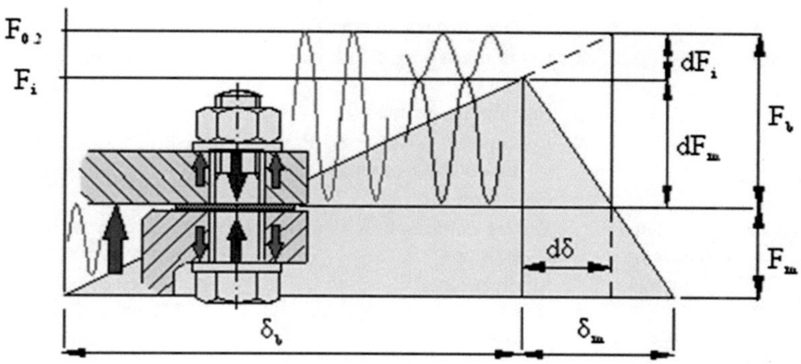

Fig. 8.8 Load diagram application of a bolted union assembly for varying loads (MIT)

F_0 = initial preload in bolt
L_1 = bolt elongation under initial preload
ΔL_2 = elongation of the joined pieces (compression) under the action of the initial preload
$c_1 = \text{tg}\ \psi_1$ = stiffness constant for bolt
$c_2 = \text{tg}\ \psi_2$ = stiffness constant for joined pieces
F_a = operational load applied to the joint
DF_1 = axial load increase in the bolt
DF_2 = load decrease in joined parts
F_1 = resultant load in the bolt after applying the operating load
F_2 = residual load binding the parts after applied operating load.

When operational loads vary, varying loads are also induced in the bolt (Fig. 8.8), which can lead to fatigue failure. Bolts or screws can fail in several ways. Table 8.6 summarizes failure modes of bolted joints, causes and consequences are highlighted.

Table 8.6 Failure modes in bolted joints

Failure mode	Cause of failure	Possible effect on the joint
Failure of the threads on the bolt	Low material quality Corrosion Fatigue	Eventual failure against overload
Failure of the bolt shank	Low material quality Operational overload Over torque	Possibility of catastrophic failure at the joint
Preload loss	Sub-torque Vibrations	Losses in the joint
Fracture of the bolt head or fractured nut	Excessive preload	Instant failure during assembly
Tearing of the threads, external and internal	Excessive gap between threads	Reduction of maximum bond strength
	Resistance difference between bolt and nut materials	Eventual failure
Loss of preload over time	Eventual overloads greater than the preload Insufficient preload Low quality of packing Rings, indentation of rings Corrosion	Loss of sealing joints Onset of shear load onto the screw Possibility of fatigue in the bolt
Grouping of threads during assembly/disassembly	Wrong combination of materials bolt/nut	Impossibility of disassembly and reuse of bolts
Premature fatigue failure	Slight preload. Excessive operational variable loads.	Possibility of catastrophic failure at the joint
Deformation of support surfaces	Excess preload	Failure by bending at the head of the bolt Possibility of loss at the joint
Indented jints at the joint surface	Excess preload	

As seen, one of the most common reasons for failure is insufficient or excessive preload. Different methods exist to tighten bolted joints, each of which has a degree of scatter in results ranging from 5 to 35 %. There are basically six main methods used to control a threaded fastener preload:

1. Control of tightening torque, vague but highly practical and widely used.
2. Tightening angle control, better accuracy, no sophisticated instrumentation is needed to assure a proper tightening.
3. Tighten by yield control, very accurate but requires sophisticated instrumentation to detect the loss of linearity of the material.
4. Control of bolt elongation, very accurate and easy to implement.
5. By thermal contraction, cumbersome and very slow in implementation.
6. Methods that apply a deformation commanded by a load applied to the bolt, hydraulic methods that depend on the accuracy of equipment and susceptible to errors in the choice of the actuators.

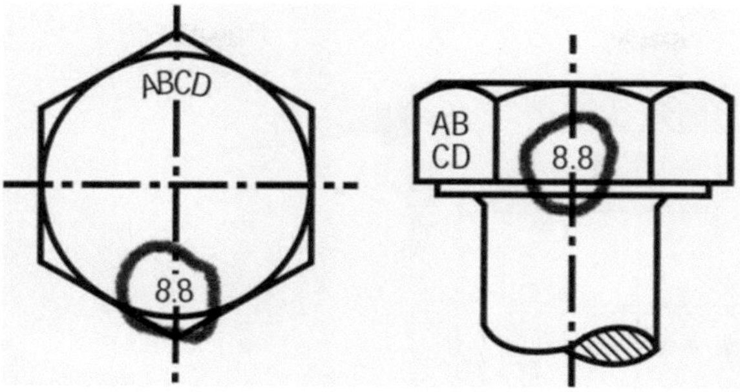

Fig. 8.9 Identification of material quality in bolts and nuts. First number indicates strength (MPa/100), second number indicates yield strength/UTS relationship (/100)

Fig. 8.10 Fatigue fracture (*left*) and ductile overload (*right*) of a threaded bolt. Note growth marks in the fatigue fracture and the large plastic deformation at failure associated with overload failures

The strength of bolts made of carbon steel is identified by symbols indented on the surfaces of nuts and heads (Fig. 8.9). The first number indicates the strength in MPa (/100) and the second the relationship between material yield and ultimate strengths ($\times 10$). Figure 8.10 shows a fatigue fracture (left) and ductile overload (right) of a threaded bolt. Note growth marks in the fatigue fracture and the large plastic deformation at failure associated with overload failures.

8.6.1 Example 8.E Failure of an Allen Type Fastening Bolt in a Compressor Piston Head

A rupture occurred in an Allen type (socket head) threaded bolt fixing the cap of a piston in a compressor at an ethylene production plant. The bolt fractured at the

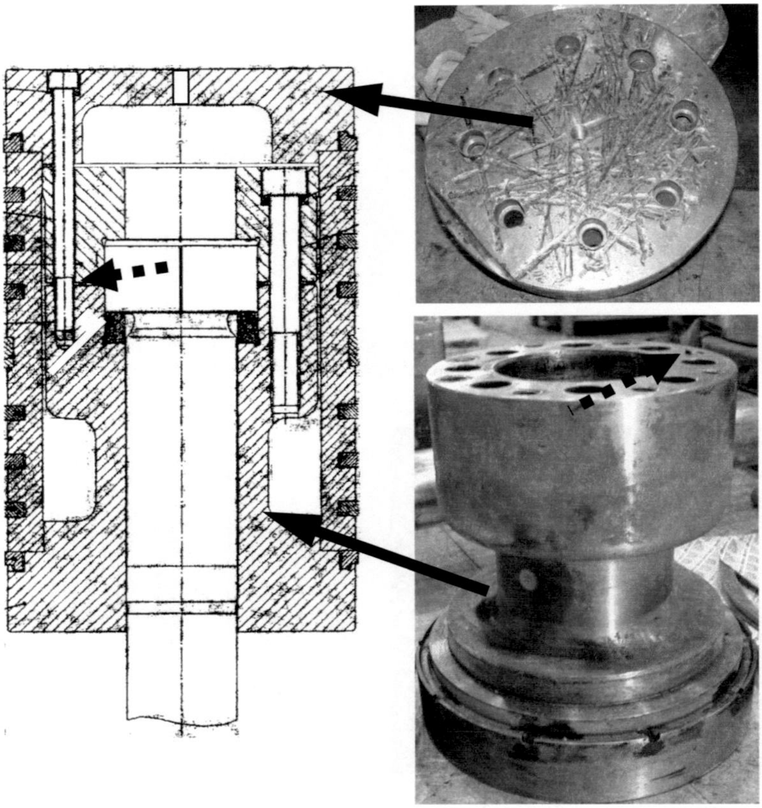

Fig. 8.E1 Scheme of piston, lower body, and cover. The position of the failed bolt is indicated

height of the first stressed thread (Fig. 8.E1). The rupture provoked the detachment of the threaded bolt from the piston lodging into the compression chamber. Severe damage was done to the piston and cylinder head.

The failed bolt, and other eight, form part of the clamping of the piston cap to the piston body. The sleeve that supports piston rings is supported between piston cap and lower piston body, the cap also forms the slot for the first compression ring. These bolts must withstand dynamic loads from the piston rings sleeve, and rings friction forces on the down stroke (Fig. 8.E2).

The failed socket head bolt is M16 $\times$ 2, grade 12.9, which corresponds (as defined before) to a high strength (1,200 Mpa) steel with a yield strength of 90 % UTS. The geometry of the thread was found to be deficient in all bolts inspected, particularly at the thread root where a rough surface from a faulty thread lamination process can be seen (Fig. 8.E3) On the other hand, quality bolts as specified, have a flawless surface finish and geometry (Fig. 8.E4).

Fig. 8.E2 Failed bolt, fracture surface

Fig. 8.E3 Deficient surface finish and geometry of a used bolt

Fig. 8.E4 Surface finish and geometry of a new bolt

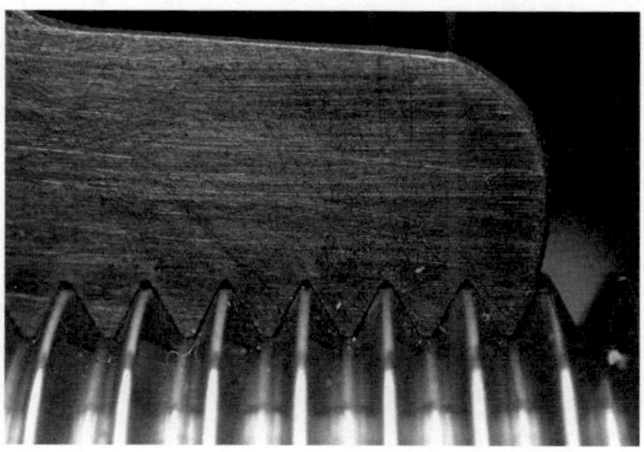

The instantaneous fracture is ductile; its surface has no apparent beach marks, and was possibly caused by an excessive preload, and misalignment between de socket and the screw head. It was recommended to review the assembly procedure and applied torque, not to reuse 12.9 grade screws, calibrate manual torque meters, and ensure proper lubrication before tightening.

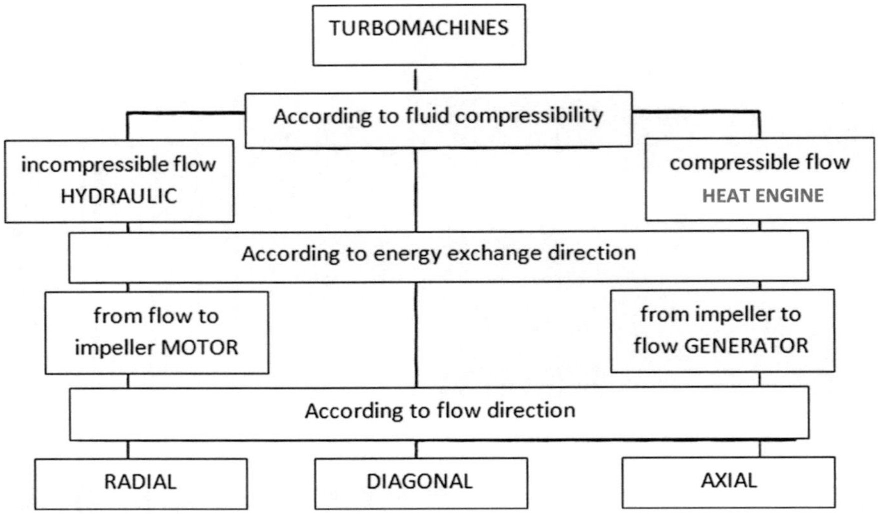

Fig. 8.11 Classification of rotating turbomachines

8.7 Characteristic Failures in Turbo Machines

A turbo machine is a machine that continuously manages fluids, which main component is the rotor through which the said fluid passes. In the process, energy exchanges between the machine and the fluid (Fig. 8.11). This exchange can be from fluid to machine or otherwise; first case are pumps, where the fluid achieves a higher velocity or pressure through the machine. The second case is that of power generators such as steam, gas, wind, or hydro turbines. Turbo machines occupy an important place in the process and energy industries, and much effort is devoted to reduce maintenance and repair time and costs.

8.7.1 Steam Turbines

Before WW2, steam turbines worked with pressures of up to 1.5 MPa and temperatures below 350 °C . Subsequently, pressures and temperatures have been rising, their design favored by the use of high-alloy steels, which dramatically improved performance. With the implementation of austenitic alloys, work temperatures rose well above 500 °C, and pressures passed 35 MPa (Table 8.7). The search for higher efficiency and the use of steam turbines in combined cycles will increase these values in the future.

Steam turbines have a very high specific power (power per unit size or weight), about ten times greater than diesel engines and with an overall efficiency of around 35 % or more. Reliability has been growing steadily, while they have shrunk in

Table 8.7 Classification of steam turbines according to working pressure and temperature

According to steam pressure at inlet of turbine:
(a) Low pressure turbine, up to 1 MPa
(b) Medium pressure turbine, from 1 to 8.8 MPa
(c) High pressure turbine, from 8.8 to 22.4 MPa
(d) Super critical pressure turbine, for superior values
According to steam temperature at inlet of turbine:
(a) Saturated steam tubine
(b) Medium temperature turbine, up to 485 °C
(c) High temperature turbine, from 485 to 565 °C
(d) Super critical temperature turbine, for superior values

Table 8.8 Failure mechanisms in steam turbines

Damage mechanism	Affected zone	Causes
Fatigue	Blades	Vibrations due to intake fluid imbalance
		Deposit accumulation
		Wear or breakage of damping elements
	Rotor shaft	Dynamic imbalances
		Mechanical damage, pitting or corrosion acting as initiator
	Bearings	Mechanical or thermal overload
Creep or thermal fatigue	Blades	Limit service hours
	Valve box	High number of stops and starts
Erosion	Blades	Droplet impact erosion
		Abrassive remains entrined by steam
	Rotor shaft	
	Bearing support	Lubrication deficiencies
		Lubricant contamination
Thermal degradation of material	Blades	Temperature above material capacity, typical in power upgrading or adaptations to combined cycle.
	Casing or volutes	Hot corrosion
		Grain resizing due to high temperature

size and increased specific power. The specific cost of installation decreases with increasing power of the unit, and efficiency increases with the temperature of steam used.

Construction of steam turbines is very robust, as they are usually fixed installations. Housing and rotor parts are heavy and sustain large mechanical and thermal stresses. The rotor is generally formed from a single block of forged carbon steel, although in some cases the rotor can have welded parts. Rotor blades are housed individually or grouped in sectors on the shaft.

The housing may be formed by different steel parts, welded or cast, and is made in two halves that meet at the midplane; thus allowing assembly of all components. The stator of the steam turbine is made of blade disks lodged in the turbine housing. The most common failure mechanisms in steam turbines are summarized in Table 8.8.

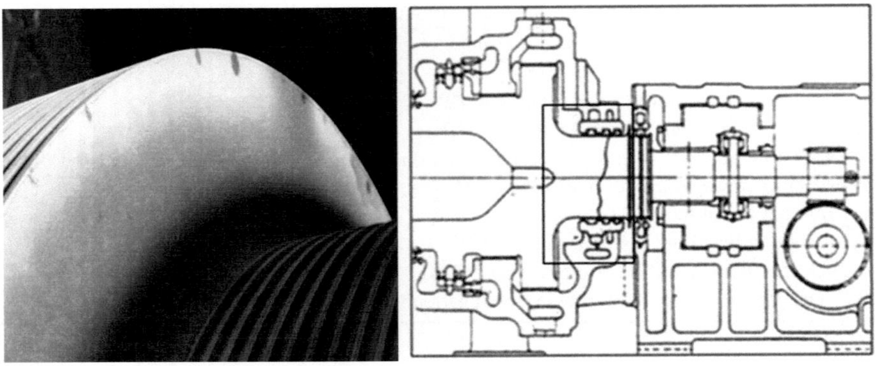

Fig. 8.F1 Cracks in the high pressure seal of the shaft

Fig. 8.F2 Detail of cracks in the high pressure seal of the shaft

8.7.2 Example 8.F Failure in the Rotor of a Steam Turbine

Cracks were detected in the area of the high pressure labyrinth seal of a steam turbine; those defects could risk the integrity of the rotor in service (Figs. 8.F1 and 8.F2).

As can be seen the cracks cover the bottom of the slots and do not propagate up to the sealing ribs. Metallographic techniques determined that these are propagating corrosion fatigue cracks, with accumulation of corrosion products at their crack tips. These cracks had probably formed before a weld rebuild of the ribs, the repair welding did not have proper penetration into the material at the bottom of the grooves.

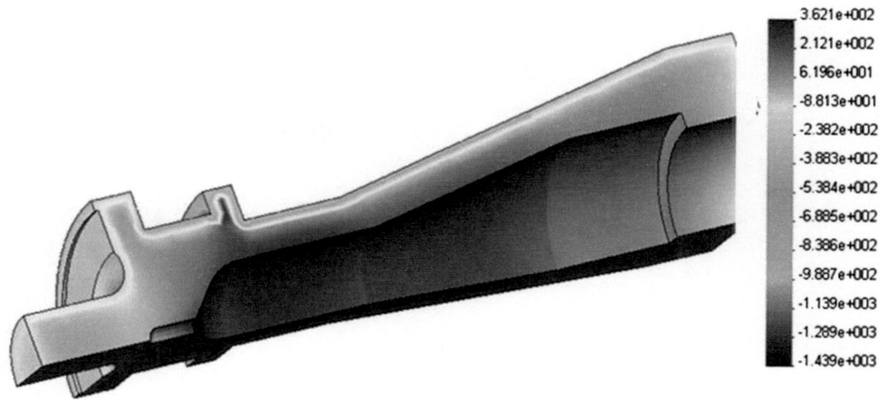

Fig. 8.F3 Modeling of thermal stresses in the start transients

Sizes of all cracks were NDT determined, and stresses due to thermal and dynamic effects in the start and stop transients we calculated (Fig. 8.F3). It was determined that the current size of the cracks did not pose a threat to the integrity of the shaft and crack propagation rates were assessed. With these data, it was recommended to continue operation and it was established a safe frequency of inspection.

8.7.3 Gas Turbines

Gas turbines have a high power to weight ratio, high rotational speed, and generate high enthalpy gases. They are conceived for direct use in propulsion and are suitable to drive all types of rotating machinery, coupled directly or through a wheel driven by exhaust gases called freewheel (Fig. 8.12).

Gas turbine combustion temperatures are of about 2,000 °C, while the maximum temperature bearable by blades is 1,300 °C. Combustion takes place with excess air in a proportion varying from 60-1 to 100-1. New developments and burner designs can decrease flame temperature to reduce emissions of nitrous oxide and other contaminants.

The materials used in the diffuser or stator vanes, rotor blades, and combustion chamber must be able to withstand high temperatures, have good creep resistance, corrosion resistance, and fatigue strength. Stability of metallographic phases at high temperatures makes them suitable for these applications. These materials are known as super-alloys and a great deal with proprietary names have developed; all of them have a phase precipitation heat treatment, which provides a gamma prime phase with Creep strength and hardness at high temperatures. At service temperatures this microstructure continues an aging

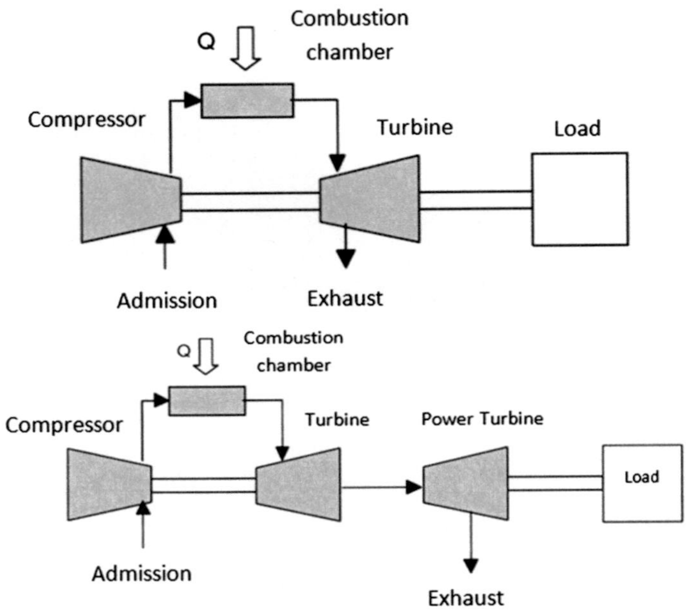

Fig. 8.12 Direct coupling and freewheel type gas turbines

Fig. 8.13 Metallographic structure of a super alloy. Its features provide high temperature stability

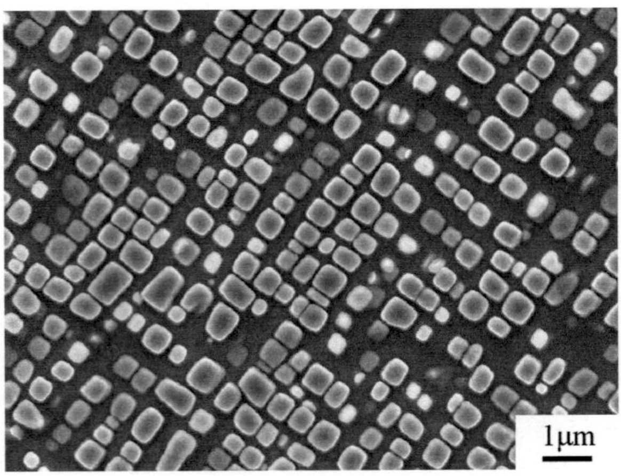

transformation, degrading the mechanical properties of the material until useful life is exhausted (Fig. 8.13).

The design of vanes, combustors, and transition pieces are performed so that these parts are cooled by fresh air and coated with thermal barrier coatings (TBC), which in turn protect from corrosion. Catastrophic failure of blades is the cause of almost half of the forced outages of gas turbines. Table 8.9 summarizes failure

Table 8.9 F1 failure modes in gas turbines

Failure type	Origin
Thermal fatigue in blades	Surface corrosion—pits
Thermal or mechanical fatigue on rotor shafts, housing	Start-stop cycles
	Emergency stops (ESDs)
Creep in blades and combustion chambers	Combustion temperature
	Flow cooling/blockage
	Engagement of the thermal barrier/erosion
Premature degradation coatings/base material of blades	Temperature-oxidation
	Corrosion, pitting
	Attack of the medium
High cycle fatigue in blades	Vibrations, stress concentrators
	Fretting/galling: at the base
	Stack up eccentricity (rotor disc series)
	Inappropriate tip windage
Thermal distortion of housing	Nonuniform heating or cooling restricted thermal expansion
	Excessive emergency stops or insufficient cool-down

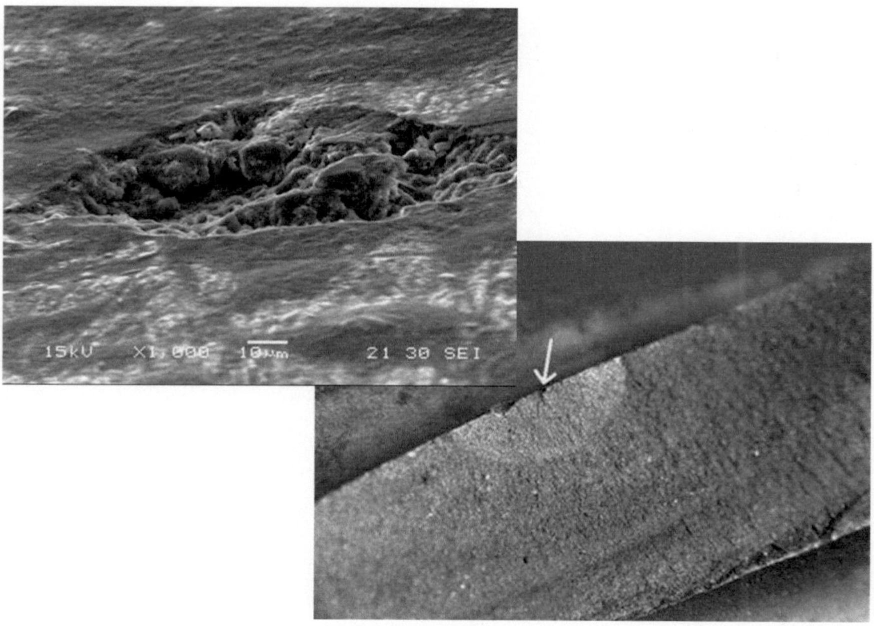

Fig. 8.14 Corrosion pit, initiating fatigue failure in a turbine blade

modes of elements of gas turbines, and their causes. Figure 8.14 shows as an example the morphology of a orrosion pit that initiated a fatigue crack, which eventually led to the failure of a turbine blade.

References

1. B.W, Smith and R.A, Grove, in *Failure Analysis and Prevention* ed. by G.W Powell, S.E Mahmoud, Handbook of ASM, vol 11 (American Society for Metals, Metals Park, 1993)
2. K.S. Ravichandran, A.K. Vasudevan, in *Fatigue and Fracture*. Handbook of ASM, vol 19 (ASM International, Materials Park, 1996), pp. 331–336
3. A. van Beek, *Advanced Engineering Design, Lifetime Performance and Reliability*. (TU Delft, Delft, 2012)
4. ASTM A370 – 12 Standard Test Methods and Definitions for Mechanical Testing of Steel Products. American Standards for Testing of Materials
5. ASTM E647 – 11e1 Standard Test Method for Measurement of Fatigue Crack Growth Rates. American Standards for Testing of Materials
6. D. Broek, *Elementary Engineering Fracture Mechanics*. (Kluwer Academic Publishers, Dordrecht, 1982)
7. P.G. Fazzini, A.A. Márquez, J.L. Otegui, P. Barcia, Cause and effect assessment after a complex failure of a large Ethylene compressor. Eng. Fail. Anal. **13,** 1358–1369 (2006). ISSN1350-6307
8. Draft API RP 585: Equipment Failure Investigation. American Petroleum Institute (2012)
9. T.L. Anderson, *Fracture Mechanics: Fundamentals and Applications*, 2nd edn. (CRC Press, Boca Raton, 1994)

Chapter 9
Failure Mitigation and Extension of Service Life

9.1 Introduction and Historical Perspective

The importance of the technologies associated with ensuring structural integrity and life extension is defined by the degree of reliability in many components of industrial plants, power plants, and civil structures. Extending the life of industrial facilities is now a key technology for many countries that are engaged in a process of integration and economic and industrial recovery.

Figure 9.1 shows a typical curve (solid line) that relates the cost of maintaining a given piece of equipment during its operating lifetime. Instead of cost, the vertical axis can represent other correlated variables such as risk of accidents, malfunctions, etc. This "bathtub" curve indicates that when a system or structure is installed and starts its operation, high maintenance costs occur due to the correction of non-conformities, project failures, and defective components. After this initial stage, a second stage develops where costs are significantly lower because, in principle, the initial problems have been resolved. The initial stage may not occur, in case of mass-produced components that have a very sleek design given by experience and/or that have undergone pre-operational destructive testing. From a certain moment, the third stage begins, where new problems arise, now motivated by the "aging" of various components. This stage progresses to the moment that costs (or risk) become so high that the structure or component loses its functionality and should be disabled.

Management analysis of the bathtub curve indicates what actions should be taken in order to reduce costs (or risks) in all phases of the life of the structure. These actions include reducing the aging stage and extending component life. Removing the third phase is associated with the detection of actual aging mechanisms, which is made through integrity assessment and life extension programs. These programs seek to bring costs back, as the dotted curve indicated at the far right of Fig. 9.1. Small cost peaks in the end of life are associated with the aging of components, their substitutions, recoveries, and reinforcements. The end of operation eventually comes up when the structure becomes obsolete in relation with new design or operation techniques, at which time it will not be worth extending its life.

J. L. Otegui, *Failure Analysis*, DOI: 10.1007/978-3-319-03910-7_9,

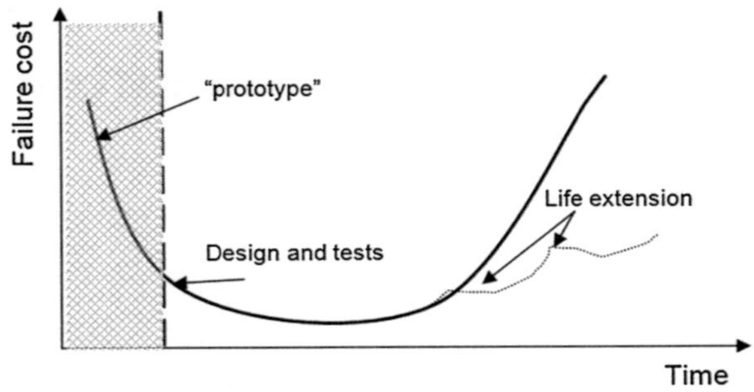

Fig. 9.1 *Bathtub curve*: cost of equipment maintenance during its operating lifetime

The systematic monitoring of aging processes in a structure and the actions taken, so that service life can be extended progressively and at low cost, is called residual life management (RLM). The RLM has proven to be the best alternative to increase short- and medium-term availability, flexibility, and reliability of industrial equipment. It is also the best alternative to increase the chances of life extension in the long run. RLM depends primarily upon inspection and monitoring of actual damage conditions in the components.

Industrial equipment should be inspected regularly for compliance with manufacturing specifications and codes. But inspections also and above all allow establishing the rate of accumulation of damage, which allows defining inspection periods and methods. The purpose of an inspection program is to ensure the integrity of the equipment, through optimum utilization of resources. In this way, safety of the plant is ensured, non-stop operation periods can be extended, in some cases deterioration rates can be reduced, and future needs for repair or replacement can be estimated.

An effective inspection program should be able to define and implement the activities necessary to detect deterioration in service of equipment before failures occur. The inspection program must systematically identify:

- What kind of damage occurs?
- Where is damage to be detected?
- How damage can be detected (Technical Inspection)?
- When or how often to inspect?

Not necessarily the inspection plan should be carried out by the responsibility of conducting inspection activities. For the preparation of the plan, an exhaustive search for information about plant processes, equipment characteristics, operating pressures and temperatures and failure history, and inspections of each component are required. In particular, we shall investigate and analyze information concerning:

- Design and manufacture
- Operation and service
- Inspection and failure history.

9.2 Failure Mode and Effect Assessment

Historically, engineers have done a good job in the design phase, to evaluate functions and shapes of products and processes; they have not always done so well when designing for reliability and quality. The failure mode and effect assessment (FMEA) is a methodology for analyzing potential reliability problems early in the development cycle, where it is easier to take actions to overcome these conditions. FMEA is used to identify potential failure modes, to determine their effect on operation and products, and to identify actions to mitigate failures.

The intention is to identify areas or components that would most likely result in failure of the assembly. FMEA defines function as a task performed by a component. For example, the function of a valve is open and close, and the failure modes are the ways in which the component may fail (the valve would fail in opening if its spring breaks, gets stuck in its guide or is kept open because of a break in the camshaft, etc.). The technique is to evaluate three aspects of the system and its operation:

- Operating and most likely early failure conditions.
- Effect of failure on performance.
- Severity of failure upon mechanisms.

The probability of failure is generally assessed in a scale of 1–10, with increasing criticality with number value. This technique is useful for evaluating alternative solutions to a problem but it is not easy to use with precision in new designs.

There are several types of FMEA, some are used more often than others; a crucial step is to anticipate what could go wrong (what if?). The development team should formulate as extensive a list as possible of potential failure modes. FMEA captures historical information for use in future product improvement. An FMEA should be done whenever a failure means harm or potential injury to the user. FMEA can focus on global system functions, in some components and subsystems, in manufacturing processes and assembly, or in service or software functions.

Changes in a process often introduce new failure modes. FMEA benefits include improved reliability and quality of the product or process, customer satisfaction, early identification and removal of potential failure modes, capturing knowledge in the organization, reducing changes and associated costs, and catalyzing technical exchange and teamwork.

The basic steps in the process for conducting an FMEA include:

1. Describe the product/process and function.
2. Create a block diagram of the product or process. This diagram shows major components or process steps as blocks connected by lines that indicate how the components or steps are connected. The diagram shows the logical relationships of components and provides a structure around which to develop the FMEA.

3. Use the diagram to define items or functions. If the articles are components, list them in a logical manner within a subsystem based on the block diagram.
4. Identify failure modes. A failure mode is defined as the manner in which a component, subsystem, system, process, etc., could potentially not meet the terms of its design. We have seen in Chaps. 4 and 8 a variety of examples of failure modes. A failure mode in one component can serve as the cause of failure in another component. Every failure should be listed in technical terms, for each step of the process. Analyzing failures that have been documented for similar products or processes is an excellent starting point.
5. Describe the effects of these failure modes. The ultimate effect for each identified failure mode must be determined. A failure effect is defined in product function or process, as perceived by the customer (internal or external). Examples of failure effects include injury to the user, inoperability of the product or process, improper operation or appearance, odor, noise, etc.
6. Establish a numerical ranking for the severity of the effect. An industry standard scale uses 1 (no effect) and 10 (very severe), such that of a failure without warning which affects both operation and safety. This ranking helps the analyst determine if a failure would be a minor nuisance or a catastrophic occurrence, and allows failures to be prioritized in order to deal with the most serious first.
7. Identify causes for each failure mode, according to the RCA tools we have seen in Chap. 7. The potential causes for each failure mode should be identified and documented. The reasons should be listed in technical terms and not in terms of symptoms. Examples of potential causes include improper operating conditions, pollution, faulty algorithms, misalignment, excessive stresses, etc.
8. Incorporate a probability factor; assign a numeric weight to each case, indicating the likelihood for a cause to occur. A standard scale uses 1 (not likely) to 10 (inevitable).
9. Identify current controls (design or process). These are the mechanisms that prevent a failure mode from occurring and ensure that the failure is detected before it reaches the customer (barriers, as they were defined in Chap. 7).
10. Determine the probability of detection by current controls, preventing customer reach.
11. Define the risk priority number (RPN). The RPN is the product of numerical grades of severity, probability, and detection. The RPN is used to prioritize actions.

$$\text{RPN} = (\text{Severity}) \times (\text{Probability}) \times (\text{Detection}).$$

12. Determine recommended actions to address potential failures that have a high RPN. These actions may include inspection procedures, test or quality selection of components or materials, reduction of normal capacity, limiting stresses or operating range, resetting the item to avoid the failure mode,

monitoring mechanisms, performing preventive maintenance, and inclusion of parts or systems redundancy.
13. Assign responsibilities and completion dates for these actions; this facilitates tracking.
14. Indicate the actions taken. After these actions have been taken, reassess severity, probability and detection, and review the RPN. Is any other action required?
15. Update the FMEA with changes in the design or process, and with all new information.

FMEA is also useful to assess whether there is an unnecessary number of components in a system, since the interaction of an assembly with another multiplies the effects of a failure. It is also useful to analyze the product and the equipment used to produce it. It also helps to eliminate excessive weakness or design complications and to identify components that can fail most likely.

9.3 Procedures for Assessing Fitness for Service

Most design codes do not address service degradation of equipment and do not take into account that previous poor workmanship or operation can be detected during inspections. Codes for fitness for service (FFS) assessment of components in industrial plants were originally created under the auspices of nuclear power plant operators. The publication of standards and best practices in several areas of technology is the result of the need to standardize techniques for evaluating the FFS of mechanical components.

The FFS assessment is defined as the ability to demonstrate the structural integrity of components that are in service and that, as a consequence of their operation, contain a certain degree of damage. The fundamental objectives of its application are:

- verify the safety of personnel as the plants become older and continue to operate,
- provide FFS procedures which are clear and technically justified, to ensure that other service providers make consistent predictions, and
- help optimize plant maintenance and operation.

The Failure Assessment Diagram (FAD) developed by CEGB R6 and originally used in the evaluation of components in the nuclear industry, has been discussed in Chap. 6.

The FAD method is also used in the British Standard BS 7910, in API Recommended Practice 579 (which will be discussed in some detail), and EXXON, CALLS, and MPC procedures. The CDF method is used by ETM and GE-EPRI (General Electric, Electric Power Research Institute, USA). These documents propose procedures with three or more levels of increasing complexity and accuracy. Each level requires different expressions for FAD or CDF to be used in the analysis. The level of analysis is determined primarily by the degree of detail of data used to define the stress–strain behavior (σ-ε) of the material, and the reliability of estimates of applied stresses and shapes of defects present in the component.

Due to its extensive use and applications, we will summarize the API STD 579–ASME FFS1: "Fitness for Service" document, which is oriented to the evaluation of piping and pressure vessels. The API 579 standard provides guidance for conducting FFS assessments using methodologies specifically prepared for equipment in the petrochemical industry and refineries, and may also apply to pressure equipment built by other recognized codes and standards.

The publication is divided into 13 sections, which are evaluated in subjects such as brittle fracture, general and localized metal loss, pits, blisters, laminations, misalignments, distortions, crack-type defects, creep and fire damage. Twelve appendices include equations for stress, pressure and thickness, and solutions for stress intensity factors, references for material properties and residual stresses, and descriptions of the various modes of failure and deterioration.

FFS assessment is a technique that measures whether a component with planar or volumetric defects can continue in service. API 579 approach uses three evaluation levels that increase complexity according to the amount of available information. Each level of assessment provides a balance between the degree of conservatism, the amount of information required, the ability of the staff doing the evaluation, and the complexity of the analysis.

- Level 1 is the simplest and most conservative; it requires few inspection data and can be performed by company personnel. Generally, this level considers the requirements established by manufacturing codes.
- Level 2 is less conservative than the previous one, requires a detailed inspection and review of the service history and must be performed by engineers with experience in the use of these tools.
- Level 3 provides the most detailed and most accurate results. Inspection and more detailed information are required; analyses are based in numerical techniques so that specialists in finite element methods or other computational tools are usually required.

The results are used to determine:

1. If the defects need to be repaired immediately before continuing the service.
2. If the equipment with defects can continue to operate normally, and when the next inspection is required (re-inspection period).
3. If the equipment needs to be replaced.

The more accurate the information collected at each level, the less need to move to a next level. Furthermore, the use of more advanced levels may not provide significant benefit if data are uncertain. It is preferable to ensure the reliability of data to work with a simpler level rather than to advance to a higher level with unreliable data. In the case where a higher level analysis is to be extended, it should be understood that additional methods of analysis of the behavior of the structure become increasingly stringent, more accurate and more expensive. Some sections of API RP 579 applicable in assessing defects are briefly presented as follows.

9.4 Evaluation of General and Local Loss of Thickness

Section 4 of API 579 refers to the evaluation of pressurized components subject to loss of material due to corrosion or erosion. Level 1 assessment is based on concepts of average thickness from which reliable results are obtained when applied to relatively uniform thickness losses. This criterion is based on the theory that when the components are subjected to internal pressure, stresses are redistributed around the weakened section, if it is small enough.

API practice 579 establishes evaluation criteria that require not less than 15 measurements of thickness. Spacing between measurements (L) should meet the following equation:

$$L = \min\left(2 \cdot t_{nom};\ 0.36 \cdot \sqrt{D \cdot t_{min}}\right) \tag{9.1}$$

where:

L = recommended spacing between measurements
t_{nom} = nominal thickness of the component
D = diameter of component
t_{min} = minimum thickness required by the standard.

The shape of the measured thickness values is analyzed based on two criteria: by simple average of measured thickness and profiles along defined planes. The use of one or another criterion is established, through a coefficient called coefficient of variation (COV) of measured values. To calculate the coefficient of variation, the following equations should be used:

$$COV = \left(\frac{t_{sd}}{t_{am} - FCA}\right) \tag{9.2}$$

$$t_{sd} = \left[\left(\frac{S_2}{N}\right) - (t_{am} - FCA)^2\right]^{1/2} \cdot \left(\frac{N}{N-1}\right)^{1/2} \tag{9.3}$$

$$t_{am} - FCA = \frac{S_1}{N} \tag{9.4}$$

$$S_1 = \sum_{i=1}^{n} (t_i - FCA) \tag{9.5}$$

$$S_2 = \sum_{i=1}^{n} (t_i - FCA)^2 \tag{9.6}$$

where:

COV = Coefficient of Variation
FCA = thickness loss expected in the future (assuming the rate of accumulation of damage will remain constant)
N = number of thickness measurements.
t_{am} = average value of the measurements.

The procedure establishes criteria based on COV for defining the acceptability of the component. Section 5 relates to the assessment of pressurized components subject to loss of material by localized corrosion/erosion and/or mechanical damage. Localized defect types are defined according to their geometry as described below:

a. LTA: loss of thickness with smooth contours and without the presence of notches. The width and length are the same order of magnitude.
b. Groove: elongated marks made by removal or relocation of material. The length is considerably larger than the width.
c. Gouge: thinning or elongated slot. Loss of thickness significantly larger than the width of the defect.

Section 6 refers to assessing pressurized components subject to loss of material by pitting. In this context, pitting is defined as localized regions of thickness loss, which can be characterized by a pit diameter in the range of plate thickness or less, and a depth less than the thickness of the plate. Level 1 is a simplified evaluation technique that does not account for the orientation of pits with respect to the direction of maximum stress. Therefore, the results are usually conservative. It is important to note that both pit depth and diameter must be carefully measured because of the variety of types of pits that can occur in service.

The extent of damage defined in FFS is the pit-couple, which consists of two pits separated by a strong ligament. The loss of thickness of each pit is shaped as an equivalent cylinder. To define a pit-couple, diameter and depth of each pit and spacing between centers are required. The occurrence of pits, their size and arrangement within the component are in practice rather random. Therefore, the inspector criterion is key when making decisions regarding sample population representing the damage to the component. Figure 9.2 shows measurements that are required to carry out the assessment.

The steps in the evaluation include:

- determination of the required thickness by standard
- calculation of depth (w), diameter (d), and distance (P) of average pit couples, and
- verification against the acceptability criteria for the service.

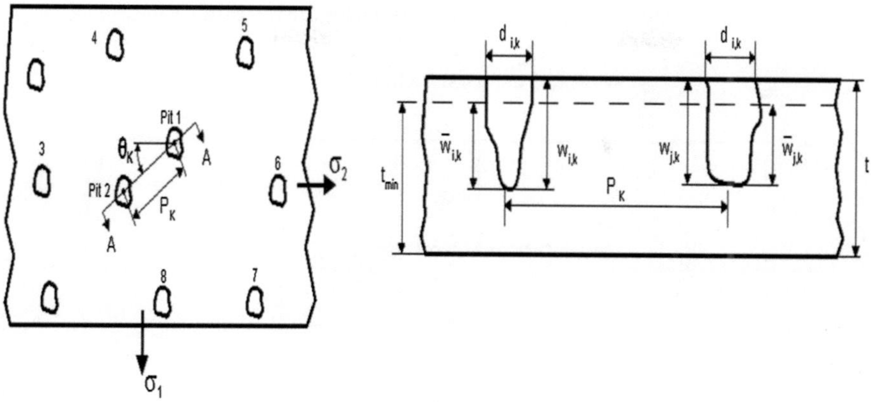

Fig. 9.2 Measurements required to carry out assessment of pitting

9.5 Evaluation of Geometrical Discontinuities and Cracks

API STD 579 recommendations in Sect. 7 involve evaluation of blisters and lamination defects, which are parallel to the plate surface or have a through thickness component. As seen in Chap. 3, steel plates used in the construction of pipes and vessels have metallographic banding which is related to nonmetallic inclusions that are elongated and parallel to the surface. When these inclusions are very large, they give rise to lamination defects, which are planar defects parallel to the surface of the equipment, or in some cases are slightly inclined. A blister is similar to a lamination, but due to the pressure of any gas trapped therein (typically hydrogen or methane), the defect swells and produces a bulge in one of the surfaces. The parameters required for assessment are outlined in Fig. 9.3.

According to Level 1, a lamination defect is acceptable regardless of size, if there is no likelihood of hydrogen absorbed in service, and the distance to any major structural discontinuity and longitudinal welds verify the following equations:

1.
$$L_{msd} \geq 1.8 \cdot \sqrt{D \cdot t_{nom}} \tag{9.7}$$

2.
$$L_{sl} \geq \max(25.4\ \text{mm};\ 2 \cdot t_{nom}) \tag{9.8}$$

In the case of blisters, as well as the equations listed above, the following requirements must be meet to be considered acceptable without repair:

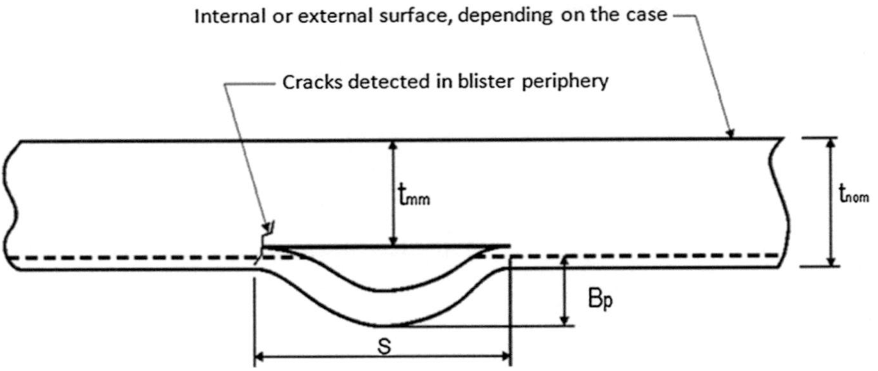

Fig. 9.3 Required parameters for assessment of blisters and laminations

3. Blister diameter is less than 50.8 mm, or $0.6 \cdot (D \cdot t_{nom}^{1/2})$.
4. The projection of the blister Bp over the surface is less than or equal to 10 % its diameter.
5. Minimum measured thickness (MMT) is greater than or equal to half the nominal thickness (t_{nom}) of the plate.
6. No peripheral cracks in the blister are directed toward the outer or inner surfaces of the component.

If the blistered or laminated component does not meet the requirements of Level 1, the damaged plate may be replaced or repaired. A blister may be removed by grinding to a smooth shape. Once grounded, the defect should be assessed according to the guidelines of Part 5 as a localized reduction of section.

Part 8 provides procedures for evaluating geometric irregularities in pressurized components associated with weld misalignment and distortion of cylindrical, conical, and spherical shells. In general, if the current geometry of the component is such that the original manufacturing requirements are met, an evaluation is not required. Exceptions include components operating in cyclic service or localized geometric irregularities, such as an indentation, a dent, or a bump. Level 1 procedures are based on the criteria of the original building codes. In Tables 8.3–8.7 of API STD 579 are specified manufacturing tolerances allowed by design codes, which are summarized in Table 9.1 for pipes and pressure vessels. Level 2 provides computational procedures for the evaluation of the irregularities described before, which do not meet the guidelines of Level 1.

Critical exposure temperature (CET) is defined in Part 3 of API STD 579 as the lowest metal temperature resulting from operation or atmospheric conditions to avoid brittle fracture. The CET may be a single temperature at an operating pressure or a set of temperatures and pressures. The CET is compared with the minimum acceptable temperature (MAT), which is the lowest temperature for a

Table 9.1 Fabrication tolerances–pipes and vessels

Item Design code Fabrication tolerance	Pipeline ASME B31.3 Requirements	Vessel ASME B&PV code section VIII, Div. 1 and 2
Out-of-roundness (cylindrical shells under internal or external pressure)	Dependent on ASTM standard according to which the pipe was designed. In general, the difference between D_{max} and D_{min} diameters should not exceed 1 %	The difference between D_{max} and D_{min} should not exceed 1 % of the internal diameter. In nozzles, this tolerance is increased to 2 %. Reference: UG-80 (a) and UG-80 (b)
Formed Heads	–	Outwards it should not exceed 1.25 % of the inner diameter. Inwards it should not exceed 0.625 % of the inner diameter. Reference: UG-81
Cylindrical shell-to-head attachment weld	–	Radial misalignment between shell and head must be lower than half the difference between the thicknesses of both components Reference: UW-13 (b)
Weld misalignment – longitudinal joints	Should there be longitudinal joints manufactured pipes should meet by default ASTM specifications for which they were designed	For $t \leq ½$ in: $e \leq t/4$ For ½ in $< t \leq 2$ in: $e \leq 1/8$ in For $t > 2$ in: $e = \min(t/16; 3/8$ in) Reference: UW-33 t: nominal thickness; e: misalignment
Weld misalignment – circumferential joints	Weld alignment shall meet the dimensional requirements of engineering design	For $t \leq ¾$ in: $e \leq t/4$ For ¾ in $< t \leq 1$ ½ in: $e \leq 3/16$ in For 1 ½ in $< t \leq 2$ in: $e \leq t/8$ For $t > 2$ in: $e = \min(t/8; 3/4$ in) Reference: UW-33 t: nominal thickness; e: misalignment
Angular weld misalignment	None specified	None specified

given material and thickness, based on its resistance to brittle fracture. MAT derives from information in mechanical design and material specification or data.

The procedures for evaluation of planar defects or cracks included in part 9 of API STD 579 are based in the FAD, which has been explained in Chap. 6, so now only a brief explanation of the methodology is given. Planar defects (or sharp notches) are predominantly characterized by a length and a depth, and are characterized as surface, embedded or through the thickness cracks. Typical examples are weld lack of fusion or lack of penetration, SCC or fatigue cracking, etc.

In some cases it is conservative, and advisable, to treat volumetric defects (such as inclusions or aligned porosities) as if they were planar defects, particularly when defect roots may contain microcracks. For a Level 1 assessment, sufficient conservatism has been added to the procedure to minimize data requirements. In brief, data required are original equipment design, maintenance and operational history, minimum specified yield strength and fracture toughness of the material, and defect geometry. However, it is limited to spherical or cylindrical components with the following further restrictions:

1. $D/t \geq 10$ and $t < 38$ mm.
2. Defects in axial or circumferential directions, away from any structural discontinuity by a distance $\geq 1.8 \cdot (D \cdot t)^{1/2}$.
3. Operating loads only due to pressure, membrane stress lower than design.
4. Carbon steel material, with allowable stress below 25 ksi, yield stress less than 40 ksi and UTS for base and weld metals less than 70 ksi.
5. Fracture toughness greater than reference K_{IC} (ASME BPVC Sect. XI).

Thus, the criterion becomes a quite restrictive condition (ensures Leak Before Break as will be defined later in this chapter). After defining loads and operating temperatures, the CET and the length and depth of the defect determine the representative figure for the case and the selection curve. The maximum allowable length of the defect is determined by entering in Fig. 9.4 with the reference temperature; this allowable is then compared with the evaluated defect.

Level 2 FFS evaluations of components containing planar defects or cracks rely on estimates of crack propagation driving force and plastic collapse equivalent stress, as used in a failure assessment diagram (FAD). The evaluation is then based on the relative location of a criticality point, depending upon stress state and defect geometry. A FAD provided by API STD 579 is shown in Fig. 9.5. The vertical axis provides an indication of resistance to brittle fracture, measured as:

$$K_r = \frac{K}{K_{IC}} \tag{9.9}$$

and the x-axis component evaluates the resistance to plastic collapse:

$$L_r = \frac{\sigma}{\sigma_C} \tag{9.10}$$

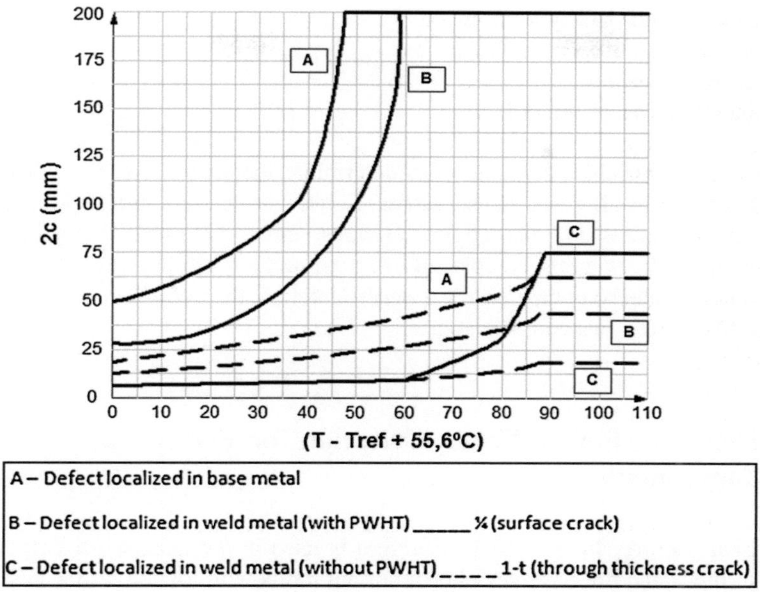

Fig. 9.4 Curves for Level 1, part 9 API STD 579

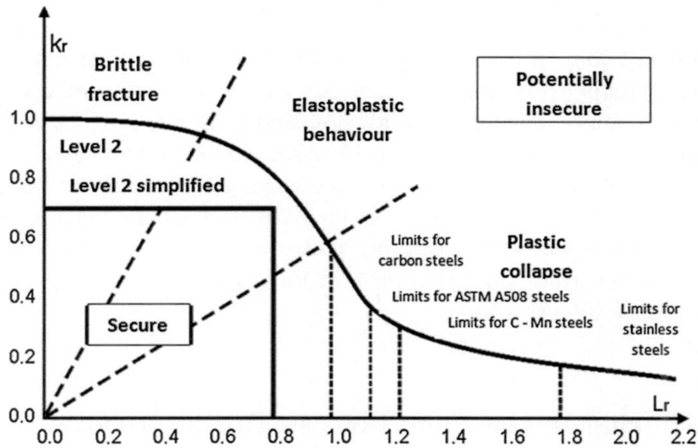

Fig. 9.5 Level 2 FAD for different materials, as defined in API STD 579

Further details of the use of this FAD can be found in Chap. 6. Fracture is expected when Kr > 1, plastic collapse is expected when Lr > 1. Typical data required for the assessment include:

- Original design data of equipment.
- Maintenance and operation histories.
- Material properties.
- Characterization of the defect.

The accuracy of this data conditions determines the accuracy of the assessment; data should be in all cases conservative. Conservative assumptions are made to determine toughness and also applied stresses, and sometimes using a simplified FAD in which the diagram is replaced by maximum allowable limits for stress and applied K, in order to compensate for uncertainties in the definition of geometry of defects from NDT inspection results.

9.6 The Leak Before Break Criterion for Pressurized Components

The "Leak Before Break" (LBB) criterion is used in the analysis of defect criticality in pipes and pressure vessels containing fluids with low or zero toxicity. If a crack propagates into the thickness of a vessel prior to fast propagation, any leaked fluid can be detected and then used to detect the failure and introduce corrective measures to prevent subsequent unstable propagation. This situation is illustrated in Fig. 9.6. LBB is the operating condition (pressure) such that for any planar or volumetric defect, there will be a leak of fluid before the onset of fracture. The determination of this pressure, completed with leak detection inspections, facilitates a safe operating limit for pipes and pressure vessels with defects.

The LBB condition depends on toughness and yield stress of the material, the axial length of the defect and the internal pressure in the pipe or vessel. Figure 9.7 shows an example of a fracture failure (blowout) prediction using the "Corlas" methodology for a 24″ pipeline built in the early 1960s. The horizontal red line indicates maximum allowable operating pressure (MAOP), the blue curve separates fracture (top) from LBB (bottom). In this example, LBB is assured if the crack in 100 mm long or less.

9.7 Monitoring Damage and Stresses During Service

Non Destructive Testing (NDT) tools to detect different damage mechanisms are well known, and their description is beyond the scope of this book. Some have been mentioned in Chap. 2, when used as part of the preliminary evaluation for a failure analysis. For the definition of an inspection plan, we must ask:

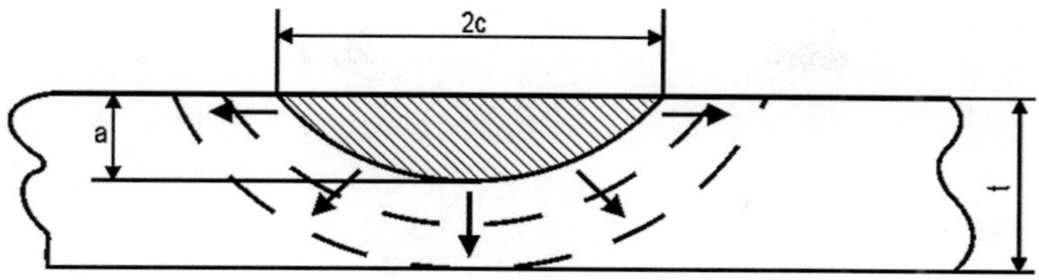

Fig. 9.6 Crack geometry for LBB

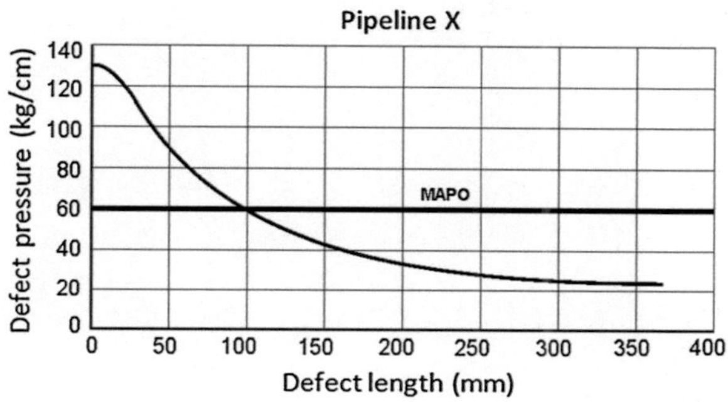

Fig. 9.7 Fracture failure (*blowout*) prediction using *corlas* methodology

- What kind of anomaly or defect are we looking for?
- What tool or methodology are we going to address?
- When are we going to do the inspection?
- How will we react to the discovery of defects?

The development of new technologies has made a mark in the industry. The engineering industry has been under the decisive influence of electronics, automation, and telecommunications, both in the operation of machinery and in industrial maintenance. The industry has new tools to effectively and efficiently operate machinery and raise inspection and maintenance to higher levels. Therefore, some recently developed techniques that increase our capacity for ensuring the FFS of mechanical components are worth noting.

Current NDT tools are based on the same principles as traditionally, but have a degree of automation and information processing that advanced significantly in the past 10 years. For example, NDT technologies have specialized in the development of methods for in-line inspection (ILI) of pipelines, see Fig. 9.8.

These tools, which travel with fluid flow inside pipes, include various types of transducers and are defined according to their purpose:

- Geometric (Caliper), which detect geometry damage such as dents.
- Magnetic (MFL and TFI) use the method of magnetic flux leakage to detect flaws and thickness reductions in tube walls. These tools are used to characterize and size external corrosion defects and internal abnormalities in welds (transverse or longitudinal), grooves, and pits. These do not detect cracks or dents, nor give good estimates of lamination sizes.

Fig. 9.8 Tool for in-line inspection (ILI) of pipelines

- Ultrasonic sensors specialize in the detection of cracks and other defects, too small to be detected by the above methods. Requires a fluid coupling, not suitable for gas pipelines unless it is used with liquid injection (batch).
- EmatScan CD (Electromagnetic Acoustic Transducer) is suitable for operation in gas lines without coupling. This method provides full coverage of the circumference and allows determining length and depth of axial cracks. Colonies of SCC cracks can also be detected.
- GEOPIG: tool with an inertial sensor to determine displacements, particularly in pipelines buried in potentially unstable soils.

Many pipelines could not be inspected by ILI, for example, those which have changes in diameter or violent curvatures. Direct assessment (DA) is a valid methodology as an alternative to ILI, generally considered when ILI or hydrostatic testing are not possible or practical: telescopic connections, small diameters, curves with small radii, valves which are not fully open, low fluid flow or low pressure.

The magnetic tomography measures surface anomalies with a magnetometer; this is an instrument (SKIF) that records changes in Earth's magnetic field above the pipeline. A software calculates the stress concentration sites and classifies them according to their magnitude. Operators interpret magnitudes and characteristics of magnetic anomalies, and locate stress concentrations. This method is still under development. Among its benefits, it should be mentioned that the method can be applied in sections, locates defects and characterizes them by their severity, does not require any preparation of the line, and does not require removing the line from

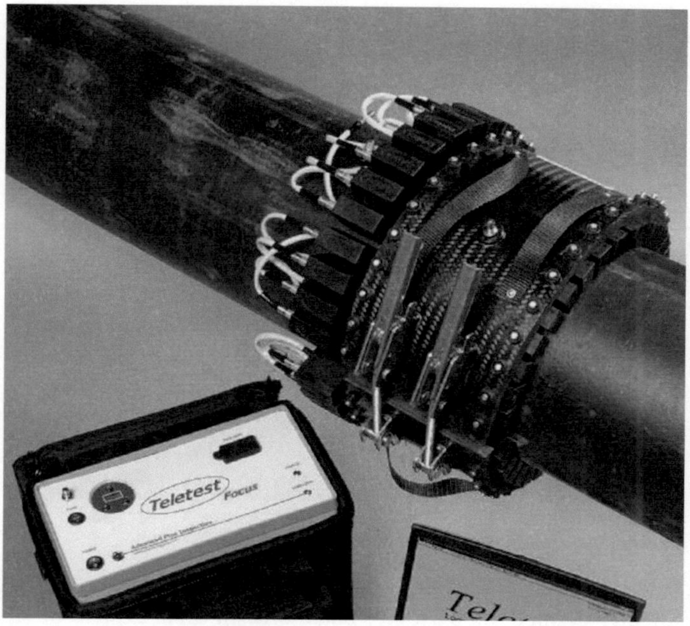

Fig. 9.9 Guided wave US test equipment

service. Among its major limitations, it is noteworthy that the methodology is not yet fully validated or approved by standards or codes, requires thorough pre-inspection assessments, and the probability of detection is less than that for the ILI techniques described before.

Another recent development is an ultrasonic method called Guided Waves Method. This is a long-range technology developed to detect metal loss in pipes. The system is computer controlled, Fig. 9.9. It is a pulse-echo system that aims to test large volumes of material from a single test point, so that it is not necessary to expose the entire pipeline. Initial application was to detect corrosion under insulation in petrochemical pipelines, but is now used in other situations where the lines are not accessible, for example in underground and air pipes, jacketed pipes, etc.

In-service measurement of strains and deformations, and from them the definition of acting stresses, involves various technologies that have also been developed recently. Methodologies developed include both in field measurement techniques and modeling strategies to determine their effects in the integrity of the structures. Figure 9.10 shows such a direct measurement of the stress state in the support of a tubular structure from a suspension bridge located in a region with strong and variable winds.

Electrical resistance strain gauges were mentioned in Chap. 5. These are small stamps that adhere to the component and send an electrical signal, which is proportional to the component deformation at the point (Fig. 5.12). These instruments measure strains experienced at a certain point in a certain direction, stresses at that

Fig. 9.10 Instrumentation of a bridge subjected to heavy winds

point are directly defined via elasticity equations. In a plane state of stress, as on the surface of a component, at least three components of the strain state must be measured in order to define the stress state at any given point.

Therefore, a general case requires placement of three gauges in different directions, close (and small) enough to be considered as being at the same point of the material. Manufacturers provide these multiple gauges in single stamps, which are called rosettes. The angles between strain gauges are typically 90° and 135°, 90° and 45°, or 120°. In case the directions where the maximum stresses are being applied at the point of measurement are known a priori, only two strain gauges may be used. In some cases it is sufficient to determine a single component of the stress state, because the relationship between it and the others is known. Such is the case of the wall of a spherical vessel under internal pressure, for example.

There is a variety of other multiple gauges on the market, varying in geometry and size, for specific applications. We have seen in Fig. 1.9 that ship hulls suffer bending loads due to waves. These are usually alternative loads, namely the hull bends in one direction and then in the other, as the hull is slid over the peaks and valleys caused by waves. Notably, the increasing size of ships has increased the risk of such failures. Today it is relatively common to place strain gauges in ships to check real time stresses in the hull. In these cases, the measurements are made in a longitudinal direction, since these stresses are known a priori to be those that may cause overloads.

Electrical strain gauges, however, have certain limitations when required for use in adverse conditions, for example, submerged, buried or at elevated temperatures. In these cases, other types of transducers are used such as Vibrating Wire Strain Gauges (VWSG). These are then used to monitor hulls, pipelines, and

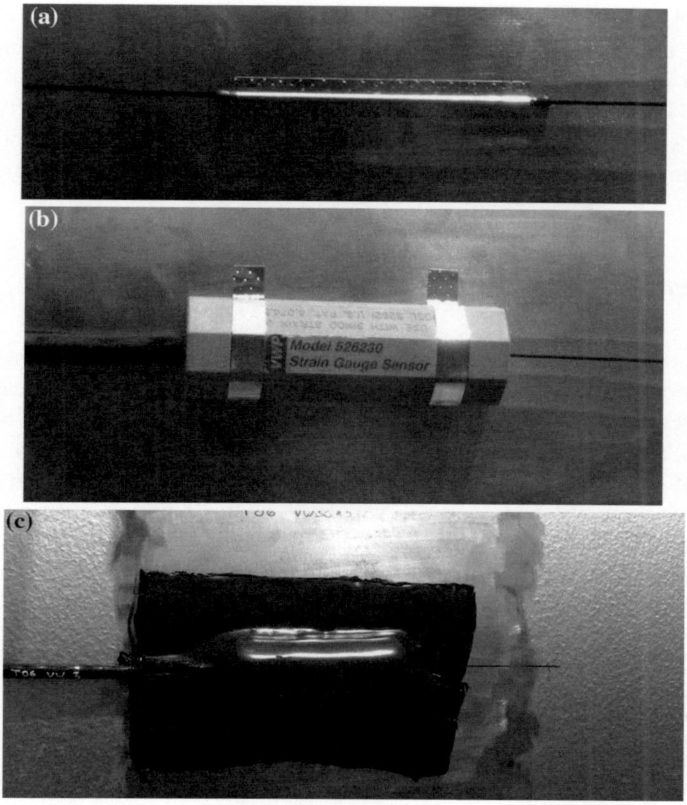

Fig. 9.11 **a** VWSG in component **b** Electromagnetic sensor **c** Protection

other buried structures. Installation requires sanding a flat surface, then soldering the VWSG, Fig. 9.11a. Similar to the string of an instrument, the resonant frequency of the wire increases as it is strained (stressed). Being attached to the component, the frequency of vibration of the wire is proportional to the stress state in that direction. To measure it, an electromagnetic sensor is placed on the strain gauge (Fig. 9.11b). This sensor is usually permanently placed, so that protective layers are placed (Fig. 9.11c).

Three VWSG are usually installed on the wall of the pipe. Having established the maximum allowable strain limits, the pipe should be intervened as soon as these limits are reached. Data can be manually or remotely collected. Periodic gauge readings are required to interpret and decide actions, so that. data collection usually includes an electronic data acquisition system. Data are taken at user-programmable intervals.

The magnitude of the axial stress and its influence on FFS are a fundamental part of the integrity strategies for buried pipes. Several failures of buried pipes, especially in girth welds, are related to the effect of soil movements. These movements may be dramatic (landsides) but are more frequently slow (soil creep), and so are difficult to detect. In unstable soils (for instance in rain forests), soil creep can cause pipe failures over a few years; other failures occurred after more than 50 years of operation. User companies have begun to implement mitigation plans that include measurement and monitoring of deformations in the pipe and adjacent soil, and the application of pipe-soil interaction models (see Example 6.A in Chap. 6).

9.8 Strength Tests Using Instrumented Indentation

Ensuring the integrity of components subjected to mechanical stress depends largely upon knowing the real properties of the materials they are made of. In many cases, transfer of assets from one firm to another, incomplete records or erroneous administrative processes have led to operators unfamiliar with the equipment and materials they are responsible for. This forces them to make assumptions, usually conservative, which increase operating costs. To take a component out of service involves process disruption, loss of income and operations that are usually costly, so that taking a sample of material for laboratory tests is often impossible, or very expensive.

There are nondestructive tests to characterize chemical composition and hardness of a material. API STD 579 presents in Annex F a table that gives a rough equivalence between hardness and UTS (tensile strength) for carbon and low alloy steels in annealed, normalized, and quenched and tempered conditions. At high and low values of hardness, the relationship is not linear, data are very scattered and linear correlations do not fit properly (see Fig. 5.11). This means that while there is a relationship between hardness and UTS, the estimation of this relationship has large margins of error and variability.

Automated ball indenters (ABI) represent a substantial advantage over conventional techniques, since they reduce the error in determining tensile strength and most importantly allow estimating yield strength. The automatic indentation test involves introducing an indenter tip made of known geometry made of a much harder material than the material being tested and while recording maximum load and diameter of imprint. We have seen in Chap. 5 that the elastic–plastic behavior of a material is defined in terms of equivalent stress and strain. This relationship is an invariant, depends only into the material (and to a lesser extent on temperature), independent from the type of test used to determine it (stress, compression, torsion, bending, indentation).

When the indenter is spherical, there are equations to transform the mean pressure and the depth of the imprint into equivalent stress and strain values. These equations are a function of the strain hardening coefficient (n), the yield stress (Fy), and the linear elastic modulus (E). With at least three points of the curve incorporating concepts of linear elasticity, it is possible to find the best fit curve to these points that is precisely the stress–strain curve of the material. Figure 9.12 shows an example of load—imprint depth curves obtained in ABI tests.

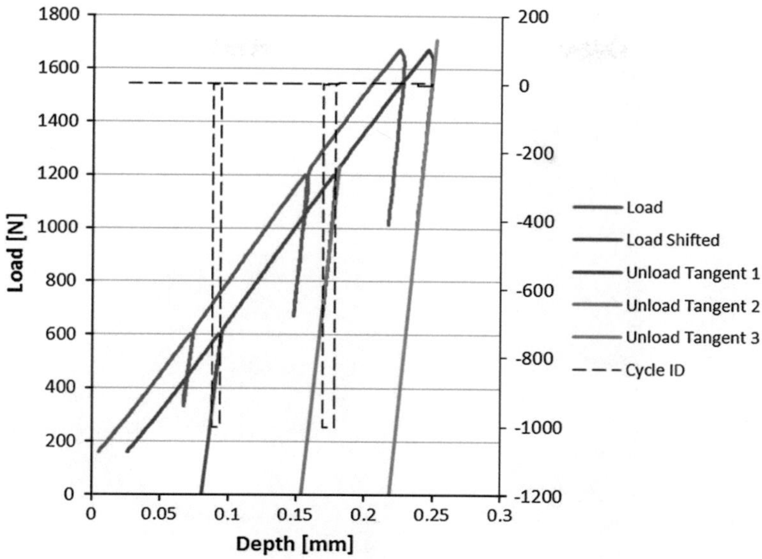

Fig. 9.12 Example of load–displacement *curves* during an ABI test

Figure 9.13 shows one such an ABI indenter. The machine has a stepper motor and a gearbox to allow fine speed control. The box is connected by a clutch to a power recirculation ball screw. This converts the rotary motion of the motor into a linear displacement which permits the introduction of the tip in the material to be tested. The control unit defines indenter penetration according to parameters set by the operator, which allows testing with different loading rates, in a different number of cycles under varying mean and maximum loads. Readings are stored and then processed to obtain the points of the stress strain curve, which finally determine yield strength, UTS, and strain hardening coefficient.

The accuracy of the instrument to determine yield strength is about 5 %. The algorithm is designed so that any errors are by default, that is, measured strengths are lower than or equal to the actual result in a destructive tensile test.

9.9 Monitoring of Machines

Predictive maintenance is a technique for predicting the future point of failure of a machine component, such that the component can be planned to be replaced prior to failure. Thus, equipment downtime is minimized and lifetimes of components are maximized. This technique involves measurement of various parameters which show a predictable relationship with the life cycle of the component. Examples of such parameters are bearing vibration, temperature, electrical connections, the insulation resistance of a motor coil, etc.

Fig. 9.13 ESYS10 ABI testing. Indenter (*left*), equipment (*right*)

Predictive maintenance requires, first, a historical perspective of the relationship between the selected variable and component life. This is accomplished by taking readings (for example, the vibration of a bearing) at periodic intervals until the component fails. Figure 9.14 shows a typical curve resulting from plotting the variable (e.g., vibration) as a function of time. In the example, the bearings should be replaced when the vibration reaches 32 mm/s.

Once determined the feasibility and desirability of predictive maintenance of a machine or unit, the next step is to determine the physical variables that are indicative of the condition of the machine. The aim of monitoring is to obtain an indication of the condition of the machine.

By monitoring we refer to the measurement of a physical variable that is considered representative of the condition of the machine, and its comparison with values indicating the state of deterioration. By automating these techniques, the meaning of monitoring has been extended to the acquisition, processing, and storage of data. The fundamental objective of monitoring is to prevent catastrophic failures. If the machine stops automatically when values indicative of condition become dangerous, then the machine is protected. But monitoring also allows troubleshooting, i.e., defining specific problems. Knowing the rate of damage progression, the ultimate goal of estimating how much longer the machine can

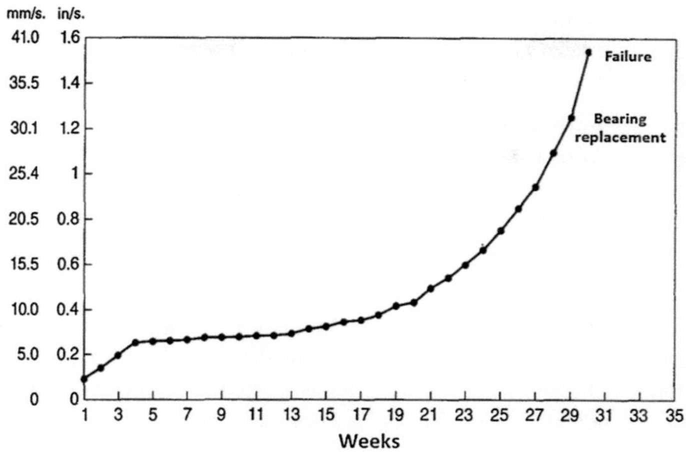

Fig. 9.14 Degradation curve for a machine component

operate without risk of catastrophic failure is achieved. This has led to the application of predictive or symptomatic maintenance through vibration analysis, oil analysis, wear control, etc. There are several techniques used for preventive maintenance, among which are the following.

9.9.1 Vibration Analysis

The interest of preventive and predictive maintenance in mechanical vibrations comes from the interest that a vibrating element in a machine has to prevent failures that vibrations might bring in the medium term. The main interest for maintenance should be identifying the predominant vibration amplitudes in the machine element, determining causes for that vibration, and correcting the problem they represent. Consequences of mechanical vibrations are increased stresses and strains, energy losses, noise pollution in the workplace, wear of materials, and the most feared: fatigue damage of materials.

Vibration parameters are:

- Frequency: The time required to complete one cycle, in vibration studies it is measured in cycles per second or hertz (Hz).
- Displacement: is the total distance that the vibrating element describes, from end to end of its movement.
- Speed and acceleration: as related to the previous.
- Direction: vibrations can occur in three linear and three rotational directions.

Test methods allow to measure oscillations in solids (vibration) and fluids (pulsations). Mechanical vibration is usually divided into two types: free, when a system vibrates at its natural frequency after an instantaneous excitation, or forced, when a system vibrates at an excitation frequency or a certain frequency spectrum. Here are the most common reasons why a machine or an element of it can vibrate:

- Vibration due to imbalances (rotating machinery).
- Vibration due to lack of alignment (rotating machinery).
- Vibration due to eccentricity (rotating machinery).
- Vibration due to failure of bearings.
- Vibration due to problems in transmission gears and belts (clearances, lack of lubrication, friction, etc.).

There are also integrity monitoring techniques for static structures that involve measuring vibrations. The existence of structural damage leads to the modification of vibration modes. These modifications are manifested as changes in modal parameters (natural frequencies, mode shapes, and damping values). The natural frequency is a sensitive indicator of the structural integrity. Thus, periodic natural frequency analyses allow evaluating the structure with reduced cost, since frequencies are easily available. Furthermore, the use of dynamic data is convenient because accelerometers can be easily attached to any structure, unlike displacement transducers (think of bridges to be tested when they pass over a river valley or a highway).

Changes in modal parameters may not be the same for each mode, because these changes depend on the nature, severity, and location of the damage. Tests developed at different times allow monitoring the structural condition over time. Responses are obtained from different types of transducers, while the excitation can be purposely caused or from natural service. However, to cause artificially transformed vibrations leads to relatively expensive tests.

9.9.2 Lubricant Analyses

These analyses are executed depending on the need:

1. Initial analysis: done on product or equipment in which doubts arise and allow corrections to product selection or changes in operating conditions.
2. Routine analysis: applies to large equipment or those considered critical, in which a sampling frequency is defined. The main objective of the analysis is to determine wear and contamination levels in the oil, among others.
3. Emergency analysis: This is done to detect any irregularity in the equipment and/or lubricant in cases of:
   - Contamination with water or solids (filters and faulty seals).
   - Use of an incorrect product.

This method maximizes reduction of operating costs, service life of components with minimum wear and lubricant use, while minimizing effluent generation. Studies in each sample allow detecting the following factors affecting a machine:

- Wear items: iron, chromium, molybdenum, aluminum, copper, tin, lead.
- Particle count: determining cleanliness, ferrography.
- Contaminants: silicon, sodium, water, fuel, soot, oxidation, nitration, sulfates, nitrates.
- Additives and lubricant conditions: magnesium, calcium, zinc, phosphorus, boron, sulfur, viscosity.
- Graphical records and history: trends over time.

Thus, by implementing widely researched and experienced techniques, and latest technology, dramatic decreases can be achieved in lost production time due to mechanical breakdown, wear of machines and their components, man-hours devoted to maintenance, and overall use of lubricants.

9.9.3 Acoustic Emission Analysis

This method studies the sound waves produced by equipments that the human ear do not perceive. Almost all mechanical friction, electric arcing and pressure or vacuum leaks generate ultrasound in the approximate range of 40 kHz. This is often quite useful in predictive maintenance. Sound waves are short and attenuate rapidly, thus reflections are small and noise does not interfere with detection. Ultrasonic measurement is complementary to vibration measurements. The application of ultrasonic analysis is especially useful in the detection of failures in rotating equipment rotating at speeds below 300 RPM, where vibration measurement is not an efficient technique.

Passive ultrasonics is a technology that allows capturing the ultrasound produced by various sources. With proper detection technology, acoustic emission methodologies allow detection of friction in rotating machines, mechanical defects, valve leaks, fluid leaks in vessels, loss of vacuum, electric arcs, and in general, verifying the integrity of sealed joints in enclosures.

The high directionality of an ultrasound wave can quickly and accurately detect the location of a failure. This method is used in static components to detect lack of thickness, cracks, and other defects that generate high stress concentrations. The element producing acoustic emission is, in this case, the plastic deformation of the material in the overstressed area. Therefore, it is required that the defect is "active" during the measurement. This is ensured when the component is subjected to higher than normal stresses, for example, by increasing pressure in pressurized components. Placing multiple sensors along the component, the position of a defect is determined by triangulation; an analysis of the acoustic signal "fingerprint" allows assessing the type of defect. However, once the failure is found, other NDT techniques (ultrasound, X-ray) are normally used to measure and characterize the defect more precisely.

9.9.4 Analysis by Infrared Thermography

We have seen that some of the damage mechanisms to components and structures are associated with high temperatures (creep and some corrosion and wear mechanisms) or sudden temperature changes (transients), both in time and in space (thermal fatigue, etc.). Therefore, a detailed map of temperature distribution is, in many cases, a fundamental tool for ensuring the integrity of both static components and machines. The vast majority of problems and failures in the industrial environment, whether mechanical, electrical, or manufacturing, are preceded by changes in temperature that can be detected by monitoring the temperature with an Infrared Thermovision system. Thermographic analyses are based on obtaining the surface distribution of temperature in the component or system, with which a temperature map (thermography or thermogram) is obtained.

Thermography is a technique that allows accurate temperature measurements without physical contact with the object being studied. Human eyes are not sensitive to the infrared radiation emitted by an object, but thermal imagers, or thermo-vision, are able to measure the energy with infrared sensors, thus "seeing" in these wavelengths. This allows measuring the radiant energy emitted by objects, and thus determining the temperature of the surface at a distance, in real time, without contact.

This technique allows detecting any hot or cold spots and related abnormalities. Thermographic studies are now widespread for testing machines and systems related to heat engines and power generation and distribution systems, chemical processes, etc. The study of energy distribution systems can alert of energy losses that are caused by poor insulation, a break, or a bad gear.

Thermal imaging cameras are portable devices and their application is defined by the range of temperatures that they can track. The most common, for lower temperatures, apply to viewing areas with potential failures with terminal boards, transformers, etc., and for monitoring and preventive maintenance of motors, bearings, and stress insulators. Creep assessment applications require a higher range of temperatures, and thus more expensive equipment.

9.10 Estimation of Residual Life Against Damage Mechanisms

We have seen (Fig. 9.1) that failure rates, or the cost associated with them, grow as the component or system ages. This is because almost all in-service degradation mechanisms are synergistic, that is, as damage grows, the rate at which it grows is also higher. Examples of this can be found in wear (see Fig. 9.14), fatigue crack propagation (see Fig. 6.D3), microstructural degradation by creep, SCC crack propagation, etc.

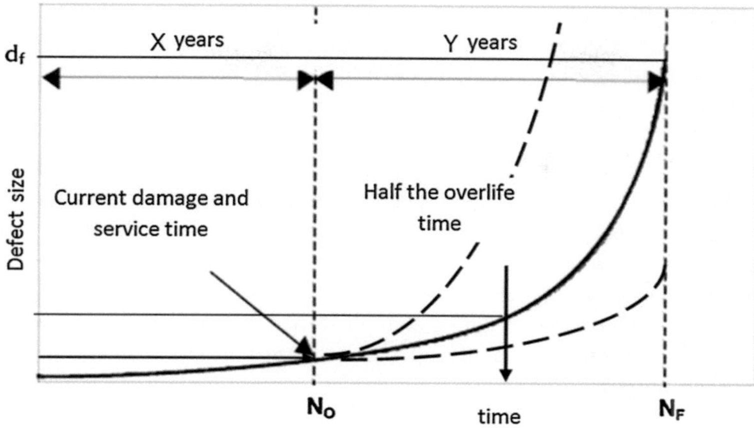

Fig. 9.15 Remaining life estimation

Predicting the time during which a component may perform safely requires an integration of damage propagation, between an initial damage (defined, for example, as the largest defect that could have escaped detection during pre-service inspection), and the damage that would lead to failure. We have seen in Chap. 6 how to perform this process for the case of constant amplitude fatigue crack propagation (Sect. 6.10 and Example 6D).

That was a relatively simple example, in which the component had been subjected to cyclic stresses of magnitude approximately constant throughout life. This allowed integrating the fracture mechanics equations that model the propagation of the fatigue crack, thus predicting component life with good accuracy. Similarly, increasing vibrations in a bearing (Fig. 9.14) shows a predictable increase rate only if the bearing is subjected to a service system (defined by loads, rotation speed, temperature, lubrication conditions, etc.) that is kept approximately constant.

Another example of remaining life estimation is shown in Fig. 9.15. The current magnitude (time N_o) of damage is defined by d_o, and d_f is the allowable damage (to be reached in time N_f). To avoid a high probability of failure, the component must be repaired or replaced at some point before the Y years to N_f are met. Note in Fig. 9.15 that at half the total survival time the damage "d" only represents about 25 % of the allowable damage. The rate of accumulation of damage increases in the later stages of the process; so defining the remaining life based on damage size would lead to unsafe situations. If the failure is the breaking point of a bearing or a steam leak, then we can approach N_f, but if the failure involves a blowout due to fast crack propagation, then mitigation should be carried out when the defect is only a small fraction of its critical size.

An additional problem, which further reduces the reliability of these predictions, appears when there are sudden changes in operating conditions. Important changes in the slope of the curve occur, for example, if the amplitude of cyclic stresses increases or decreases with respect to historical magnitudes. This leads to life curves as indicated in dashed lines in Fig. 9.15. It is obvious that the accuracy of predictions decays, and the allowable service time before repair should probably be reduced.

Fortunately, the variation of the amplitude of fatigue cycles is to some extent predictable, and is related to historical changes. We close this chapter with an example of remaining life prediction under these conditions, in which fracture mechanics techniques defined in Chap. 6 are extended with the introduction of the concept of average annual damage. This concept states that even in cases of varying loads, it is possible to define an annual load spectrum that is repeated over time, so that the problem is analyzed based on what fraction of the total life is spent in each period.

9.10.1 Example 9.A Fatigue Life Prediction in a Pipeline

In-service pressure cycles in a pipeline are generally not of constant amplitude, so that estimates of fatigue life for an intended (design) or past (integrity) service require the definition of a load spectrum (in this case, pressure vs. time). The spectrum is broken into blocks of pressure cycles of different amplitudes; there are several counting methods for doing this. Of the accepted methods for counting cycles, as indicated in ASTM E 1049, the most widely used is the so-called Rainflow method. This method optimizes the detection of large cycles in broader load spectra. Numbers of stress cycles of different amplitudes are defined this way, either from historical records or from predictions of future service of the specified or similar equipment,

Stepwise integration is used then to assess fatigue damage done by each of the cycle amplitude blocks, using Miner's rule for variable amplitude cyclic stresses. The Miner criterion simply sum all damages done by each block of cyclic stress amplitude. The damage produced during 1 year, for example, is the inverse of the number of years of total life calculated for a given block. The total damage in 1 year is the sum of damages produced by all the blocks into which the stress spectrum was subdivided. In this way, the cumulative fatigue damage into the component in 1 year of operation is estimated, and from this value, calculation of the remaining life in years is straightforward:

$$\sum \frac{n_i}{N_i} = C \tag{9.11}$$

where $\mathbf{n_i}$ is the number of cycles to which the component is subjected to a stress amplitude $\boldsymbol{\sigma i}$, $\mathbf{N_i}$ is the number of cycles required to cause failure at that stress amplitude and C is a constant (usually considered equal to one). The value of $\mathbf{n_i}$ is a material constant that can be obtained from S–N curves available in the literature or from the integration of the Paris equation, as seen in Chap. 6. Most design or integrity assessment codes include evaluation criteria that provide fatigue curves for steels and other common materials (see for example Table 6.3).

To estimate the fatigue life of a variable amplitude-loaded component, initial and final defect depths are first defined, as we saw in Sect. 6.10. After performing cycle count and grouping blocks of equal pressure amplitude, only those pressure amplitudes that produce fatigue propagation, that is, stress cycles larger than ΔK_{th} (threshold ΔK, Sect. 6.10) are taken into account. Larger stress cycles obviously generate higher propagation rates, but frequencies of occurrence of these large cycles are usually small. Then the Miner criterion of damage summation is applied, and fatigue life in years is then estimated.

The reliability of these estimates depends on maintaining the assumed operating conditions in the future. If there is a substantial increase in frequencies or amplitudes of cycles (especially those of large amplitude), life predictions should be updated periodically.

Let us assess as an example the fatigue life of a 12″ diameter, 4.78 mm thick steel pipeline with 25 years of operation. Figure 9.A shows the distribution of number of pressure cycles as a function of their amplitudes, obtained with the Rainflow counting method in a 12-month period of pipeline operation. The detection limit of the nondestructive testing tools, i.e., the largest defect depth that may have gone undetected, is 2 mm. The critical size of a crack in the outer surface (i.e., that would cause a blowout at maximum operating pressure) was calculated in 3.32 mm. Let us simply consider a constant crack shape ratio $a/c = 0.1$. Fatigue propagation properties for this material are defined in Table 6.3.

From that table, fatigue threshold is $\Delta K_{th} = 2$ MPa $m^{1/2}$. What pressure cycles should not be considered?. For an average crack depth of 2.67 mm, the minimum amplitude of pressure cycles below which fatigue crack propagation does not occur is $\Delta P_{th} = 6.6$ Bar (threshold cut in Fig. 9.A).

As seen in Fig. 9.A, there are a limited number of cycles of large amplitude, which are the most harmful. To predict the number of total life cycles for the hypothetical crack under study, the crack propagation Eq. 6.14 between initial and final crack depths is now integrated. Thereby determining the number of cycles required to propagate the crack from 2 to 3.32 mm, considering the effect of each of the amplitude blocks separately. The inverse of the estimated life is the damage to the pipeline due to the cycles in each block. Then cumulative damage in the pipeline in 1 year of operation is estimated using the Miner criterion. Again, the inverse of this value is the total fatigue life in years.

Table 9.A shows the results of calculations of the average propagation rate of fatigue cracks for each of the pressure amplitude blocks. Calculations were performed for the two extreme conditions defined in Table 6.3: base material and ERW material. Larger propagation speeds rates are predicted for the weld. To reduce the length of the table, and as just an example, only rows for smallest and largest amplitudes are included in the table.

The last two rows show the total fatigue damage (as a fraction of total life) predicted in a year of operation, and the remaining life of the pipeline. Estimated

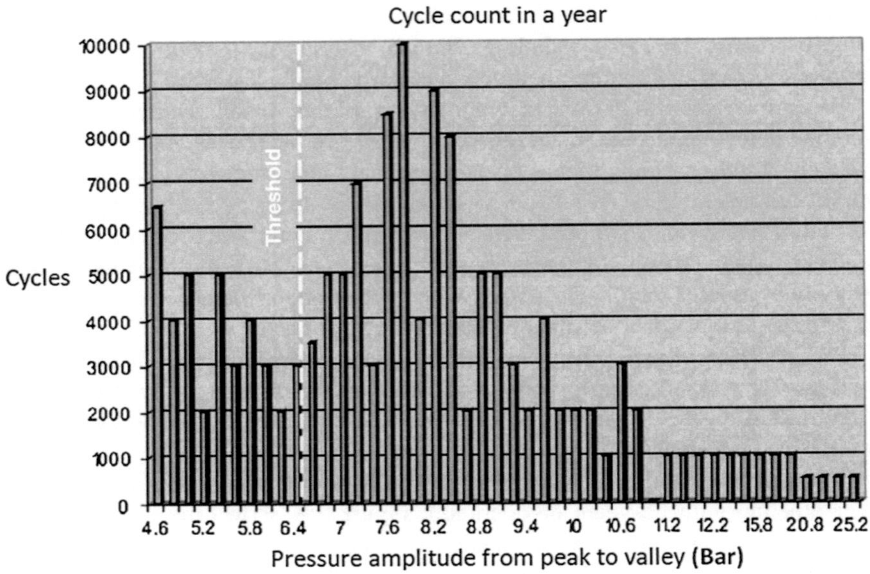

Fig. 9.A A distribution of amplitudes of pressure cycles

Table 9.A A Remaining fatigue life, to propagate a crack from 2 mm deep to 3.32 mm

Delta P (Bar)	Cycles per year (n)	Welding		Base material	
		Cycles until break (N)	Damage in a year	Cycles until break (N)	Damage in a year
6.6	3,500	7.47E+08	4.687E–06	5.79E+09	6.050E–07
6.8	5,000	6.43E+08	7.773E–06	4.84E+09	1.034E–06
7	5,000	5.56E+08	8.986E–06	4.06E+09	1.230E–06
7.2	7,000	4.83E+08	1.448E–05	3.43E+09	2.039E–06
20.8	500	2.40E+06	2.082E–04	5.90E+06	8.468E–05
23	500	1.45E+06	3.441E–04	3.23E+06	1.548E–04
25	500	9.58E+05	5.221E–04	1.96E+06	2.553E–04
25.2	500	9.20E+05	5.433E–04	1.87E+06	2.678E–04
Sum of damage			2.691E–03		1.026E–03
Predicted remaining life [years]			371.57		974.53

life is around 370 years. This is not such a high value as it seems: code rules for buried pipelines state that calculated life must be at least three times the desired life. Therefore, this pipeline would be able to operate safely against this type of damage for about 120 years.

References

1. J.L. Freire et al., Aspectos Generales de la Evaluación de Integridad y Extensión de Vida de Equipamiento Industrial y Estructuras. Doc. No. 1. Proyecto PROMAI. Impr. CNEA, Buenos Aires. 1996
2. J.L. Otegui, E. Rubertis, Cañerías y recipientes a Presión. Eudem, ISBN 978-987-1371-18-1. 2008
3. ASME Boiler and Pressure Vessel Code, Section V, *Nondestructive Examination*. The American Society of Mechanical Engineers 2004
4. R6, Assessment of the Integrity of Structures Containing Defects. British Energy Generation Report R/H/R6 Revision 3 1999
5. SINTAP, Structural Integrity Assessment Procedures for European Industry. Project BE95-1426. Final Procedure, British Steel Report, Rotherham 1999
6. R.A. Ainsworth et al., An overview of the European flaw assessment procedure SINTAP and its validation. Int. J. Press. Vessels Pip. **77**, 869–876 (2000)
7. BS 7910:1999, *Guide on Methods for Assessing the Acceptability of Flaws in Fusion Welded Structures* (British Standard Institution, London, 1999)
8. API RP 571, *Damage Mechanisms Affecting Fixed Equipment in the Refining Industry* (American Petroleum Institute, New York, 2003)
9. API STD 579, *Fitness for Service* (American Petroleum Institute, New York, 2007)
10. T.L. Anderson, *Fracture Mechanics: Fundamentals and Applications*, 2nd edn., (CRC Press, Florida, 1994)
11. J. Barsom, S. Rolfe, *Fracture and Fatigue Control in Structures* (Prentice Hall, New Jersey, 1987)
12. API RP 1111, *Design, Construction, Operation, and Maintenance of Offshore Hydrocarbon Pipelines (Limit State Design)*, 3rd edn., (American Petroleum Institute, Washington, 1999)
13. J.C. Wachel, J.D. Tison, Vibrations in reciprocating machinery and piping systems. Engineering Dynamics Incorporated – Proceedings of the 23rd Turbomachinery Symposium, Texas A&M University, USA, 1994
14. I. Franco, *Mantenimiento Predictivo* (Venezuela, 2004), www.monografias.com
15. R.C Rosaler, Manual del Ingeniero de Planta. Mac-Graw-Hill/Interamericana de Editores, S.A. de C.V (2002)
16. L. Bittel, J. Ramsey, *Enciclopedia Del Management* (Ediciones Centrum Técnicas y Científicas, Barcelona, 1992)
17. ASTM E 1049-85 (reapproved 1990), *Standard Practices for Cycle Counting in Fatigue Analysis* (American Society for Testing and Materials, USA, 1998)
18. J.L. Otegui, M.D. Chapetti, J. Motylicki, *Fatigue assessment of an E.R.W oil pipeline*. Int. J. Fatigue, ISSN 0142-1123, **24,** 21–28 (2002)
19. P. Von Versen, (2012) www.npd-solutions.com

Chapter 10
Consequences of a Failure Analysis

10.1 Writing Reports After a Failure Analysis

A well-written report that clearly explains what happened, how it happened, and why it happened is valuable and serves many purposes. There are several formats for reporting the results of an investigation. The easiest format is a simple narrative, where the investigator simply describes the investigation in chronological order. A narrative report works fine when the investigation involves only some issues and the evidence is simple. However, it can be difficult for a reader to imagine the event reconstruction when there are many facts to consider. The line of reasoning becomes entangled when test results and lab reports, eyewitnesses who have to be correlated with each other, and several scientific principles must be interconnected. Thus, sometimes some answers are shown before questions have been fully developed.

Alternatively, an investigation report can be prepared as an academic paper, full of technical jargon, equations, graphs, and copious reference notes. This type of report, appropriate for research reports or for conclusions of a particular investigation by experienced professionals, is often insufficient. To determine what type of format to be used, it is important to first consider who will read the report. The audience of a failure investigation report includes the following:

- Claims adjusters (insurance): The adjuster will use the report to determine if a claim is under the terms and conditions of the insurance policy. If potential subrogation is suspected, adjusters would forward the report to the company attorney for evaluation. Subrogation is a type of lawsuit that an insurance company files to a third party that might have had something to do with the cause of the loss, to recover at least part of what is to be paid for a claim.
- Authorities: they are certainly interested if the report is related to a crime, criminal negligence, or violation of the law. The report may become part of the body of evidence for the prosecution. They can compare the facts and findings of the report against what can be collected and either repeat or check any lab work to confirm the conclusions and observations that are discussed in the report. Authorities may take possession of any evidence stored during the

J. L. Otegui, *Failure Analysis*, DOI: 10.1007/978-3-319-03910-7_10,

preparation of the report. If evidence or information indicates that there was a crime, in most cases there is a legal and professional obligation to bring this to the attention of the authorities. The only exception that could apply to this obligation is when the principle of attorney-client privilege is invoked. If this is the case, carefully clarify obligations to the registered attorney in this matter.

- Lawyers: they examine every line and every word used in the report. Often, lawyers give meanings to words or phrases that the investigator never intended to give. The legal meaning may be different to the engineering meaning. Engineers are generally poor writers, lawyers live by it. Despair language skills often provide lawyers on both sides a variety of possibilities for reinterpreting the engineer's report in the sense that they need.
- Technical experts: technical experts who read the report have knowledge and skills similar to those of members of the investigation team. Experts on the other side will challenge every facet of the report that is detrimental to their client and try to show that the report is a sham and worth nothing. Any research standard will of course be shown to be incorrect, or misapplied, not as good as those used by the other experts. A common technique used to discredit a report is to segment it into tiny pieces, none of which, considered individually, is harmful. This technique is designed to disconnect the interrelationships of the various components and destroy the context and the general sense.
- The same author: years after the end of the report the investigator may have to address the issue again. Court cases can take several years and the expert should address his own report to remember the details of the case.
- Judge or jury, most jurors get uncomfortable with equations and statistical data. Since high school, their main sources of new scientific or technological knowledge have been television shows and popular publications.
- Administrators: people who run the company want to understand what is in the report and use the report as a basis for corrective action to prevent something similar from happening again.
- Other professionals: in related industries, who wish to avoid such events in their plant or facility.

To meet these diverse audiences, the report format should be consistent with the method of the investigation pyramid noted in Chap. 7. The report shall include:

- *Report Identification*: This includes the title and date of report, the names and addresses of the author and customer, and any identifying information, such as file or case number, and the date of the failure.
- *Purpose*: This is a concise statement of what the investigator is trying to accomplish. It is usually a single sentence or short paragraph.
- *Background*: This part of the report sets the stage for the rest of the report; contains general information about what happened to cause the reader to understand what is being discussed.

- *Analysis*: This is the section in which the investigator explains how different events relate to one another.
- *Results*: This is a list of all factual findings and observations relevant to the investigation. The organization of events is important. Listing the general observations and conclusions first and then more detailed elements is a useful technique.
- *Conclusions*: In just a few sentences, maybe even one, summarize findings. The conclusion must be said clearly, unequivocally, using the indicative verbal form.
- *Observations*: Sometimes during the course of investigation information is obtained in other matters that may affect public safety and welfare.
- *Appendices*: Detailed calculations, specific tests, or relevant data but extensive to report are to be included in appendices. Calculations and data put here ensures that the general reading flow of the report is not interrupted for readers who cannot follow the detailed calculations or simply are not interested in them. And for those who wish to immerse themselves in the details, these are available for review.
- *Attachments*: This is the place to post photos and descriptions, excerpts of standards and codes, lab reports, and other related items which are relevant but are too large or which direct insertion into the body of the report is inconvenient.
- *Executive Summary and Table of Contents*: When the report is long and complex a summary and/or a table of contents may be added at the front of the report. The executive summary, which is usually a few paragraphs and no more than one page, notes the highlights of the investigation, including the conclusions. A table of contents indicates the organization of the report and allows the reader to quickly find the sections and items he might want to review.

10.2 Failure Analysis in Potentially Litigious Cases

Recent trends in the law are creating a growing number of lawsuits against individuals and organizations that handle materials involved in litigation. The legal obligation begins when the investigating team or person is handling or performing tests on a sample and causes or permits any degradation, loss, destruction, or alteration of material or documentary evidence. This type of obligation can be a significant problem when pursuing analysis, tests, or even storage of materials that are or may be involved in litigation. The concept not only applies to physical evidence, but also photographs, documents, computer records, or any evidence that a person has the duty to preserve.

It is common for independent forensic consultants to protect and maintain the materials involved in litigation. These people are becoming the focus of lawsuits. The legal obligation may be imposed even if they followed all instructions and have made appropriate tests. The negligence claim could be generated not only by the directly involved, but also by any other party who may have been adversely affected by any loss or damage of materials, or even the loss of the opportunity to be present and observe the document, examination, or destructive testing.

ASTM issued a standard (E860) that is applicable to this type of examination. Some of the most important requirements of this standard include the following:

1. Material involved is what is or could reasonably be expected to be the subject of litigation.
2. Specific actions are required if any testing, examination, removal or other action alters the nature, status or condition of the evidence, and prevent or limit further examination.
3. The analyst must provide photographic or other evidence documented before any examination.
4. If test conditions can alter or limit other evidence or future tests, it is required that the person performing the test notifies the client of this fact and recommend him to notify all other potential stakeholders. It also requires the analyst to advise the client that all other stakeholders are given the opportunity to witness and videotape any test, disassembly, inspection, etc.
5. At the end of the test, all the evidence must be returned or retained, unless all parties agree for it to be disposed.
6. Tests should always be delayed until all concerned have been notified, except when the delay adversely affects the quality of the results due to changes in the state of the evidence.
7. The rules also require full documentation of all testing, examination, removal, or other actions. Under the requirements of this standard, the analyst might be "negligent," even if following client's instructions, if these instructions do not conform to the standard.

The analyst should have a clear understanding of his specific role and the nature and extent of any examination or test to be carried out. If the customer does not provide this in writing, the analyst can send a written description of what is understood the requested actions are and ask the customer to confirm this understanding in writing. If the analyst doubts whether the customer is the actual owner of the material in question, or has the authority to request a review, he must raise it to the client. The analyst may need to get additional specific authorization from the actual owner.

If the client is not aware of proper procedures, the analyst must advise him of the rules and requirements. Although the services may be requested by a large corporation or an insurance company, it is best to discuss these issues openly and not assume that the person with whom the analyst is dealing with is aware of the proper procedures.

The analyst shall inform the customer of the nature and extent of any destructive testing, and ask to provide proper notice to all interested parties (such as those required by ASTM standards). The analyst should seek a written guarantee of his client who claims that this has been done before starting any destructive testing. When other potentially interested parties are involved in the study, protocols will be agreed in order to determine the precise order and mechanism of testing the evidence, as well as document the agreement between the groups with regard to procedures and how others can participate in the test.

The procedures for inspection of materials must include labeling or identification of all relevant parts, documenting the chain of custody during transport, and storage of such materials. If the materials involved are potentially degradable in a short period of time (evaporation, oxidation, etc.) the necessary instructions and authorization must be acquired to proceed with the appropriate test. Material must be photographed often to document its condition, as we saw in Chap. 3: (a) upon receipt, (b) before any test, (c) during any removal or review, (d) at the conclusion of any test. Material conditions should also be documented at the time which it is no longer in the analyst control and is delivered to some other group.

In potentially litigious cases, the analyst must have professional liability insurance or the like and determine if he has adequate protection against potential claims. For cases involving substantial amounts, analysts could have at least one person present during the examination and tests to verify and document the condition of the material.

The analyst must develop forms that define the reception of materials, the scope of the assignment, notification of destructive tests, demand for the customer to notify third groups, storage, and shipment or disposal of the material. An experienced attorney can be helpful at this point. When asked to prepare reports of incidents at issue, or expert opinions, the analyst should consult the requirements of ASTM E-1020 and ASTM E-620. Confidentiality is extremely important in cases involving litigation. As much as the findings can be academically interesting, it is risky to share or discuss these findings, even on an informal basis. In case of questions about who could be informed of findings, the analyst should clarify it in writing with the client.

The analyst will determine if the client wants a written report, whatever the outcome. In some cases, the client may want to report findings orally, and then decide the need or magnitude of the written report. The analyst has a duty to report findings accurately and completely to the client. However, in some cases, the client may not want to have a written report, he will define the nature and extent of the documentation. The analyst should be aware that all documents, correspondence, files, notes, records, etc., may be required in a lawsuit and shown to a judge. The analyst must choose carefully before discarding preliminary documents or notes, even if they have already been fully incorporated in the final version of the report.

When storing materials, written documentation must be kept well. It is important to establish a chain of custody and to show that no unauthorized person may have had access to the material. It is advisable to store the material in a labeled container with an inviolable seal. In particularly contentious cases it is preferable to return the material to the customer after completing the examination. Dispose only when it is clear that no possible use of the material is required in the future; no material should be disposed of without customer verification, in writing.

10.3 Learning or not Learning from Accidents, Here the Question

Confucius, a Chinese positivist philosopher said some thousands of years ago that man has three ways of learning (Fig. 10.1):

- First, by meditation, this is the noblest.
- Second, by imitation, this is the easiest.
- Third, through experience. This is the most bitter.

Applied to risks of failure:

- Meditation includes systematic techniques such as Hazard and Operability Study (HAZOP) and design calculations.
- Imitation includes learning from the experience of others, as described by design codes, codes of practice, books, and reports of accidents.
- Experience waits until we've had an accident ourselves.

Samuel Coleridge described History as "a lantern on the stern" that illuminates the risks through which the ship has passed, instead of illuminating those risks hiding ahead. The loss prevention expert Bill Doyle said that for every complex problem there is at least a simple, plausible, and wrong solution. The tools defined in this book aim to ensure that failure causes remain truly enlightened, and to establish appropriate preventive measures.

Since its first public announcement, Murphy's Law is usually summarized as "What can go wrong will go wrong," and has spread rapidly to all technical cultures. The spirit of the phrase implies the principle of defensive design; anticipate errors that the end user will likely commit. The final user (Murphy's assistant in the historical version discussed in Chap. 1) could choose how to connect the cables. When chosen incorrectly, the sensors did not do their job properly. In most user-oriented well-designed technology, incorrect connections become difficult. For example, the 3.5-inch disk used in computers would not fit into the drive unless it is oriented correctly. In contrast, the 5.25-inch previous disks could be inserted in a variety of orientations that could damage the disc or the drive. But again, compact discs allow being inserted upside down.

Corrective action should at least remove the immediate causes and should eliminate or mitigate the negative effects of the root causes. When there are multiple causes, no budget, or is difficult to determine what should be corrected:

- A quantitative analysis can be used to determine the total contribution of each cause to the undesired result.
- A concurrent diagram may be used to place causes in order of importance.
- Causes that contribute most to the undesirable outcome or adverse effects must be removed or mitigated to minimize the risk.

Fig. 10.1 Learning methods

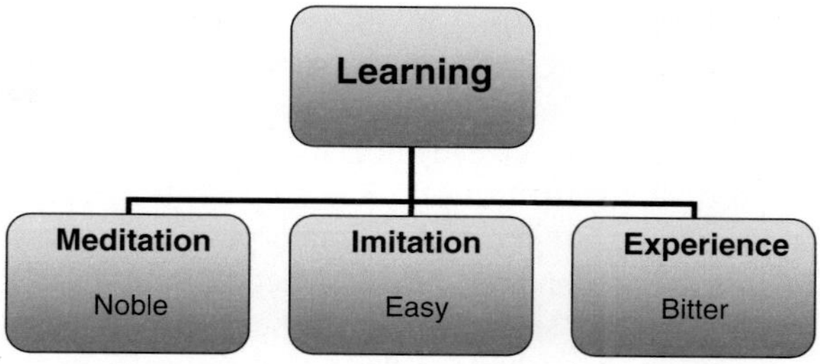

In Chap. 7, we addressed some of the recommendations included in the standard DOE-NE-STD-1004-92, a guide for root cause analysis in the nuclear power industry. This Standard also addresses and defines some guidelines to ensure that corrective measures are derived from an effective RCA. Specified remedial programs include:

- Emphasis on the identification and correction of problems that may affect the performance of individuals and teams, including the assignment of qualified personnel, corrective actions, and monitoring to verify the effectiveness of corrective actions.
- Development of administrative procedures to identify resources and assign responsibilities.
- Development of a work environment that demands accountability and allow correction of obstacles to the reliable performance of individuals and teams.
- Development of a work environment that encourages the voluntary reporting of deficiencies, errors, or omissions.
- RCA training programs, training of staff and management levels to recognize and report incidents, including in particular the identification of significant and generic problems.
- Development of programs to ensure the prompt investigation following an incident or identification of significant reductions in production.
- Adoption of a mechanism for identification and classification of the mechanisms and factors that pose problems, particularly those with global implications.

R. K. Noon mentions in his quite attractive book "Scientific Method, Applications in Failure Investigation and Forensic Science" four common reasons why corrective actions resulting from a failure investigation sometimes fail:

1. *The tail wags the dog*: If the analyst anticipates that some corrective actions will be more palatable to the customer or the authorities, he can begin, more or less consciously, to lead the rest of the investigation to reach a conclusion with the palatable actions. He may seek evidence to support a cause that leads to desirable actions, while the evidence that could distort that evidence is not actively looked upon. This evidence could even be discarded as irrelevant or unnecessary, or simply ignored because there is much evidence to support favored causes.

2. *Lipstick on a corpse*: In this case, the analyst has successfully investigated the event of failure, the cause is supported by numerous tests, but he is aware that there will be more consequences than just corrective actions. Faced with this dilemma, some investigators try to minimize the negative findings and highlight positive results. Instead of using plain language and objectively describe what happened, less precise and more positive language is used.
Instead of defining terms as deficient or insufficient to describe a process, phrases like "less suitable" can be chosen. Because it is a report about causes of an event, it is assumed that its main goal is to be a practical report, based on documenting what went wrong and telling how to fix it. However, a secondary objective is introduced, which is to convince the reader that the failure event and its underlying cause are not as bad as the reader might think. Corrective actions work best when they are concrete and specific. Diluted or minimized causes, however, often lead to diluted or minimized corrective actions. By giving a good "spin" to the problem, management or authorities may not properly support what needs to be done to fix the problem. In other words, the report fails to convince its audience that the failure event is a serious problem that deserves serious attention.
3. *Elementary, my dear Watson, elementary*: Almost no one can resist to speculate on what caused a failure, in the same way that fans of mystery books or movies often try to guess clues to identify the murderer at the end of the story. It is certainly fun and may even be a point of pride, if our initial estimate proves to be correct. However, such assumptions can jeopardize the integrity of an investigation. In particular, where:

   - The conjecture is made by a manager involved in the investigation process.
   - The entire plant has an authoritarian, chain of command style of organization.
   - Management culture rewards correctness and has zero tolerance about mistakes.

   The deficiency of this scenario is that the analyst would favor some hypotheses instead of trying to rebut them (Chap. 7). Once the cause is presumed found the collection of important evidence is suspended. Why waste resources when we already have the answer? The effect of not seeing the obvious due to being focused on a specific task or activity, such as the famous test of a Gorilla in a basketball field, is sometimes called "inattention blindness."
4. *Dilution of the solution*: This situation is usually unique for large organizations with many administrative processes, with individual departments whose priorities are aligned with business priorities. Investigation detected an undesired situation or event, adequately identified the causes, developed an action plan, and provided senior management, which approved and supported the plan and provided sufficient resources to run it. But the same directors and departments that created the problem in the first place are now in charge of solving it; the work is likely to be faced with the same attitude about its importance, and will

probably end with the previous result.
To successfully overcome this obstacle, the timing of the corrective action plan should be carried out by a higher authority than those involved in the execution of its various parts. This person should have the authority to gather and prioritize resources to ensure that all parts of the plan are realized on time and are properly coordinated, in the manner of a Project Manager.

10.4 On Human Error

In the introduction to this book, we assumed that all failure-related causes are uniquely human. Even obvious physical causes can be considered a result of human error, but this error was committed at another time and/or in another company. However, this does not mean we have to blame someone for having been the cause of a failure.

The Human Performance Evaluation Process (HPEP) is the method used to assess how people's actions contributed to the cause of an event. For example, if an event depended on the actions of a person to succeed, and instead the event was a failure:

- Did anyone not act?
- Did anyone act but made the wrong decision?
- Was the person fatigued, stressed, or error prone?
- Was the person properly trained to act properly?
- Did the person have the skills to make the right decision?
- Was someone or something distracting the person from doing the right thing?
- Did the person have any motives to allow failure, increase its probability through inaction or promote it?
- Is there a motive, means, and opportunity if the failure was deliberate?
- Was the person physically able to do the right thing?
- Did the person have the correct tools or elements?

Many accidents have been attributed to human error, usually from someone low-ranking that could not blame someone else down. Managers and designers seem nonhuman, they do not make mistakes. In the process industries this attitude is much less common now than before. Many companies now realize that almost every accident is due to a faulty direction and human error is no longer accepted as the cause of any accident. In the accidents at *Chernobyl* nuclear plant, the *Herald of Free Enterprise* ship and some aircraft (in Germany and Tenerife), those above blamed those below. Airplane pilots, although very experienced, are in a similar position to the operators: they have no one below them to blame. Human error is not useful when it does not lead to action, and we tend to say that someone should be more careful. Nobody is deliberately neglectful. T.A. Kletz (1993) divides "human error" into four groups that require different actions:

- **Errors**: someone does not know what to do (or, worse, thought he knew but was wrong). We must improve training or instructions or, better yet, simplify the work.
- **Violations**: someone knew what to do but decided not to. Why the work should be done in the right way must be explained, we do not live in a society where people follow the rules just because someone says so. Occasionally, we must verify that the correct methods are in use. Incorrect methods can be followed for months or years before an accident occurs (as in Clapharn Junction train crash, due to defective light signals). How many times have you exceeded the speed limit without being arrested? It is always advisable to simplify the work, if the safe way to do is difficult, people will find an easier, unsafe way.
- **Incompatibility** (between work and skill of the person): someone thought he knew what to do and tried to do it but it was beyond his physical or mental ability, perhaps beyond any skill. In some cases we may have to change the person, but it is usually better to change the work situation, i.e., change design or operation methods for the plant or equipment.
- **Distractions and attention spans**: someone knew what to do, he thought he could do it and did it, but incorrectly. These are like the distractions of everyday life, but more serious in their consequences. Human nature makes them inevitable and there is little we can do to prevent them, but we can reduce stress and distraction. It is usually more efficient to change the work situation so that there is less opportunity for mistakes or errors that have less serious consequences.

Types of human error can be illustrated by a hypothetical accident that happened because someone did not open (or close) a valve. It could be a mistake: he did not know that he was expected to close the valve. It may have been a violation: he knew he was expected to close the valve but decided it was unnecessary or, in extreme cases, would not bother. It may have been an incompatibility: someone thought to close the valve but found it was too difficult. Finally, it could have been a lapse of attention: someone thought to close the valve but he was busy or under stress, and forgot about it.

In Appendix H of the document DOE-NE-STD-1004-92, as discussed in Chap. 7, a list for human performance evaluation is carried out in more detail. Four stages or steps are defined in the performance of a person:

(a) Detecting the information
(b) Understanding the information
(c) Selecting the action
(d) Implementing the action.

The operation and management of facilities and equipment and related procedures and documentation are part of the working environment to be evaluated for each of these steps. These are common problems that should be considered:

- Cognitive Overload
- Cognitive under-use/boredom

- Intrusion of habits
- Lapse of memory recall
- Spatial disorientation
- Preconceptions
- Tunnel vision or lack of picture
- Ignorance
- Poorly made hypotheses
- Instinctive actions
- Uncoordinated actions and thoughts
- Insufficient attention applied
- Shortcuts to complete the work
- Complacency and lack of perceived need for concern
- Confusion
- Misdiagnosis
- Fear of failure and its consequences
- Tiredness and fatigue.

Where high-risk requirements are very sensitive to nonconformity, each of these factors of human performance should be considered in order to achieve a high degree of reliability. These factors also should be considered in the design of control systems and operator training.

If accidents are due to failure of management to provide adequate training, instruction, supervision, equipment, or operation methods, should managers be punished after an accident? Public, encouraged by the media, seems to believe that there must be someone to blame for every each accident. However, managers need, like the rest of us, training, instruction, and supervision, and like the rest of us, do not always succeed. Often managers and above do not adequately provide because they do not understand they could do more to prevent accidents. Most management schools do not include safety in their curricula and most management books do not mention it.

10.5 Human Errors at Different Levels, the Chernobyl Case

10.5.1 The Worst Nuclear Disaster

The V.I. Lenin nuclear power plant was 18 km from the city of Chernobyl, in Ukraine, Fig. 10.2. The well-known event, the reason for these lines, is the most serious nuclear accident and one of the largest environmental disasters in history. The plant had four reactors with capacity to produce 1,000 MW each, the accident thwarted the completion of two reactors that were under construction. The design of these reactors did not meet some safety requirements as imposed on all civil

Fig. 10.2 Nuclear power plant, city of Chernobyl

nuclear reactors at the time. The most important of unfulfilled requirements is that it lacked a containment building, those spheres so obvious in most nuclear plants.

One day in April 1986, during a test that simulated a power failure, a sudden increase of power in reactor 4 caused overheating of the nuclear reactor core, which ultimately led to the explosion of hydrogen accumulated inside. The amount of radioactive and/or toxic gases released was 500 times what the atomic bomb released in Hiroshima in 1945. It caused the direct death of 31 people and forced the Soviet government to evacuate 116,000 people, causing international alarm upon detection of radioactivity in 13 European countries

The heroics of firefighters during the first three hours of the accident prevented the fire from spreading to the rest of the plant. Army helicopters dropped 5,000 tons of a mixture of sand, clay, lead, dolomite, and neutron absorbing boron on the core. The work on the containment reactor avoided a second explosion, with dramatic consequences that could have left Europe uninhabitable. After the accident, 600,000 people participated in a massive process of decontamination, containment, and mitigation. The international community funded the decommissioning of the plant, completed in December 2000, including a "sarcophagus" to isolate the reactor.

A report to the International Atomic Energy Agency studied the causes of the accident. As we shall see, there was a series of human errors. The team that operated the plant on Saturday April 26, proposed a test with the intention of increasing the safety of the reactor. To do this it was required to calculate for how long the steam turbine power would continue generating after loss of power supply from the main reactor. The emergency cooling pumps, in case of failure, required a minimum of power to get going (until diesel generators went on) and plant technicians were unaware of whether, once cut the flow of steam, turbine inertia could keep pumps running.

To conduct this experiment, technicians would not stop the chain reaction in the reactor to prevent a phenomenon called poisoning by xenon, a neutron absorbing gas. While the reactor is in normal operation, so many neutrons are produced that absorption is minimal, but when the power is too low or the reactor is stopped, the amount of Xe135 increases and prevents the chain reaction for a few days. The reactor can only be restarted when the Xe135 disintegrates.

Operators inserted control rods to reduce reactor power up to 30 MW. With such a low level, automatic systems would stop the reactor and therefore operators disconnected the power regulation system, the emergency core coolant system and all mechanisms of automatic reactor shutdown. These actions, as well as removing the central computer that prevented prohibited operations, were serious multiple violations to the Nuclear Safety Regulation of the Soviet Union.

At 30 MW power, reactor power was increased to prevent xenon poisoning by taking out control rods, but with the reactor about to go out, operators manually removed too many control rods. Of the core's 170 boron steel bars, safety rules always require a minimum of 30 bars down; this time eight were left. With disconnected emergency systems, the reactor experienced an extremely rapid rise in power that operators did not detect on time.

Four hours after the experiment started, people in the control room began to realize something was wrong. When operators wanted to lower control rods back into emergency mode, rods did not respond because they were deformed by heat. Rods were disconnected to allow them to fall by gravity. In a few seconds, the power increased almost 100 times above nominal. The light water coolant was not able to extract that huge amount of heat and vaporized in a fraction of a second, producing a steam explosion. There were loud noises and then came the explosion caused by the formation of a cloud of hydrogen within the core that blew the 100 ton reactor roof, causing a fire on the ground and a giant emission of fission products into the atmosphere. The reactor was destroyed. At the core, exposed to the atmosphere, graphite burned, fuel and other metals became an incandescent liquid mass. The temperature reached 2,500 °C.

10.5.2 Levels in the Chain of Human Errors

While the effects on people and the environment were huge, what concerns us now are those aspects related with the chain of human errors. The Chernobyl accident occurred in a period of decline of the Soviet Union, and is explained as a result of it. The gross errors in the Five-Year plans and the excessive rigidity of the prevailing economic system caused a rapid decline of the economy. For example, 10 years before the PC central committee had decided not to prioritize industrial automation technologies just not to increase unemployment (curiously, the word "robot" comes from the Russian word for work). At the same time, the loss of political power in a centralist regime produced a relaxation of discipline and will in most of the society, especially in countries forced to belong to the union, as was the case of Ukraine.

The human causes of the explosion were of three types (A) design errors, (B) management failures and mistakes made by the operating staff, and (C) lack of means to mitigate the impact.

A. Design errors: RMMK reactors were difficult to control at low power and any tendency toward a chain reaction was rapidly amplified. In other reactors any increase in chain reaction is automatically stopped due to the design of the reactor core. The Chernobyl explosion occurred during a low-power test, i.e., a time in which the reactor was unstable. Russian nuclear engineers knew of this instability; Soviet authorities had been warned before the accident, but warnings fell on deaf ears.

Control rods in a RBMK reactor are inserted slowly. While in other reactors it takes 2 s, complete insertion in a RBMK requires 20 s, too slow to prevent runaway when operating in an unstable mode. Control rods are boron carbide with a cover of graphite; when the rod begins to be inserted, graphite increases reactivity. This dangerous phenomenon had been observed 3 years before the incident in a similar reactor in Lithuania, then a part of the USSR too.

The neutron moderator consists of 600 tons of graphite. When very hot graphite enters in contact with air, it bursts into flames. In Chernobyl, the graphite fire vaporized radioisotopes within the reactor and dispersed them into the atmosphere along with the smoke. RBMK reactors do not have structural containment or a system for filtering exhaust gases.

B. Mistakes made by the operating team: six errors have been identified. Two permanent operation rules were violated: not operating below a power level, and never with less than 30 control rods fully inserted into the core. Another mistake was not following the test procedure, and deliberately taking out of service three safety mechanisms: one for emergency water injection, and two for emergency stop.

If any of these six mistakes had not been committed, the explosion would not have happened. But why did they do this? It is clear that the operators were not properly trained and did not understand the dangerous nature of their actions. It is easy to blame the operation team for the disaster, but they were doing their job with the training they had received. That training was insufficient and totally inconsistent with the lack of passive safety features in the RBMK reactor design. Not knowing much about the behavior of the reactor core, operators were unable to appreciate the implications of the decisions they were making, and their situation was even more dangerous because the test was being done at low power and in violation of standing orders.

In addition, operating instructions and specific instructions for the test were incomplete and inaccurate. A detailed examination of what happened in the few hours and minutes leading up to the explosion would have shown that this would happen.

C. Lack of means to mitigate the impact: Nuclear problems were well known in the USSR since the 1950s. Soviet scientists had excellent knowledge in this field, and had made useful recommendations to the authorities which unfortunately were ignored. Soviet biologists knew how the iodine settles in the thyroids and the importance of the simple and efficient protection provided by the stable iodine. An emergency plan for protection in case of a nuclear accident was developed as far back as 1964. This plan included all measures now universally accepted: stay indoors with windows and doors closed, distribute stable iodine, temporarily evacuate the threatened population, prohibit the consumption of probably contaminated food, move cattle to uncontaminated pastures, etc. Each action was accompanied by criteria for the level of radioactivity that justified its application.

This emergency plan was approved by the Minister of Health of the USSR December 18, 1970, more than 15 years before the Chernobyl disaster, but remained a dead letter. A new plan was introduced in 1985, but was rejected by the Minister of Nuclear Engineering in September, just 7 months before the accident. The reason provided was that an accident of this nature and requiring these measures was "impossible in the USSR." In the absence of an emergency plan, basic precautions around Chernobyl were implemented with delay:

- instructions to stay indoors with windows and doors closed (after 36 h);
- a ban on the consumption of fresh milk, fruits, and vegetables produced locally (after 7 days);
- the distribution of stable iodine (the U.S. offer was rejected);
- the provision of protective clothing and respirators to firefighters and recovery workers (for long unavailable).

Elementary precautions that should have been taken immediately after the failure were unknown to local authorities and to the direction of the plant. They had no emergency plan to intervene, or to administer stable iodine, or medical supplies, or protective clothing, not even the instruments to measure radioactivity.

10.6 Business Impositions and Political Decision-Making

Similar flaws continue to exist, because it is humans who handle the plants and, more importantly, humans define the social, economic, and political environments in which the plants must operate. Most scientists and technicians are honest and open. Others, who are competent and recognized as such, are strongly motivated by personal interests, and scientific objectivity succumbs to lack of courage to stand for what is technically correct. Managers and directors accept or even encourage politicians in making questionable and even dangerous decisions; the struggle for influence often replaces the technological debate.

We saw in the Sect. 10.5 that relevant knowledge developed by Soviet scientists did not spill to the medical and nuclear communities. The combination of outdated technology and weak discipline largely explains human failings that caused the accident. The Chernobyl accident exposed the once-proud Soviet Union (which had led such remarkable spatial developments in the 1960s) to international humiliation. Although the reformist Mr. Gorbachev had replaced Mr. Chernenko (a bureaucrat whose greatest merit was to have been a contemporary of V. Lenin) in office, decadent middle managers were eagerly looking for a scapegoat. That was how the plant manager was arrested. No wonder, considering the magnitude of the damage caused and a Stalinist regime in the background. If this book has left an imprint, the reader will agree that this would be counterproductive. The release of the plant manager, a year after the fall of the Soviet Union, would therefore not be surprising.

In the Cold War, which at times threatened to get hot, the look at the production of plutonium from RBMK reactors imposed a sense of urgency to the Soviets. Scientists and engineers worked to produce weapons-grade plutonium, as much and as quickly as possible. On the other side of the war, in the "West," the strategies were somewhat different. Political propaganda, however, was overwhelming on both sides; all should contribute to the success of the "cause." In Chap. 7 we talked about the Challenger accident, which occurred the same year as the Chernobyl disaster. Then we talked about technical issues. We return to this case now, to address some points that show a remarkable similarity with the Soviet case.

Ronald Reagan, then U.S. President, was a mediocre actor, but exceptional communicator. Two years earlier he had proposed that a school teacher be part of the shuttle crew, in order to regain public interest in the (American) conquest of space. He succeeded. Millions of people around the world were excited about the live broadcast from the Challenger. The teacher McAuliffe was chosen for her empathy to communicate with students; her work would be to take a science class from space. Mission 51L, like others before it, had been postponed several times due to technical failures and adverse weather conditions. However, this time the flight was to leave according to the interests defined by government propaganda. Mr. Reagan would use the presence of the teacher in space in his annual State of the Union speech, which was already scheduled. By then, the Challenger should be in space.

Recall that the rubber rings used to seal the impeller rocket failed during take off due to low temperature. How much did the boss's order contribute that cold Tuesday for NASA technicians to give the nod to take off, with an even lower temperature than in other cases, in which they did not hesitate to postpone the launch? The same media that had idolized the teacher, summarized his version of the tragedy months later: The Challenger crew paid with their lives the indolence of NASA technicians who ignored warnings about the vulnerability of Rubber rings to extreme cold, clearly identified by the manufacturer of the solid rocket boosters. To analyze what is worse for a technical person involved in a failure, a prison sentence in Siberia or condemnation by the American media, lies beyond the capabilities of the author of this book.

10.6.1 Example 10.A Incident on a Water Pumping Plant

The Root Cause Analysis of a failure during the decommissioning of a pump in a pumping station for drinking water yields some interesting results. The plant is over 50 year old. It is underground, located under a public square. The plant has six pumping lines in a pump room, located between a very large water storage tank and the command room, the latter slightly above the level of the pump floor, Fig. 10.A1. Maintenance was in the process of changing internal drive valves. At the time of the incident, Line 3 had been out of service for the last 10 days.

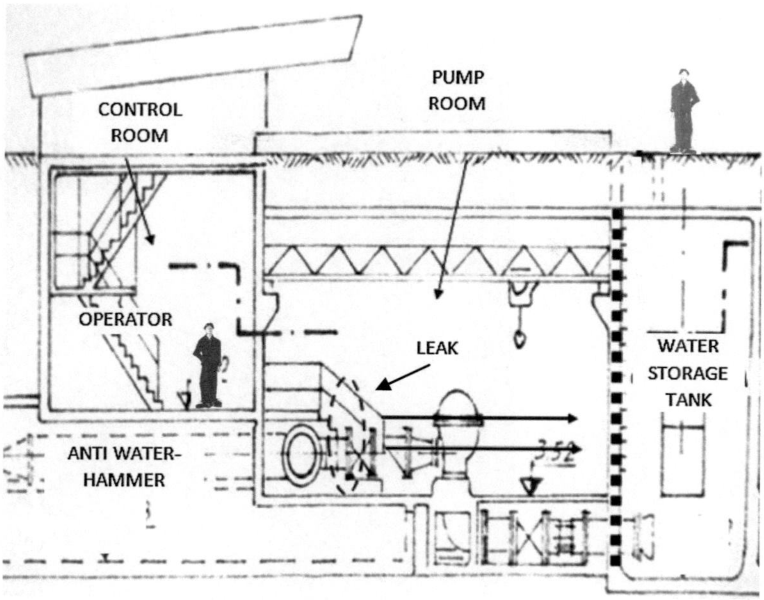

Fig. 10.A1 Leak location in the pump room, note positions of control room, and storage tank

The main output pipe of this pump was disassembled, and the discharge and locking valves had been removed. The plant had the following important characteristics when the incident occurred:

- The main and drive lock valves in Line 3 were disconnected.
- The external impulsion valve in Line 3 was closed, but no blind flange was placed.
- Electrical control fuses for pump 3 were placed, and the switch at the main electrical panel was left on.
- The control room and the pump room are both below the level of the water distribution lines.

To meet water consumption, the night of the event the three enabled (4, 5, and 6) lines were in operation, with valves strangled. Around midnight it was decided to decommission line 6, for which it was required to gradually close the valve for line 6, and simultaneously open valves for lines 4 and 5, to compensate pressure.

Inadvertently the drive for valve in line 3 was released, causing injection of downstream water from the other lines and distribution pipelines. This probably occurred when the operator was instructed to fully open the eponymous valve for line 4. Both control buttons are within a short distance, in the same control table, see Fig. 10.A2. A striking feature is that the operator failed to timely correct his mistake, even when he saw the water filling in the pump room, and had more than 15 min until the water reached the level of the control room.

Fig. 10.A2 Buttons in control table

The operator called his superior, located in the company headquarters (some 5 km away), warning of "a tank failure." Figure 10.A1 highlights the position of the front wall of the tank; water is likely to have collided against the wall and would have appeared as to be coming out of any cracks therein. As complete collapse of the tank wall could indeed have some risk of immediate flood, operators were instructed to turn off motors and cut the electrical power. Which they did, then left the control room. The water continued to flow, and eventually completely flooded pump and control rooms. Electrical transformers were overloaded; adding to a service outage that affected a significant number of people for over one week.

Opening of power keys and removal of electrical fuses to ensure isolation of line 3 should have been mandatory before starting any maintenance. However, the existing procedure only indicated opening the keys, but not the removal of the fuses. The keys were either not open or closed again before the incident. The valve in line 3 was left closed but no blind flange was placed to stop any leakage, as it was not specified in the maintenance procedure. Nor were placed any warning signs such as "out of service" or similar.

Root causes of the incident can be grouped as:

- Causes inherent top the plant.
- Causes inherent to training of operation and control staff.
- Causes inherent to maintenance and construction.

One group of causes is associated with the other. Thus, an organizational effort to add value to the workplace and bestow prestige to the activity of operators would result in better staff, which will be better able to take hard training; a virtuous cycle of improvements of this sort would also results in improved organizational climate.

RCA findings indicate that most promising actions are those aimed at improving the working environment and motivation. The city-operated company is strongly divided between staff, professionals, and "political" managerial levels, designated by different administrations throughout its 30 years history. This situation resulted in a bad relationship among all different parties. The authorities in charge at the time of failure were viewed with suspicion by senior managers, who in turn did not receive sufficient cooperation from subordinates. These problems were added to the usual vertical division between the three components of the business: operations, maintenance, and engineering.

Because of its location in a public place, the control room could not rise, and be well protected from any consequences of the event that operators must control. Being a highly visible public service, the event was magnified by the media. Fortunately in this case, the rapid supply of water by alternative sources to those affected ultimately resulted in an improvement in the image of the company.

10.7 The Rules are Relaxed: The Case of the Macondo Oil Well

We have seen that the errors of the past (own and others) have made us learn. It is known that the memory of each person is selective; for example, most end up remembering only the good things in any stage of our lives or wherever we used to live. Despite our best efforts, this also happens with organizations: measures for the safe operation of equipment and processes are often victims of their own success. The absence of accidents is piercing the perception of risk, makes forget bad experiences and take increasing risks.

In the spring of 2010, the oil well blowout that led to the explosion of the offshore oil rig Deepwater Horizon, its sinking and the subsequent oil spill in the Gulf of Mexico took everyone by surprise. In part because few remembered that something similar had happened in the area. The failure caused the death of 11 workers, and spillage of a billion gallons of oil into the sea. Due to the position of the platform in the Gulf of Mexico, shared by the United States, Cuba and Mexico, the oil spill could spread an extremely wide area. It is estimated that cleaning operations succeeded in eliminating 75 % of the oil released. The rest was rapidly degraded; apparently bacteria that consume oil are numerous in the Gulf of Mexico, so that the impact on marine organisms and regional economies was lower than expected.

The Deepwater Horizon was a dynamic positioning semi-submersible ultra-deepwater oil rig built in 2001, Fig. 10.3a. The purpose of the tower was to drill oil wells on the seabed, moving from one place to another as required. The rig could accommodate a crew of up to 130 members, and had “e-drill,” a monitoring system that allowed receiving real-time information of the drilling process and information on maintenance and errors. Once a particular drilling is finished, the extraction is carried out by another rig. In September 2009, Deepwater Horizon drilled the deepest oil well in history, in the Tiber field, with a depth of over 3,000 m in 1,250 m of water.

Deepwater Horizon sank in April 2010 while working at the BP Mississippi Canyon in the block known as the Macondo prospect. Its fall was a result of an explosion that had taken place two days earlier, and caused one of the largest oil spills, Fig. 10.3b. The oil spill, mixed with a small amount of methane, could not be controlled due to the difficulty of sealing the leak in the pipe on the seabed. The underwater photo in Fig. 10.3c shows the magnitude of the leak, after one of several unsuccessful attempts to blocking. Several attempts to seal the well pipe failed. First an iron bell was tested, and then, in late May, with the injection of heavy mud and cement, a procedure called “top kill”. In July, BP placed a new cap to stem the leak.

In the “top kill” procedure mud (drilling fluid) is pumped heavily, until the weight of the liquid exceeds the oil pressure. The “junk shot” involves injecting objects like golf balls to clog the blowout prevention mechanism in the valve stack on top of the well. BP inserted a tube over one kilometer long on the broken riser to divert some oil to a drill ship on the surface, about 1,500 m above the wellhead. The tube diverted about 22,000 barrels of oil for nine days, but closed after the operation began “top kill”. BP built a four-story containment dome, intended to control most of the leaks. In the dome, the team discovered that the opening was obstructed with a mixture of water ice and gas. The dome was placed on the seabed, 650 m away from the leak. BP conducted drilling relief wells to plug the leak. Each time one of the new wells intersected the existing well, heavy drilling mud was injected, followed by cement.

Finally in August, BP managed to seal the Macondo with the “static kill.” The method is similar to the “top kill” that had not worked in June. But this time the engineers pumped in mud more slowly and at less pressure due to the new bell installed on top of the well, which had reduced the oil flow. After the “static kill,” BP proceeded to seal the well from the bottom.

BP’s internal investigation of the Macondo incident attributed many of the root causes to issues that could be classified as quality problems. Studies revealed mechanical failure, faulty design and operational controls, poor management of contractors, and decisions by the technical staff. Here we focus on two causes: the one that caused the original eruption of oil, and the failure of the containment system.

The cause of the eruption of oil is related to human error, of course. The well cementing process, once completed, can be left safely, until a new system is installed for production. Verifying correct cementing involves a leakage test,

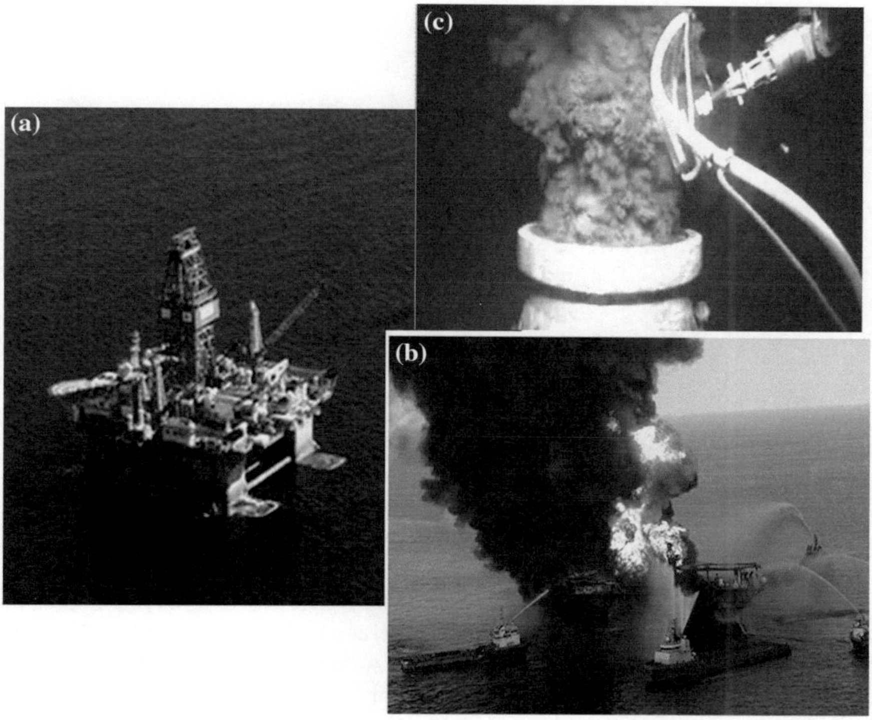

Fig. 10.3 **a** Deepwater Horizon **b** Platform fire **c** Subsequent uncontrolled leak

which must ensure that the pressure exerted by the oil and gas does not increase during a certain time. This time the head of the test okayed without having met all the requirements to ensure success. This decision was in the context of a great pressure to meet the schedule, a shift change, and exhausted staff. On 20 April the leaders approved the mud to be replaced with seawater, which was not heavy enough to prevent gas that had been seeping into the well to finally reach the platform.

Intended to stop any leak is the "blowout preventer," see Fig. 10.4. This is a valve at the wellhead on the seabed, with jaws designed to collapse the pipe and seal the well in case of emergency. This time, however, several attempts to activate the mechanism failed. A barrier was exceeded.

It is remarkable number of explanations attempted for this failure. According to the U.S. Congressional investigation, the valve had a leak in the hydraulic system and also a failed battery. According to BP's vice president in charge of the drilling, the blowout preventer was not connected to the correct pipe.

The possibility that the line of communication between the tower and the control unit of the blowout preventer on the seafloor could have been cut by the explosion was also evaluated. It was also said that the blowout preventer did not

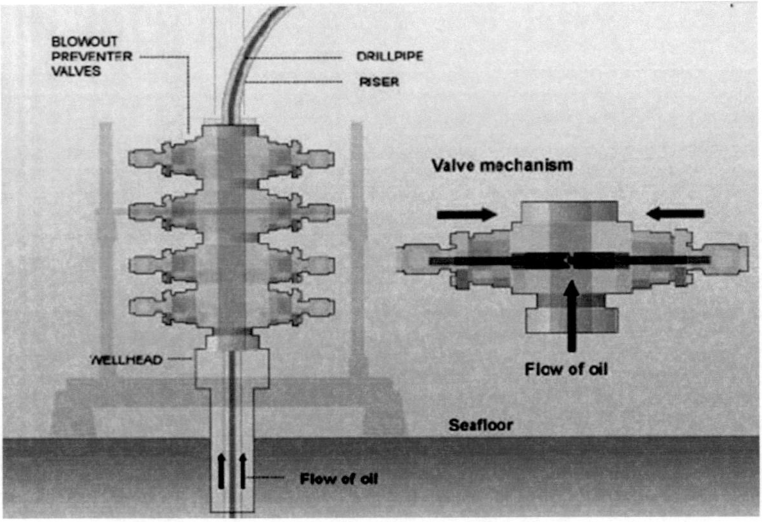

Fig. 10.4 Schematic of blowout preventer

match specifications by Transocean (rig operator), the cutting arms were supposedly not designed for such a strong pipe. Actually, what happened is that the jaws worked and sectioned the tube. But the tube was not in the middle of the jaw, just on one end, and jaw crushing failed to seal it completely. The eccentricity of the tube could be due to buckling of the pipe when compressed by the forces produced by the oil and gas leak.

There had been a history of similar failures in oil wells, which are now returning to light. Three decades earlier, in 1979, the "Ixtoc I" semi-submersible drilling rig experienced a confined loss of pressure in an exploratory well just over 50 m of water; its subsequent explosion generated an oil leak for almost a year. In total, more than 3 million barrels (600 million liters) of crude oil were spilled into the Gulf of Mexico and its surroundings. Immediately, and as a result of that accident, the oil industry began to operate with greater awareness to avoid failure in wells, and so took extra precautions in operation. However, over time and with a growing successful record of oil drilling in the Gulf, platforms and operations in oil wells were neglected, and this was the basis for the Deepwater Horizon explosion and subsequent environmental catastrophe.

It was of no accident that these two unfortunate events occurred 30 years apart, highlighting the span of one generation of engineering and technology memory for any industry. During this period, we can expect successful periods interrupted by faults, and depending on when he enters, a young engineer can be more or less conscious. This awareness tends to dominate the design and operational behavior for a period, but eventually a paradigm of success tends to suppress a paradigm of failure and an atmosphere of overconfidence, complacency, and arrogant relaxation prevails until a new failure provides a new wake-up call.

Moreover, this failure is particularly timely. At the moment the Macondo spill occurred, the U.S. was about to lift the ban on new exploration in the Gulf of Mexico, in force at the time. Immediately after the explosion, this prohibition was again revised. The U.S. Department of the Interior ordered a 6-month moratorium on new drilling in more than 150 m of water. The moratorium suspended work on 33 wells, and was later overturned by courts. Finally, the ban was lifted, to the delight of almost everyone, large energy consumers in developed countries as well as inhabitants in more frugal countries with large oil reserves, that could be the unwitting hosts of next military adventures to secure oil sources.

10.8 Underestimating Consequences of Natural Events, the Fukushima Event

We said in Chap. 1 that natural disasters (earthquakes, floods, etc.) or "acts of God" cannot be considered human errors. But the consequences of these natural events on the failed component are human errors. A failure means someone did not properly estimate the probability or intensity of such an event, and did not make the necessary adjustments to the design and construction of the component or structure. We will now see a recent example that quite dramatically illustrates this situation.

The Fukushima I nuclear power plant was a set of six nuclear reactors located in the village of Ōkuma, with a total capacity of 4.7 GW. Fukushima I was one of the 25 largest nuclear power plants worldwide. Fukushima II was the first nuclear reactor built and managed independently by the Japanese company Tepco. Fukushima III was 11 km away.

The plant was designed by General Electric and was connected to the network in 1971. Forty years later, in March 2011, after a massive earthquake and scale nine tsunami, the plant was severely damaged, people in a 20 km radius had to be evacuated. That day reactors 1, 2, and 3 were operating, while units 4, 5, and 6 were out in a periodic inspection. When the earthquake was detected units 1, 2, and 3 automatically turned off and stopped the production of electricity. Normally, the reactors can use electricity from external power lines for reactor cooling and control room, but the network was damaged by the earthquake. Emergency diesel power generation began to work normally, but stopped abruptly with the arrival of the tsunami that followed the earthquake.

A state of emergency at the nuclear power plant was declared because of the failure of the cooling systems. Vapor pressure increased, and the company released part of that as radioactive steam. Three days after, a new explosion rocked the complex due to the accumulation of hydrogen in the reactor 3. Another explosion occurred the next day in the reactor 2.

Fig. 10.5 Fukushima nuclear power plant

In April, the Fukushima accident was raised by the Japanese government to level 7 in the nuclear accident scale, equaling in severity to Chernobyl. All this after a series of explosions, dramatic increases in radiation levels in the surrounding area confirmed partial melting of at least one of the nuclei, radioactive water leaking into the sea, and successive failed attempts to lower the temperature in the compromised reactors.

The most relevant aspect of this failure is, without doubt, the drama of its consequences. Not so much by the immediate damage to people and the environment directly related to the incident, but by its medium-term economic consequences. Nuclear plants were built when Japan believed in a nuclear-powered future. Japan was the third largest producer of nuclear power, which supplied 30 % of its (huge) energy needs. They came to have 54 reactors in operation. Due to their cooling water needs, all these reactors were built along vast stretches of the sea coast. These reactors fed power lines across the country, fueling the economic heart of Tokyo and other cities around the country (Fig. 10.5).

Japan is highly seismic; during the design of the nuclear power plants, 50 years ago, the impact of strong earthquakes were obviously considered. But the possibility of the combination of a massive earthquake with a subsequent tsunami, with the inevitable flooding of coastal areas, was not taken into account. The Fukushima incident forced the government to close all plants susceptible to injury from a new combination of earthquake and tsunami. Within this plan, they had to close all of them. In May 2012, Japan closed the Kashiwazaki-Kariwa nuclear power plant, the last still operating. Inside the visitor center of the plant there was still on display a certificate from the Guinness Book of Records: "nuclear plant with higher electric power capacity in the world."

"We have coexisted with nuclear power plants and convinced that they were safe," the local mayor said during the closure of the plant, "but with the Fukushima accident we can no longer assume that they are safe. Our confidence in Government and plant operators has been shattered." At the time of writing this book, Japan, once heavily dependent on nuclear power, was facing the near future with serious energy problems. The country is affected by a recession cycle that has lasted over 10 years and after the consequences of the 2011 natural disaster Japan must now increase imports of LNG and oil. The cost to the economy is high. The dramatic increase in fuel imports helped push its trade deficit as the largest in its history.

Large retaining walls were built during 2012 and 2013 in Kashiwazaki-Kariwa and other plants, which, hopefully, will be sufficient to resist any possible tsunami. It is likely that by the time the reader reads these lines, Japan has regained much of its nuclear generating capacity. However, environmental concerns are still at issue. In August 2013, Japan's Nuclear Regulatory Authority (NRA) warned that the build-up of contaminated groundwater at the Fukushima plant was on the verge of tipping out of control. A survey by Japan's Ministry of Economy said water laced with caesium and other radioactive materials had been flowing into the Ocean at a rate of 300 tons a day for 2 years, since the crisis began.

10.9 Management of Old Plants and Equipment

The most effective implementation for prevention is long before the possible occurrence of an event. For instance, T. A. Kletz summarizes in his book "Learning from Accidents" a series of fire and explosion prevention techniques, based on the study of a large number of failures in industrial plants:

- Avoid large inventories of hazardous materials by intensification, substitution, or attenuation.
- Thoroughly inspect during and after construction.
- Install gas detectors so that leaks are discovered quickly. This does not eliminate the need for regular inspection tours. Even in plants with gas detectors, half of leaks are discovered by men.
- Warn people when a leak occurs. Those who are not required to deal with the leak should leave the area by a safe route.
- Isolate the leak by means of emergency isolation valves operated remotely.

- Disperse leaks by open construction, supplemented if necessary by steam or water curtains.
- Remove known sources of ignition. Although this may seem a strong line of defense, it is really the weakest. It takes very little energy to ignite a flammable vapor leak in air.
- Protect against the effects of the leak:
  - Fire: isolation and water spray.
  - Explosion: stronger buildings, distance (avoid nearby development).
  - Toxicity: distance (avoid concentrations of people).
- Provide means for emergencies.

Management of plant modifications:
Many accidents occur because plant and process modifications have unforeseen adverse effects. People dealing with complex systems tend to think straight. They think about the effects of their actions on the road to the immediate goal but are unaware of side effects.

Testing and inspection of protective equipment:
Many accidents would not have occurred if protective equipment had been maintained in good condition. To prevent similar accidents in the future we need:

(1) An education program to convince all levels that safety equipment must be kept in good condition and is not an optional extra that can be neglected under work pressure. An ongoing program is needed to involve people in regular discussions of accidents that have occurred. Training in loss prevention is particularly important during the formative years of a student.
(2) An audit program to ensure that equipment is in good condition. Most equipment must be tested at regular intervals: instruments monthly, relief valves annually. Managers must ensure occasional audits.

User friendly designs:
Friendly is a term coming from the field of computers and is used to describe plants that tolerate far from ideal operation or maintenance without accidents. Thus, Bhopal would not have happened if there had not been large and unnecessary storage of chlorides. What is not does not leak or explode.

We must keep reserves of hazardous materials to a minimum (intensification), use safer materials (substitution), or use hazardous materials under nonhazardous conditions (attenuation). Nuclear reactors at Three Mile Island and Chernobyl were less friendly than gas-cooled reactors and operators had less time to respond. Chernobyl was particularly hostile as it had a positive power coefficient (when it got warmer, production increased at once).

The shell cover of the failed rocket propellant in the space shuttle Challenger was not a single stretch but was made in several segments with joints between them. The O-ring on the joints was a weak point and failed. Rear turbine aircraft as the Trident are more likely to stall than those with turbines mounted in the wings.

The closing of the cargo door on the DC 10, even after modification, was hostile because someone had to look through a hole to make sure it was properly closed.

Hazard and Operability Studies (HAZOP): a lantern in the bow

Many accidents in the literature show the need for critical design review through a Hazard and Operability (HAZOP) study or similar techniques. It takes two earlier studies, one at the stage of conceptual analysis, when it is decided which process to use and where the plant is located, and another in the design phase. A conceptual study in Bhopal, for example, could have discussed the product to be made (other insecticides were available), the process used and the need to accumulate intermediates.

It is better to light the risks after they crossed than not illuminate at all when we can go back, but better still is to illuminate the risks that lie ahead. Like the chameleon, we need to keep one eye in the past and the other into the future.

Top management: the danger of ignorance

In many accidents, such as in chemical plants as Flixborough and Bhopal, in nuclear plants as Chernobyl and Three Mile Island and in railways as Clapham Junction, lack of training was a significant factor. The operators not only need training in their duties, but also an understanding of the process and the risks, so they can handle problems unforeseen in previous instructions.

A common feature in all cases is failure to learn from past experience. Sometimes knowledge is not distributed to those who need to know; frequently companies or plants, even individual managers, do not learn. People involved in an accident do not forget, but after a while others fall into place. Organizations, as we already discussed, have no memory.

An ongoing program is required to occasionally discuss past accidents or describe recent accidents, communicate in bulletins and safety meetings, include in codes of practice notes on the accident that led to the recommendations. The use of information storage and retrieval systems of information enables improved access to accident reports and recommendations. A shortly followed recommendation is to keep a "black book" or "memory book" in each control room, and a file of technical reports on accidents that have occurred, including accidents at other facilities. The black book should be required reading for newcomers; the oldest should dive into it to refresh their memories from time to time.

The culture and climate of a company affects staff behavior, the actions taken to prevent accidents and enthusiasm. Different companies, even different plants within the same company, may have different cultures. The culture depends on training of staff within the company and during their formative years as students. Changing the culture is difficult and takes time, but can be done. Each accident is different and complex. Each accident requires action at several levels: action to prevent the events that occurred immediately in advance, action to remove the risk and action to improve the steering system. The first and third are normally possible, but remove the risk is often difficult in an ancient plant.

How far should we go for adapting old plants to modern standards?
There is no simple answer to the question. Several accidents show what can happen when good procedures do not compensate for poor design. To summarize this section: if a change is almost as easy to do in an old plant as a new one, then it must be done, unless the plant is to be closed or there is another way to accomplish the same goal.

Higher risks for operators or the public in an old plant should not be accepted, but the methods to obtain the standard may be different from those in a new plant. We can for example improve software. The most modern plants often get the best managers and operators; new plants have glamour, and people like to be transferred there, even when the old plants have the greatest dependence on high standards of performance and handling.

When new rules are designed to prevent damage or loss of production and are not related to the injury of people, it is legitimate to leave old plants undisturbed; it depends on the balance between probabilities and costs. Here Risk and Operability Studies are required.

10.10 Causes and Consequences of the Most Famous Accident

A century after its occurrence, we should not conclude this chapter without some mention of the consequences of the accident that has had more significance, at least at the level of ordinary people: the sinking of the liner Titanic, on the morning of April 15, 1912. This ship struck an iceberg in the North Atlantic when she was coming to New York on her maiden voyage from Southampton. Of 2224 people on board, 1517 died. It was a very modern ship in its time, built with the latest technology.

Aside from the loss of hundreds of lives, the most interesting point is also the most amazing: the sinking of the Titanic was a good deal. Revenues from sales of books and films have exceeded the economic losses caused by its collapse. The circumstances of the sinking have been part of the myths and legends of several generations. So much so that even when we have enough evidence of the root causes of its sinking, public has not shown interest in learning. It is, by now, an almost religious matter.

The many studies that have been made, from different perspectives and from samples rescued from the more than 4,000 m deep wreck, provided information about the accident. "Or should we be surprised that there was no single cause for the sinking of the Titanic, but a chain of circumstances?" Science writer R. Corfield provides interesting data about construction defects and material problems. Yes, it sank after colliding with an iceberg, but why did it sink so quickly? The Titanic was one of the first ships that had transverse bulkheads forming sealed compartments to prevent a leak could flood the entire hull. In an

Fig. 10.6 The Titanic during construction, in a Belfast shipyard

emergency, compartments were closed with emergency gates electrically operated from the bridge. The hull was made of steel plates joint together with 3 million rivets of steel and iron.

Modern researchers have found that the hull rivets were not all the same, they did not have the same composition and they were not all placed in the same way. Those on the bow and the stern were not of the best quality, as amidships. They were placed manually, because the hydraulic press system used at the time to place the rest of the rivets could not be used where the curvature of the hull was more pronounced. These rivets also had more impurities, making them especially vulnerable to shear

Laboratory tests have shown that these rivets did not resist the impact load against the iceberg, so let the hull plates open, allowing the influx of water into the internal compartments. By design, if more than four of the 16 compartments were flooded, the ship was doomed to sink, and six flooded compartments were found by the liner designer, Thomas Andrews, when went down to look for the damage caused by the collision with the iceberg (Fig. 10.6).

Corfield mentions another fateful element that conspired against the Titanic: climate. When it is too hot in the Caribbean, the reinforced Gulf Stream favors the formation of a wall of ice, icebergs in the North Atlantic off Newfoundland, due to accentuated differences in temperature and water density between the warm Gulf and cold Labrador currents. And that was the case that spring of 1912. On the other hand, 3 months before the accident the closest approach between the Moon and Earth in over 1,400 years took place. With this approach, the tide was high. The exceptionally high tide in January could have dislodged many trapped icebergs to retake their way south, so that chances increased for any one intersecting the Titanic.

Another topic of interest is the dramatic consequence in human lives. The lifeboats were seen as a regulatory requirement, a nuisance, because this ship was "unsinkable." At this point there has been a tremendous learning. Not so much in

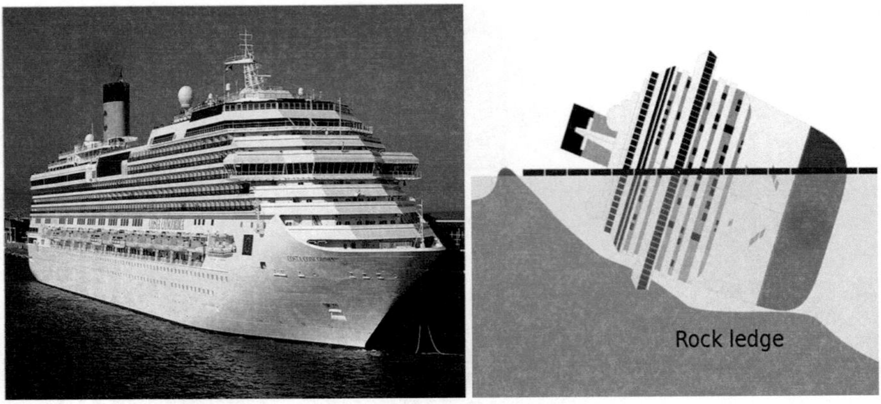

Fig. 10.7 Cruise ship Concordia during its final voyage

the design of the bulkheads of the hull. One hundred years after the sinking of the Titanic, another great liner, the Concordia, sank in shallow water, causing a few fatalities, Fig. 10.7. This ship sank, among other things, because these vessels have dispensed of watertight compartments to facilitate the transfer of passengers in lower decks. In a new twist, the criterion was: with the ship having control systems, it is "almost" impossible hitting anything. But as with all failures, there was a human error: on issues of the day, the captain approached the ship to shore beyond recommended by the charts. Days after the sinking of Concordia, media echoed the widespread disapproval to the Captain of the ship: apparently he had neglected his duties, having traveled accompanied by a lover and leaving the ship before making sure it would have been totally evacuated.

We should not be so hard on the captain: as in many other cases, a scapegoat was found. It is "normal" that officials in recreational ships are solicited by ladies wanting to increase the enjoyment of the trip. The traditions of "women and children first" and "the captain goes down with his ship," on the other hand, although reassuring ideas, contradict the evidence. When analyzing fatalities listings, it was found that survival rate among the crew is much higher than among passengers. This is logical: crew members are generally physically, mentally, and emotionally better trained to cope. Captains who have gone down with their ship, although there are honorable cases, are not the majority. And that way is not bad. To require otherwise would be to put them to death for their role, before any opportunity for root cause analyses.

At the time of writing this book, the victims had established demand upon those responsible for the ship, alluding to unsafe design. If demand thrives we might be seeing a new shift in design criteria.

References

1. H.K. Lee, *Safety Management, The Hong Kong Experience* (Lorrainelo Concept Design, Hong Kong, 1991)
2. R.K. Noon, *Scientific Method, Applications in Failure Investigation and Forensic Science* (CRC Press Taylor&Francis, Boca Raton, 2009). ISBN 978–1–4200–9280–6
3. H. Petroski, *To Forgive Design: Understanding Failure* (Harvard University Press, Cambridge, 2012
4. DOE-NE-STD-1004-92 Root Cause Analysis; Guidance Document. U.S. Department of Energy, Office of Nuclear Safety Policy and Standards (1992)
5. T.A. Kletz, *Learning from Accidents*, 2nd edn. (Butterworth–Heinemann Ltd, Oxford, 1994). ISBN 0 7506 1952 X
6. T.A. Kletz, *Plant Design for Safety—A User–Friendly Approach*, (Taylor & Francis, New York, 1991)
7. T.A. Kletz, *Hazop and Hazan*, 3rd edn. (Institution of Chemical Engineers, Rugby, 1992)
8. R.E. Knowlton, *A Manual of Hazard and Operability Studies* (Chemetics International, Vancouver, 1992)
9. T.A. Kletz, *Lessons from Disaster—How Organizations have No Memory and Accidents Recur* (Institution of Chemical Engineers, Rugby, 1993)
10. P. Garnsey, in *Understanding Catastrophe*, ed by J. Bourriau, (Cambridge University Press, Cambridge, 1992), p. 145
11. Gie Asset Integrity. Procedure for handling specimens in potentially litigious cases. (In spanish). Argentina, www.giemdp.com.ar. (2010)
12. T.A. Kletz, *An Engineer's View of Human Error*, 2nd edn. (Institution of Chemical Engineers, Rugby, 1991)
13. Jacques Frot. Causes for the Chernobyl event, Nov. 2000
14. Dr. Fairlie, Ian y Sumner, David (2006), Abstract of Report TORCH, Green Group., European Parliament, Berlin, Brussels, Kiev [28 August 2006]
15. Current evaluation of the Chernobyl reactor accident release (Nuclear Energy Agency of OCDE Feb. 1996). Consulted 13/8/2010
16. AIMPGN: Health Effects of Chernobyl—20 years after the reactor catastrophe
17. United States Environmental Protection Agency. Deepwater Horizon Response April 2010. Download KML Geographic Data
18. A.S. Brown, A system approach to safety (a report recommends ways to avoid a repeat of the Macondo Well blowout). Mech. Eng. J. pp. 42–44 ASME (2012)
19. R. Corfield; The role of physics in the sinking of the Titanic. Phys. World (2012)

MIX
Papier aus verantwortungsvollen Quellen
Paper from responsible sources
FSC® C105338

If you have any concerns about our products,
you can contact us on
ProductSafety@springernature.com

In case Publisher is established outside the EU,
the EU authorized representative is:
Springer Nature Customer Service Center GmbH
Europaplatz 3, 69115 Heidelberg, Germany

Printed by Libri Plureos GmbH
in Hamburg, Germany